Fundamentals and Applications of Additive Manufacturing

Through nine chapters covering software, hardware, solid-based, liquid-based and powder-based 3D printing processes, this textbook provides a comprehensive but easy-to-understand and application-oriented guide to the fundamentals and applications of 3D printing.

Readers are guided through various topics in a structured and logical manner that takes them from an initial comprehensive discussion of the topic to specialized chapters on advanced areas. The authors compare additive manufacturing with conventional processes; introduce computer-aided design; explore data preparation techniques, including 3D printing interfaces; and provide information regarding STL files, model slicing, tool path generation, 3D printing material, solid-liquid-powder-based 3D printing processes, post-processing techniques, advancements and future trends in 3D printing. The book also discusses printing accuracy, precision and tolerance and open-source data preparation software such as Fusion 360 and TinkerCAD. To ensure readers' comprehensive understanding of the different printing methods, the book discusses solid-, liquid- and powder-based 3D printing processes and their principles, workings, applications, post-processing techniques and future trends. Images and descriptive figures effectively illustrate the concepts and processes throughout, aiding in the understanding and retention of the concepts and processes. Every chapter includes learning outcomes, discussion topics, self-check exercises and multiple choice questions to help teachers and students in assessing their learning.

The broad coverage and engaging discussion format make this an ideal textbook for undergraduate and postgraduate students and an accessible reference for enthusiasts with elementary knowledge. In guiding readers from the basic concepts through to individual methodologies and printing techniques, it is invaluable to any reader who aims for a career related to any related application and industry.

Fundamentals and Applications of Additive Manufacturing

Hardware, Software, Methods, Materials and Future Trends

Sharad Kumar Pradhan, Ankit Nayak,
Surendra Singh Thakur, Vishal Francis
and Aniket Nagargoje

CRC Press
Taylor & Francis Group
Boca Raton London New York

CRC Press is an imprint of the
Taylor & Francis Group, an **informa** business

Designed cover image: @Shutterstock

First edition published 2026
by CRC Press
2385 NW Executive Center Drive, Suite 320, Boca Raton FL 33431

and by CRC Press
4 Park Square, Milton Park, Abingdon, Oxon, OX14 4RN

CRC Press is an imprint of Taylor & Francis Group, LLC

ISBN: 9781032898049 (hbk)
ISBN: 9781032912752 (pbk)
ISBN: 9781003562344 (ebk)

DOI: 10.1201/9781003562344

Typeset in Palatino
by Newgen Publishing UK

Contents

Preface

Additive manufacturing (AM), which is also known as 3D printing, is the technology that is changing the way we understand, create and manufacture products in different domains. AM has come a long way from being a simple tool for creating prototypes and has rather been perfected to become a manufacturing optimization strategy that ensures the best customization, sustainability and efficiency. These days, its use is not limited to borders of traditional applications but ranges through areas such as aerospace, healthcare and life sciences, automotive, construction and consumer goods, among many others.

This book seeks to provide substantial insight into the world of additive manufacturing and its diverse applications through exploration. The book is structured in such a way that it captures both the new learners as well as the practitioners who wish to update themselves with the recent developments of the industry.

In preparing this book, we have strived to strike a balance between theory and practice. The book comprises nine chapters. The first chapter provides a comparison between subtractive and additive manufacturing processes, including an overview of all the established 3D printing approaches. The next chapter delves into the hardware and software aspects of 3D printing. This chapter focuses on the components of a typical 3D printer, outlines the specifications of various printer types, and illustrates the assembly of an open-source fused deposition modelling (FDM) printer. Chapter 3 guides learners about the initial step required in any 3D printing technique that is to develop a digital model through a case study for understanding the typical steps involved in the creation of a digital model using computer-aided design (CAD) software. Chapter 4 focuses on the critical preprocessing stage of 3D printing, where the digital model is prepared for the printing process by integrating CAD and computer-aided manufacturing (CAM). Topics covered in this chapter include the generation of support structures for complex parts, ensuring successful prints, and the slicing process, where the CAD model is segmented into layers and converted into G-code. Once the learner is ready with an overview of various 3D printing processes, different components of a typical 3D printer and associated software, and the skills to develop a digital model and its G-code file, the next step is to select suitable a 3D printing material. Chapter 5 equips learners with the knowledge needed to select the most suitable material for a given 3D printing process, ensuring optimal results. The next three chapters deal with the printing of this digital model using solid-, liquid- and powder-based 3D printing processes providing a detailed overview of their working principles, applications and other technological aspects, illustrating how AM is revolutionizing manufacturing processes, enabling complex geometries, reducing material waste and opening new avenues. Post-processing techniques are essential for enhancing the quality, functionality and overall performance of 3D printed components. The last chapter provides an in-depth overview of the various post-processing, inspection and testing techniques used in solid-based, liquid-based and powder-based 3D printing processes.

The book also explores the troubleshooting, advancements and future trends in connection with solid-, liquid- and powder-based 3D printing processes, including issues related to scalability, cost, material limitations and quality control. In doing so, it aims to foster a balanced perspective that celebrates the achievements of AM while acknowledging the ongoing efforts needed to fully realize its potential.

Every chapter includes learning outcomes, discussion topics, self-check exercises and multiple choice questions to help teachers and students in assessing their learning.

Because of its extensive coverage and its engaging discussion format, this is a particularly appealing textbook for both undergraduate and postgraduate level students and an easy book for amateurs with basic understanding. We are pessimistic that students, researchers and engineers who are interested in applying AM to the current challenges and invent new ideas will find this book useful. This book has been written with the hope that the ideas presented in it will lead to new research and further development of this interesting area.

All Authors
2024

Acknowledgements

This book would not have been possible without many individuals and organizations' support, guidance and inspiration. First and foremost, we extend our deepest gratitude to the researchers, engineers, practitioners, enthusiasts and hobbyists in the field of additive manufacturing whose available work has laid the foundation for this book.

This book is based on the classroom and lab sessions that we have all conducted at our institutes and on massive open online learning platforms for students and enthusiast. The experience gained through all these sessions, the questions raised by learners, the assignment submitted by them, the interactions, all motivated us to write this book on 3D printing and its applications.

We are particularly indebted to our fellow faculty members, industry personnel, practicing engineers, technicians who provided valuable insights, constructive feedback and encouragement during the writing process. Their expertise and dedication have greatly enriched the quality and scope of this book.

We are also grateful to the publishers and CRC editing team for showing confidence in our work, their professionalism, patience and meticulous attention to detail and suggestions in bringing this work to realization.

Special thanks go to our family members, and Mr. Rishabh Sood, who relentlessly cheered us from the start of our book writing efforts.

We are grateful to Professor Prashant K. Jain (IIITDM Jabalpur), Professor Richard Su, and Professor Jane Pu (OMFS, The University of Hong Kong) for their encouragement and thoughtful support.

Finally, I would like to acknowledge with gratitude the support extended by all the authors. This book would not have been possible without academic support and the discussion offered by them.

It is our hope that this book will serve as a meaningful contribution to the ever-evolving field of additive manufacturing.

About the Authors

Sharad Kumar Pradhan is Director of THDC Institute of Hydropower Engineering and Technology, Tehri, India (on deputation) and is a Professor Mechanical Engineering at the National Institute of Technical Teachers' Training and Research (NITTTR), Bhopal. Before joining NITTTR Bhopal he was a Scientist at "Central Mechanical Engineering Research Institute" (CMERI), Durgapur, (WB), India. He attained his Ph.D. from the Indian Institute of Technology, Delhi, India.

Ankit Nayak is an Assistant Professor at UPES, Dehradun, India. Prior to this, he was a Postdoctoral Fellow in Oral and Maxillofacial Surgery at the University of Hong Kong. He obtained his Ph.D. in 2020 from the Indian Institute of Information Technology, Design & Manufacturing (IIITDM) Jabalpur.

Surendra Singh Thakur is a Senior Project Engineer at 3D Engineering Automation LLP, Pune, India. He obtained his B.Tech from the National Institute of Technology (NIT) Surat, followed by his M.Tech and Ph.D.

Vishal Francis is an Associate Professor in the School of Mechanical Engineering at Lovely Professional University, India. He obtained his PhD from the Indian Institute of Information Technology, Design and Manufacturing, Jabalpur, India.

Aniket Nagargoje is an Assistant Professor at MIT Academy of Engineering, India. He obtained his BE from Savitribai Phule Pune University in 2014, his M.Tech from GGV, Bilaspur in 2017, and his Ph.D from Indian Institute of Information Technology Design and Manufacturing, Jabalpur in June 2022.

1

Overview of Additive Manufacturing Processes

This chapter presents an overview of additive manufacturing processes with a comparison to subtractive manufacturing processes. It also outlines the different additive manufacturing processes and their working principle. So far, seven types of additive manufacturing processes have been identified and characterized. The chapter comprises the following main sections:

After reading through the material, self-check exercises, objective questions and discussion topics presented in this chapter, the learner will be able to do the following:

(a) Differentiate between subtractive and additive manufacturing processes.
(b) Outline the various 3D manufacturing processes.

1.1 Additive vs Subtractive Manufacturing Processes

Additive manufacturing (AM), also referred to as additive layer manufacturing (ALM), is the industrial term for 3D printing. This technique involves creating physical objects from a digital model in a computer-controlled process. It includes depositing materials like metal, plastic and ceramic in solid (wire/sheet), liquid or powder form in thin, two-dimensional layers to construct a three-dimensional object. Presently, seven types of additive manufacturing processes have been identified and defined.

1.1.1 Additive vs Subtractive Manufacturing Processes

AM or ALM is the industrial term for 3D printing. 3D printing is the technique of creating physical items from a digital model. The manufacturing

DOI: 10.1201/9781003562344-1

processes are categorized as formative, subtractive and additive. In formative manufacturing processes, there are forming, casting and joining processes where the product is either created with the help of predefined shapes of die/punch/mould or standard pieces are joined to create it. Forming and casting processes are generally used when there are a greater number of parts to be manufactured, which justifies the cost of die/punch/mould. Subtractive manufacturing removes material in the form of chips or particles from the stock to create an object based on drawings or digital models. In contrast, an additive manufacturing process is a computer-controlled technique that includes depositing materials, such as metal, plastic and ceramic, in solid/liquid/powder form in thin two-dimensional layers to create a three-dimensional item. It melts or fuses powder or cures liquid polymer materials to build components based on a computer-aided design (CAD) model in a layer-by-layer manner.

Generally both subtractive and additive manufacturing methods utilize CAD drawings/models for building components but in conventional subtractive manufacturing processes, the products can also be made from production drawings created using hand drawing instruments.

Additive manufacturing involves melting wire or fusing powder/sheets or curing liquid polymer materials to create components based on a CAD model using 3D printers, while subtractive manufacturing removes material (in the form of chips or particles) from the stock to produce an object based on drawings or models using lathe, milling, drilling, grinding machines, etc. A third group of manufacturing process is called the formative method, which requires a die or mould or standard sections for creating objects. This method includes forming, casting and joining/welding processes. Figure 1.1 illustrates the basic principle of all these three approaches and Table 1.1 gives a comparison between additive and subtractive manufacturing processes.

Figure 1.2 depicts the differences between the processes, highlighting the waste produced during additive and subtractive manufacturing processes.

1.1.1.1 Application of 3D Printing

3D printing is utilized by a variety of industries, providing numerous benefits and applications:

- Manufacturing: enables rapid prototyping and production of complex, custom parts.
- Dental: used for creating dental prosthetics and orthodontic devices.
- Healthcare: facilitates the creation of medical devices and patient-specific surgical guides.

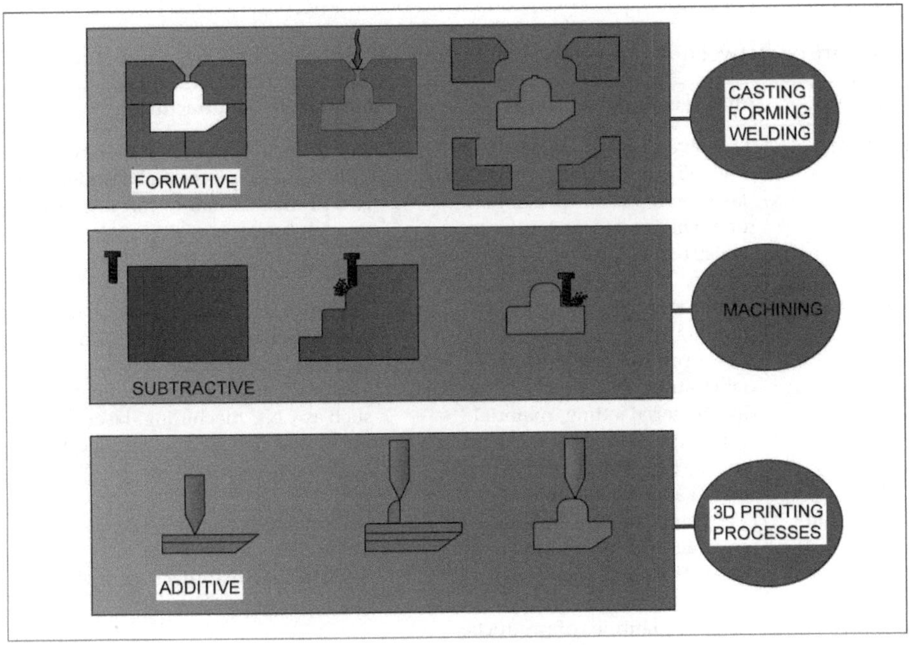

FIGURE 1.1
Basic difference between formative, subtractive, and additive manufacturing processes.

- Jewellery: allows for intricate, customized designs and production.
- Biomedical implants: offer the ability to produce patient-specific, complex implants.
- Pharmaceuticals: enables personalized drug dosages and unique drug delivery systems.
- Custom-fitted personal products: allows for tailored products, such as footwear and eyewear.
- Educational materials: enhances learning experience by hands-on models and visual aids.
- Replacement parts: provides on-demand production of spare parts for household items.
- Civil construction: showcases potential for sustainable, efficient buildings and infrastructure.

By its versatility, 3D printing continues to expand its reach and impact across these various industries, driving innovation and customization.

The technique has a wide range of applications, from prototypes (visual and functional) to end-use parts as presented in Figure 1.3.

TABLE 1.1

Comparison between Additive Manufacturing and Subtractive Manufacturing

Features	Additive Manufacturing	Subtractive Manufacturing
Process	It involves adding material in a layered manner, by placing each layer on top of the previous one. This process is similar to constructing a house by stacking bricks.	Subtracting (removing) material from a bulk material piece can be understood as a process similar to carving a cave or sculpture.
Equipment	This process uses digital manufacturing techniques such as 3D printing and additive fabrication, which include methods like material jetting, material extrusion, binder jetting, directed energy deposition, powder bed fusion and sheet lamination.	This process uses traditional machining processes (lathe, milling, drilling, shaping, grinding machines) and advanced metal cutting techniques, such as CNC machining, laser cutting, EDM, plasma cutting, waterjet cutting and abrasive processes.
Production	These processes are suitable for creating prototypes and small batch production, making them ideal for testing designs and producing limited quantities of products.	These methods are best suited for mass production, allowing for efficient and cost-effective manufacturing of large quantities of products.
Additional Equipment	Post-processing systems are essential for certain printers to ensure that the final product meets desired quality and appearance standards. These systems typically involve curing, finishing, and cleaning and painting processes, which help to enhance the strength, durability, and aesthetics of the printed parts.	Different tooling, fixtures, robots, material handling systems, coolant systems, tool changers and waste removal equipment may be required.
Complexity	Complex designs using natural geometry can easily be created through 3D printing.	It is suitable for mechanical designed parts but may not be suitable for nature-inspired or natural designs (bone, etc.)
Materials	Most of the materials are plastic. Metals, ceramics, plasters, graphite, carbon fibre, nitinol, polymers and paper are among other materials used. Material is required in the form of wire, powder, sheets and liquid.	Metals, ceramics, plasters, graphite, carbon fibre, nitinol, polymers and paper are some other materials used. Material is required in the form of stocks, ingots and billets.

1.1.2 Overview of 3D Printing or Additive Manufacturing Processes

The encompassing realm of additive technologies features the following:

- Material extrusion 3D printing is the process of layer-by-layer material deposition through a nozzle or aperture, involving materials (wire form), such as plastic, nylon, rubber, etc.

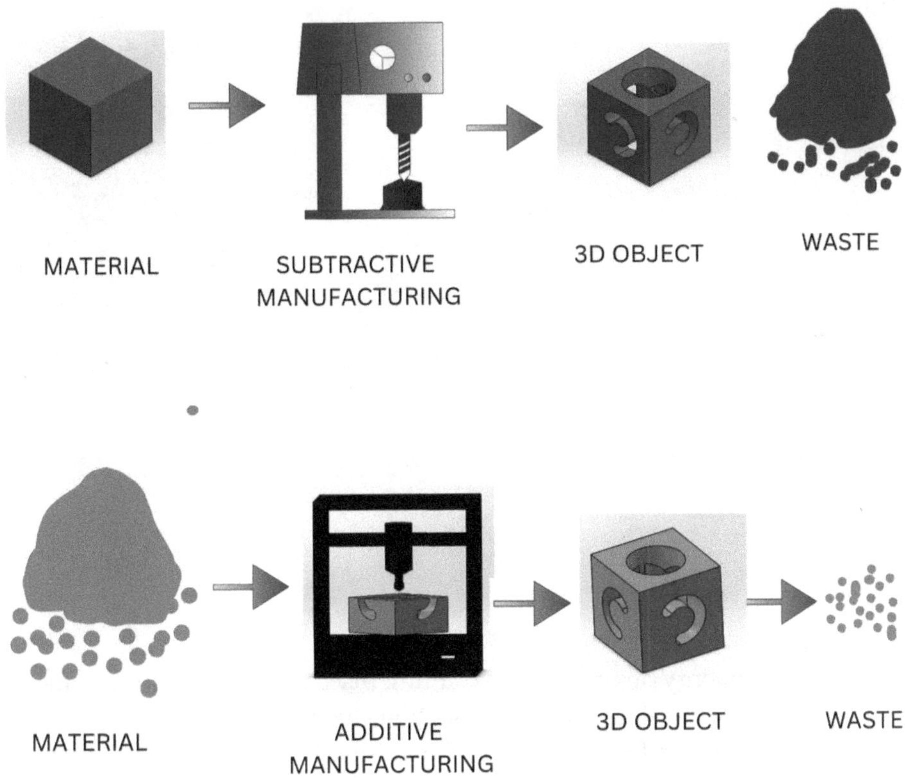

FIGURE 1.2
Comparison of waste/scrap after performing additive and subtracting manufacturing processes.

- Vat photopolymerization 3D printing involves the selective curing of predeposited photopolymers (liquid material) through light-activated cross-linking, resulting in the creation of stable and intricately connected structures.
- Powder bed fusion 3D printing selectively fuses specific areas within a powder material bed using thermal energy, with materials including metals, polymers and fibres.
- Material jetting 3D printing is a technique where carefully deposited droplets (liquid material) combine to form intricate objects using a variety of materials, such as photopolymers, wax, sand and poly-jetting substances.
- Binder jetting 3D printing involves the precise application of liquid bonding agents to powdered material, utilizing materials like metal, plastic, ceramic and sand.

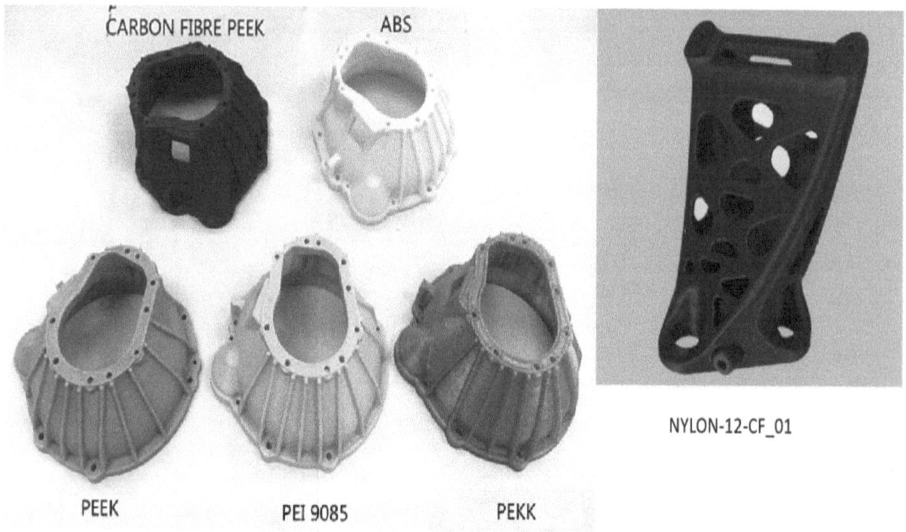

FIGURE 1.3
Application of 3D printing and materials.

- Directed energy deposition 3D printing fuses deposited materials using concentrated heat energy, with materials including wire and powder.
- Sheet lamination forms a model by bonding sheets of materials together, with materials such as metal, paper, wood and plastic.

All seven 3D printing approaches are shown in Figure 1.4 along with their micro version manufacturing as the eight options.

1.1.2.1 Material Extrusion 3D Printing Process

A fused deposition modelling (FDM) 3D printer functions by depositing melted filament material (wire) onto a printing platform layer by layer, ultimately producing a final product. The printer utilizes digital design files that are given to the machine and then transformed into physical dimensions. To prepare an object for printing, its CAD file must be converted into a format that the 3D printer can recognize, usually an .STL format (accessible in most CAD software through the export tab). FDM printers use two types of materials: modelling material for the end product and support material, which acts as a temporary scaffold throughout the printing process.

To start the printing process, a spool of thermoplastic filament is inserted. Once the printer reaches the desired nozzle temperature, it pushes the filament through an extrusion head and nozzle. The nozzle warms the filament to its glass transition temperature before extruding it onto a base, sometimes referred to as a build platform or table. The controller guides and coordinates

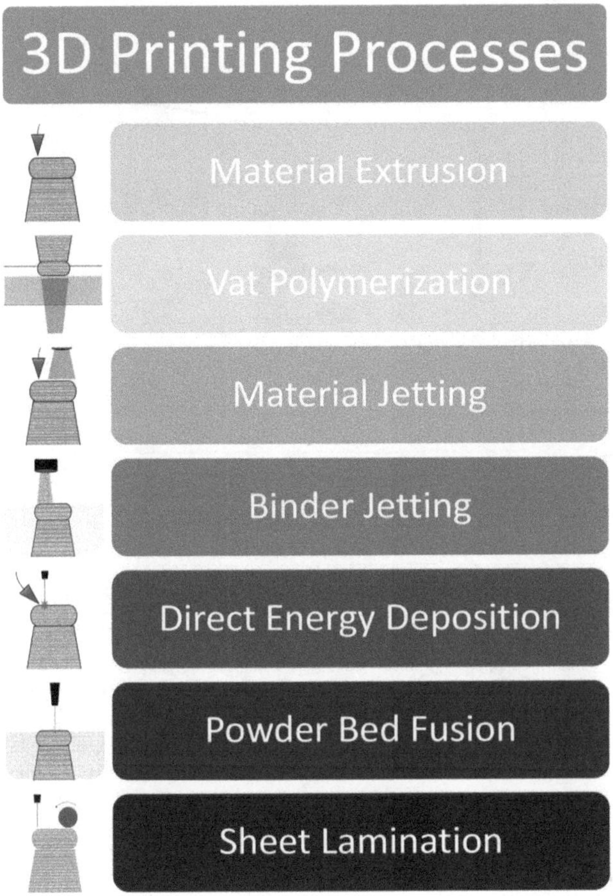

FIGURE 1.4
Seven 3D printing processes.

the nozzle and base (as per the coordinates received from the G-code file) during printing based on the deposition path, also known as the tool path. Figure 1.5 shows the schematic diagram of the process.

FDM is a technique of depositing material in thin strands, layer by layer, along a specified route. When a layer is finished, the build platform moves down, or the nozzle (printer head/extrusion head) is raised by one layer thickness to maintain the gap between the deposited layer and the nozzle, and the process starts again with the next layer. The success of FDM relies on strong layer adhesion. As molten thermoplastic passes through the nozzle, it presses against the previously printed layer due to gravity. The combination of high temperature and gravitational force causes the layer to cross-link the polymer, allowing it to fuse with the layer beneath it. This process continues until the final component is completed.

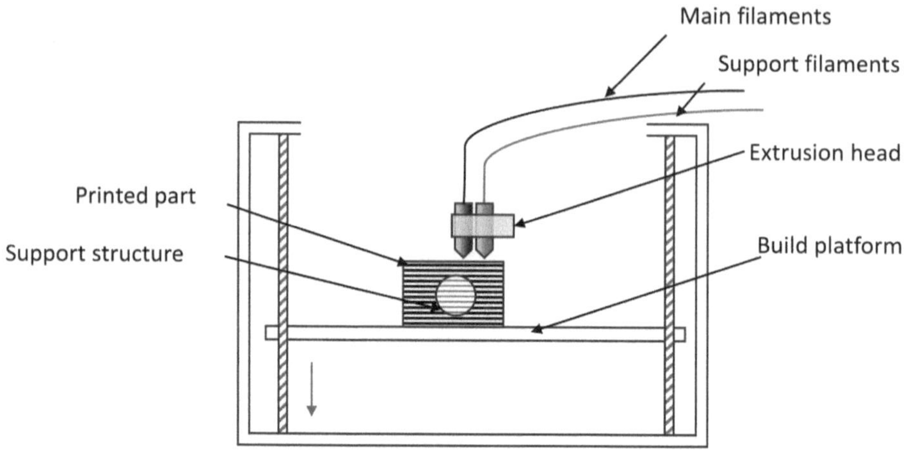

FIGURE 1.5
Fused deposition modelling 3D printing process.

After removing the object from the FDM printer, the support materials are typically removed by soaking them in a water-and-detergent solution or by physically breaking off the thermoplastic supports. Additional finishing techniques, such as sanding, milling, painting and plating can be utilized to enhance the functionality and aesthetics of the object.

Types: fused deposition modelling (FDM is a registered trademark of the process developed by Stratasys Inc.) It is often called fused filament fabrication (FFF) if the filament is used as the material.
Materials: plastic filaments encompass a variety of materials such as PLA, ABS, PET, PETG, TPU, nylon, ASA, PC, HIPS, carbon fibre, and more. Innovative materials containing metal fillers are now available, which are incorporated within the extruded plastic "rods".
Dimensional accuracy: ±0.5% (lower limit ±0.5 mm).
Common applications: electrical housings, jigs and fixtures, investment casting patterns, custom enclosures, functional prototypes, architectural models, medical models, art sculptures, automotive components, aerospace parts, etc.
Strengths: lowest costing 3D printing method, wide range of materials.

1.1.2.2 Vat Polymerization 3D Printing Process

Vat polymerization, a captivating 3D printing method, utilizes a concentrated light source to selectively harden photopolymer resin in a container. To solidify a thin layer of liquid plastic, light is concentrated on specific areas (as per the coordinates received from the G-code file). This process is carried out layer by layer until the desired 3D object is produced. Stereolithography (SLA), digital light processing (DLP) and masked stereolithography (MSLA) are the three most commonly used vat polymerization methods. Although

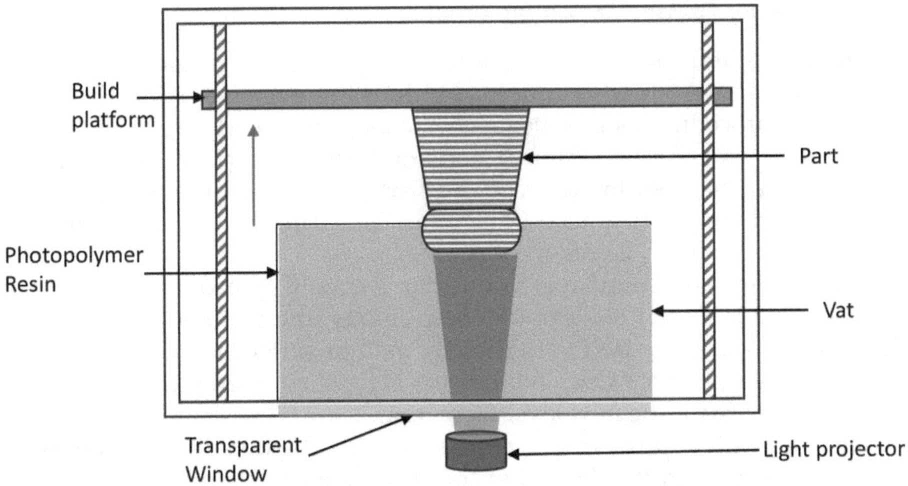

FIGURE 1.6
Vat polymerization 3D printing process.

each of these 3D printing techniques is based on the same fundamental concept, the main difference lies in the type of light source used to cure the resin. Understanding the variances between different light sources is crucial in determining the optimal choice for a particular application.

Photopolymers, a unique form of resin, are utilized as the printing material for all vat photopolymerization printers. These special substances have the exceptional capability to rapidly bond and solidify when they are exposed to specific light wavelengths. This process, known as photopolymerization, is the underlying principle of this 3D printing technique.

The majority of vat photopolymerization printers employs a container or vat to contain the liquid photopolymer, with the build platform partially submerged in the liquid. By using information from a CAD file, the printer directs a light source to selectively (as per coordinates received from the G-code file) harden the liquid photopolymer, transforming it into a solid layer. Once this layer is formed, the build platform is reimmersed in the remaining resin, and the process is repeated for the subsequent layers. This sequence continues until the entire design is meticulously brought to life, layer by layer. Figure 1.6 shows the schematic diagram of the process.

Types: SLA, MSLA µSLA and more.
Materials: photopolymer resins (castable, transparent, industrial, biocompatible, etc.).
Dimensional accuracy:±0.5% (lower limit ±0.15 mm or 5 nanometres with µSLA).
Common applications: injection mould-like polymer prototypes; jewellery casting; dental applications.
Strengths: smooth surface finish, fine feature details.

1.1.2.3 Powder Fusion 3D Printing Process

Like other methods of additive manufacturing, the process starts by creating a 3D CAD file, which is then uploaded into the machine. After the file is loaded, the procedure usually begins by heating the powder bed to maintain a consistent temperature. Printing actually begins when an energy source is applied (as per coordinates received from the G-code file) to powdered material, such as plastic, metal or ceramic. This causes the particles to fuse together and gradually create the desired object.
During the process, a thin layer of powder is evenly spread across the build surface using a recoater blade or roller. An energy source, such as a laser or an electron beam, is then used to selectively melt or sinter the material needed for that particular layer. Once the process is finished, the build plate moves down to prepare for the next layer formation.

In the 3D printing process, a layer of powder is added and then fused using a laser beam to form the desired shape. This sequence is repeated until the entire object is produced. The unfused powder serves as support and casing for the completed product, ensuring its structure and integrity. Following the printing process, the extra powder is eliminated to reveal the final product. Figure 1.7 shows the schematic diagram of the process.

Types: selective laser sintering (SLS), selective laser melting (SLM), electron beam melting (EBM), direct metal laser sintering (DMLS), multi-jet fusion (MJF).
Materials: thermoplastic powders (Nylon 6, Nylon 11, Nylon 12, etc.), metal powders (steel, titanium, aluminum, cobalt, etc.), ceramic powders.
Dimensional accuracy: ±0.3% (lower limit ±0.3 mm).

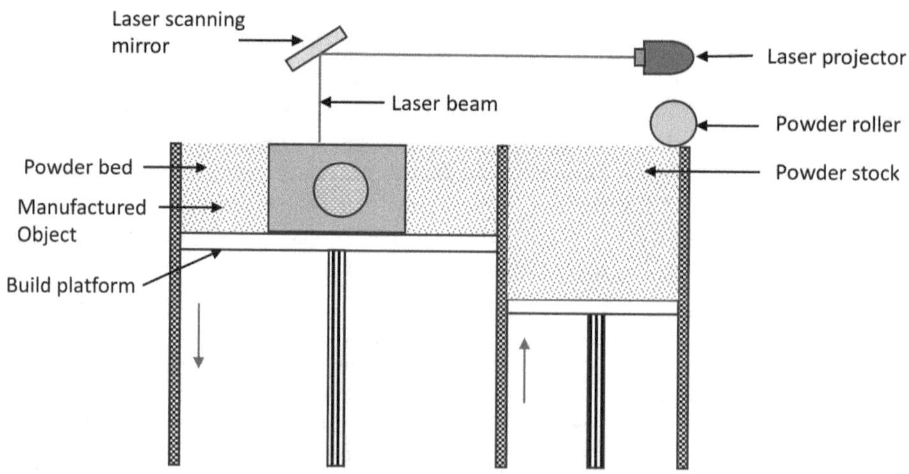

FIGURE 1.7
Powder fusion 3D printing process.

Common applications: functional parts, complex ducting (hollow designs), low-run part production.
Strengths: functional parts, excellent mechanical properties, complex geometry.

1.1.2.4 Material Jetting 3D Printing Process

To start this 3D printing process, it is necessary to first create and upload a 3D CAD file to the machine. A print-head, similar to the ones found in traditional inkjet printers, then releases small drops of photosensitive material onto the build platform according to the coordinates received from the G-code file. When exposed to ultraviolet (UV) light, the material solidifies in specific areas.

During the printing process, the material undergoes either partial or complete curing, depending on the specific needs. After each layer is formed, the build platform lowers to allow for the next part of the model to be printed. This process continues, layer by layer, until the entire object is successfully created.

The material jetting process starts with generating and uploading a 3D CAD file. Next, a print-head deposits photosensitive material onto the build platform. This material solidifies under UV light based on the coordinates received from the G-code file. After each layer is partially or fully cured, the platform moves down to make room for the next layer. This process is repeated until the final object is finished. Figure 1.8 shows the schematic diagram of the process.

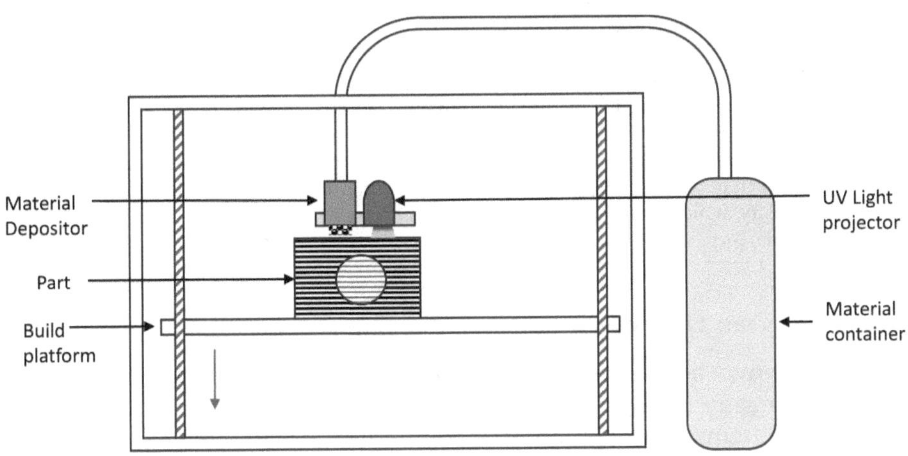

FIGURE 1.8
Material jetting 3D printing process.

Types: material jetting (MJ), drop on demand (DOD).
Materials: photopolymer resin (standard, castable, transparent, high temperature). It includes rigid (various colours), flexible (rubber-like from 30 to 90 Shore A), transparent (lens/window quality) and biocompatible options.
Dimensional accuracy:±0.1 mm.
Common applications: full-colour product prototypes; injection mould-like prototypes; low run injection moulds; medical models.
Strengths: best surface finish; full colour and multi-material available; exceptional print resolution, ideal for printing intricate models with fine detail and thin sections. It is relatively fast compared to SLA.
Weaknesses: limited to proprietary photopolymer resins. Additional post-processing steps are commonly needed to achieve a cosmetic finish.

1.1.2.5 Binder Jetting 3D Printing Process

Binder jetting is a type of 3D printing that utilizes print-head technology similar to material jetting. In material jetting, building components and their support structures are deposited, whereas in binder jetting, liquid glue is added to a powder surface. This additive manufacturing technique involves depositing small amounts of binder material onto a powder surface through a print-head (as per coordinates received from the G-code file) to construct an object layer by layer. The resulting item is solidified and strengthened within the powder bed before being extracted, and any excess powder is removed using compressed air. Binder jetting is commonly used to produce complex geometries, prototypes and small production runs using a variety of materials, including metals, ceramics and polymers. Figure 1.9 shows the schematic diagram of the process.

Types: metal binder jetting, sand binder jetting
Materials: sand, polymer, or metal powder: stainless/bronze, full-colour sand, silicia (sand casting); ceramic–metal composites.
Dimensional accuracy: ±0.2 mm (metal) or ±0.3 mm (sand).
Common applications: functional metal parts; full-colour models; sand casting.
Strengths: low cost; large build volumes; functional metal parts; superb colour reproduction; fast print speeds; support-free design flexibility. Supports a surprisingly wide range of materials from stainless steel, bronze and aluminum, through ceramics, to sand and plaster.
Weaknesses: mechanical properties are not as good as metal powder bed fusion. Generally has high porosity, which may require additional steps like infiltration or sintering to improve strength and density.

1.1.2.6 Directed Energy Deposition 3D Printing Process

DED, also known as directed energy deposition, is a sophisticated 3D printing method primarily utilized for adding material to existing components or for constructing items from scratch. It is commonly used in industrial settings, such as for repairing damaged turbine blades or propellers.

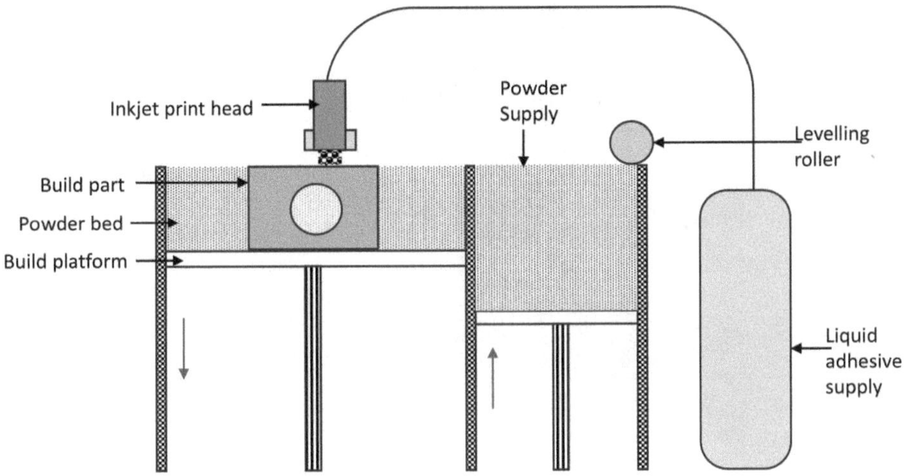

FIGURE 1.9
Binder jetting 3D printing process.

Similar to other powder bed fusion (PBF) technologies like laser powder bed fusion (LPBF) or electron beam melting (EBM), DED uses a focused energy source, such as a laser, electron beam or plasma, to melt the material. The main difference lies in the delivery of the material itself, which is either in the form of wire or powder. As the material exits the nozzle, it immediately melts, allowing the creation of intricate structures.

The process continues with the layers solidifying (as per coordinates received from the G-code file) until the object is either newly formed or repaired. Overall, DED is a versatile and efficient method for repairing and building parts in various industrial applications. Figure 1.10 shows the schematic diagram of the process.

Types: laser engineered net shaping (LENS); electron beam additive manufacturing (EBAM); cold spray.
Materials: metals, in wire and powder form.
Dimensional accuracy: ±0.1 mm.
Common applications: repair of high-end automotive/aerospace components, functional prototypes, and final parts.
Strengths: support structures rarely required; mixing metals; ability to work in three dimensions.
Weaknesses: poor surface finish requires post-processing; expensive.

1.1.2.7 Laminated Object or Sheet Lamination 3D Printing Process

Sheet lamination is a 3D printing method that involves layering ultra-thin sheets of material. Initially, a thin layer of material is rolled out or placed on

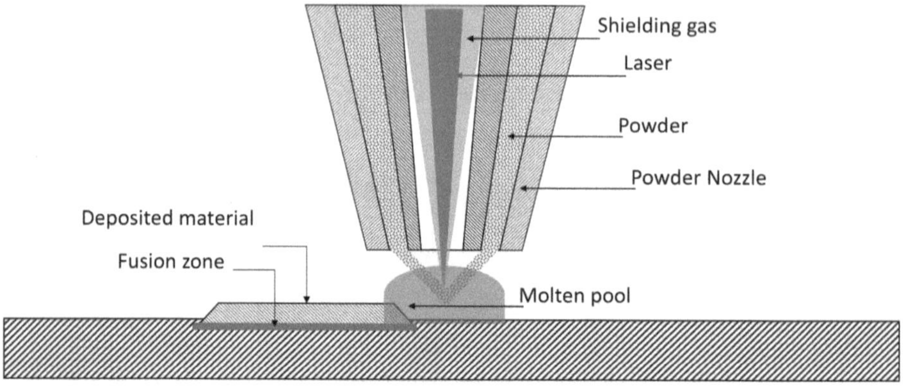

FIGURE 1.10
Directed energy deposition 3D printing process.

the building platform. The method for bonding the layers varies depending on the type of sheet lamination used. For selective deposition lamination (SDL) and ultrasonic additive manufacturing (UAM), the layers are first glued together, then the 3D shape is cut. On the other hand, computer-aided manufacturing of laminated engineering material (CAM-LEM) cuts each layer to shape (as per coordinates received from the G-code file) before bonding them together. This process is repeated until all layers reach the full height of the object. After the layers are completed, the print block is removed, and any unwanted outer edges are trimmed to reveal the final 3D object. A variety of technologies, such as heat and sound, can be utilized to fuse the material layers together. Figure 1.11 shows the schematic diagram of the process. Despite sheet lamination producing items with less additive precision than other AM processes, it offers the advantages of cost-effectiveness and shorter manufacturing time. This enables product designers to create low-fidelity prototypes using readily available, low-cost materials.

Types: laminated object manufacturing (LOM), ultrasonic consolidation (UC).
Materials: paper, polymer and metal in sheet forms.
Dimensional accuracy: ±0.1 mm.
Common applications: non-functional prototypes, multi-colour prints, casting moulds.
Strengths: low cost; quick production possible; composite prints.
Weaknesses: low accuracy; more waste; plenty of post-production work required for parts.

In summary, according to differences in formation and processing of the materials, 3D printing technology can be divided into seven types, as depicted in Figure 1.12.

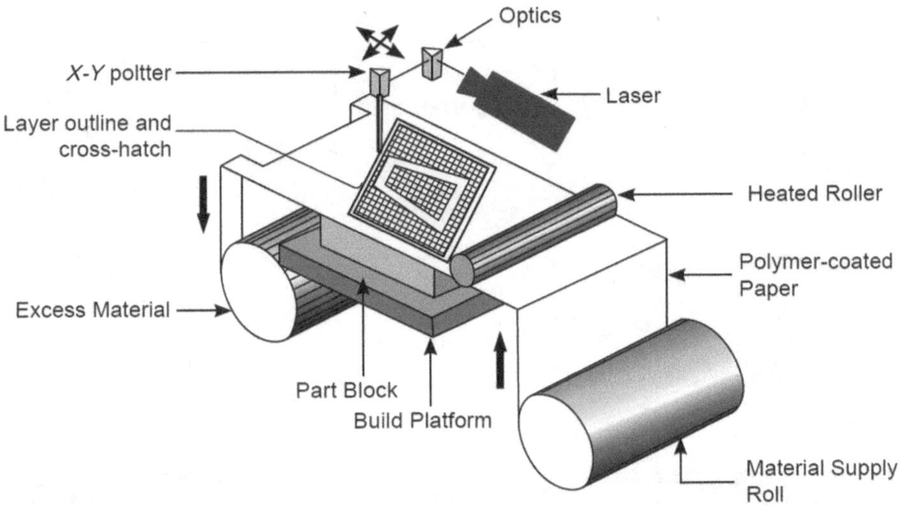

FIGURE 1.11
Sheet lamination 3D printing process.

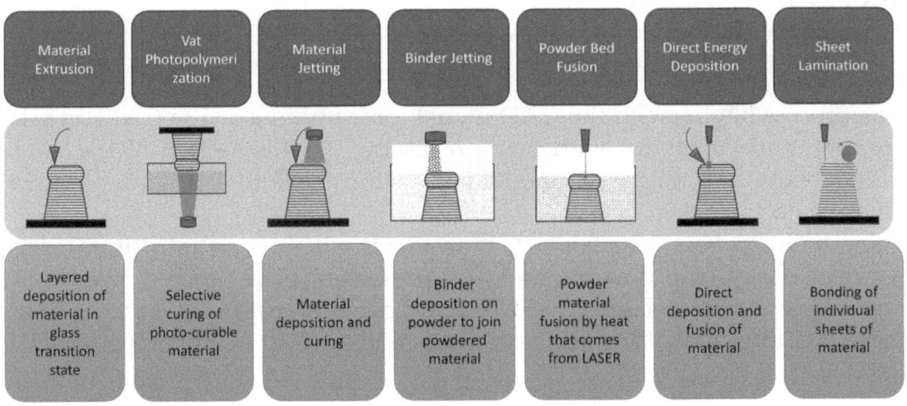

FIGURE 1.12
Outline of seven types of 3D printing technologies and their functional processes.

1.2 Typical Steps of 3D Printing

3D printing allows us to fabricate physical models of any geometry provided as a CAD file, typically in STL format, to the 3D printing system. The series of steps required to create a final product is known as the process chain.

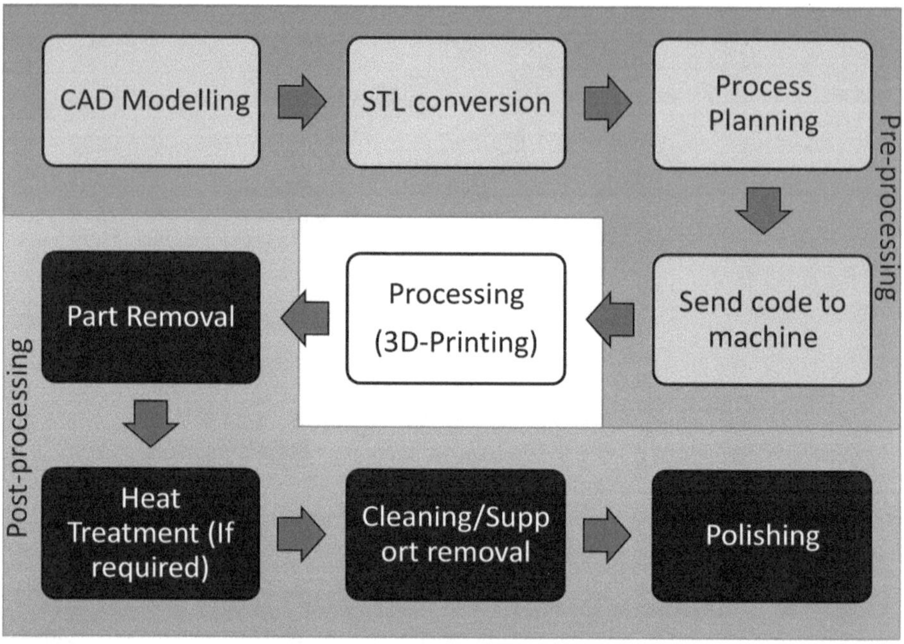

FIGURE 1.13
Process sequence of 3D printing.

This process begins with the CAD model, progresses through CAM and 3D printing, and concludes with post-processing. In this section of the book, we will explore how a CAD model is transformed into a tangible product, covering each stage in detail.

The entire process chain of 3D printing is categorized into three parts, i.e., preprocessing, processing and post-processing. The process chain of 3D printing is shown in Figure 1.13. Detailed discussions on the process chain based on individual 3D printing processes will be dealt in with in Chapters 6, 7 and 8.

Step 1 – There are three different methods, as previously mentioned, that can be used to create the solid model of the product we aim to print with a 3D printer.

Creating a CAD model is the first step in 3D printing. The design must be provided in digital format to process planning software. To achieve this, various CAD modelling programs or reverse engineering tools are used to develop a solid model of the required part. As previously mentioned, the CAD model for 3D printing must have a completely closed boundary.

TABLE 1.2

Native CAD File Formats

Format	Extensions
AutoCAD	.dwg
CATIA V4	.model
CATIA V5, V6	CATPart, CATProduct, CGR
Creo	.prt, .asm
NX, Unigraphics	.prt, .jt,
Pro/E	.prt, .asm
Inventor	.ipt, .iam
MicroStation	.dgn
SolidWorks	.sldprt, .sldasm
Solid Edge	.par, .asm, .psm

Step 2 – We have to convert the CAD file that is primarily in native file format into neutral file format.

- STL (or stereolithography) is the most widely used file format for 3D printing. (Chapter 4, Section 4.2, will present more information on this format.)
- STL can be substituted with .OBJ and .3MF files. There is no colour information in these file formats.
- File formats like .X3D, .WRL, .DAE and .PLY are required for full colour 3D printing.

The native file formats (refer to Table 1.2) are primarily determined by the specific CAD software you choose to use. By utilizing these native formats, you gain the ability to fully explore and take advantage of all the detailed features and functionalities the program offers, thereby enhancing your design capabilities and overall experience.

On the other hand, to enable file sharing between users working with different CAD software, neutral file formats (refer to Table 1.3) have been created. The table shows some examples of these neutral formats.

Most CAD software allows you to save or export models in various formats, including STL. For instance, in SolidWorks, any part can be saved as an STL file, as shown in Figure 1.14. SolidWorks also offers advanced options to adjust the resolution and choose between binary or ASCII formats for the STL file. Naturally, higher-resolution STL files will take up more memory to store. On pressing the OK button, the number of triangles, file size and file format will be displayed as shown in Figure 1.15.

Step 3 – The STL file is then sent to the process planning software, which first checks its quality and repairs any errors if needed. The software

TABLE 1.3

Neutral CAD File Formats

Format	Extensions
STEP	.stp, .step
STL	**.stl**
IGES	.igs, .iges
3D PDF	.pdf
Parasolid	.x_t, .x_b
VRML	.vrml
X3D	.x3d
COLLADA	.dae
DXF	.dxf

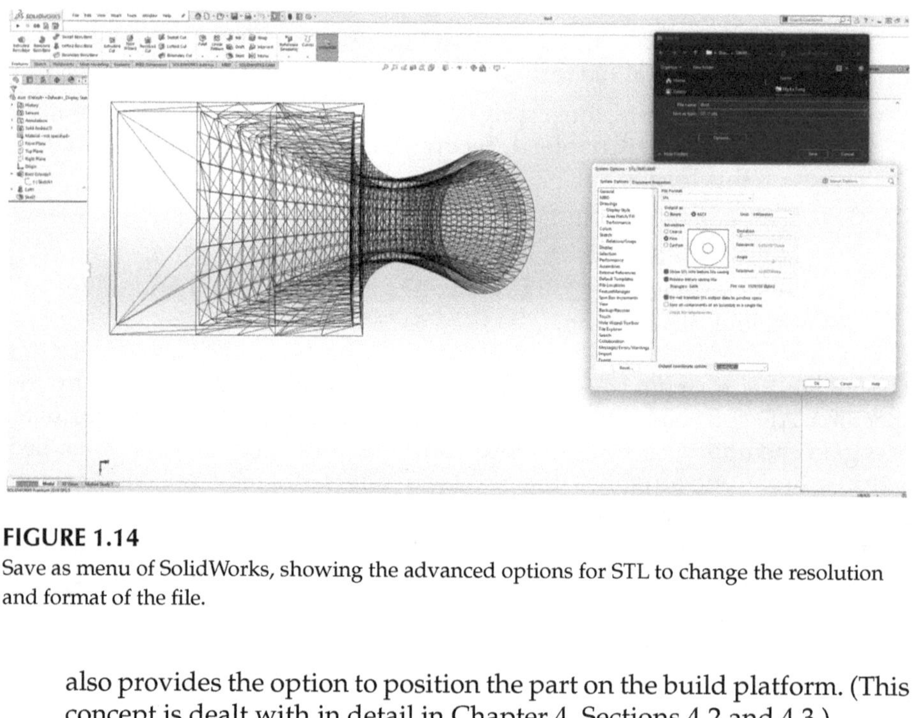

FIGURE 1.14
Save as menu of SolidWorks, showing the advanced options for STL to change the resolution and format of the file.

also provides the option to position the part on the build platform. (This concept is dealt with in detail in Chapter 4, Sections 4.2 and 4.3.)

The process planning software allows for the simultaneous printing of multiple copies of the same part on the platform. Additional settings, such as part orientation, size scaling, support generation, and, in some machines, colour, can also be configured at this stage. Machine-specific process parameters must be set at this point, and these parameters vary based on the machine type and the 3D printing method used. Common parameters include layer thickness and infill density, which are essential in almost all 3D printing

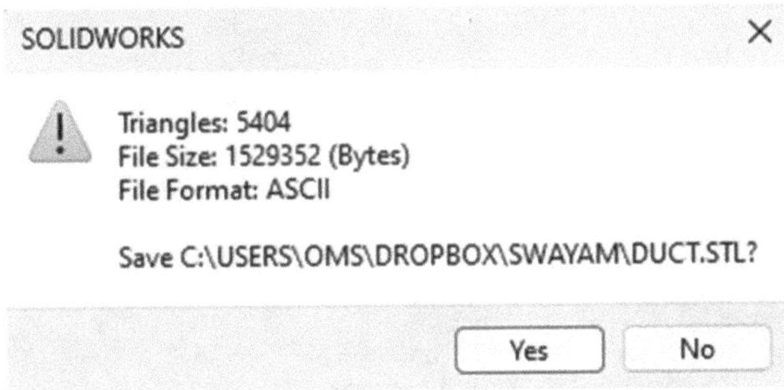

FIGURE 1.15
SolidWorks message box showing the number of triangles, file size, and format.

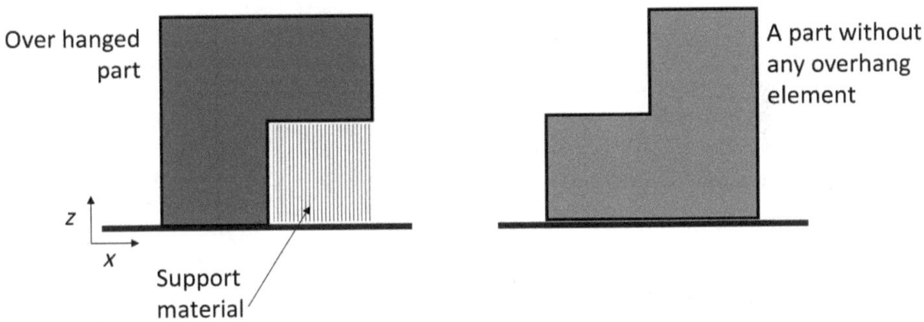

FIGURE 1.16
Same part with two different print orientations.

scenarios. Layer thickness relates to print resolution, while infill density determines how much material fills the interior of the part. It is important to note that increasing these parameters will lead to longer print times, higher material consumption, better surface finish and more strength in the product printed.

Print orientation plays a crucial role in affecting print quality, time and material usage. As illustrated in Figure 1.16, the same part can be positioned on the build platform in two different ways. In the first orientation, an overhanging feature is created, which requires support material for successful printing. On the other hand, the second orientation allows the part to be printed without the need for any support material.

Printing support structures consumes both time and material, while also increasing post-processing efforts. In some 3D printing processes, part strength in the build direction is significantly impacted by the bonding strength between layers. Therefore, carefully considering the print orientation is crucial for optimizing print quality, reducing material usage and enhancing part strength.

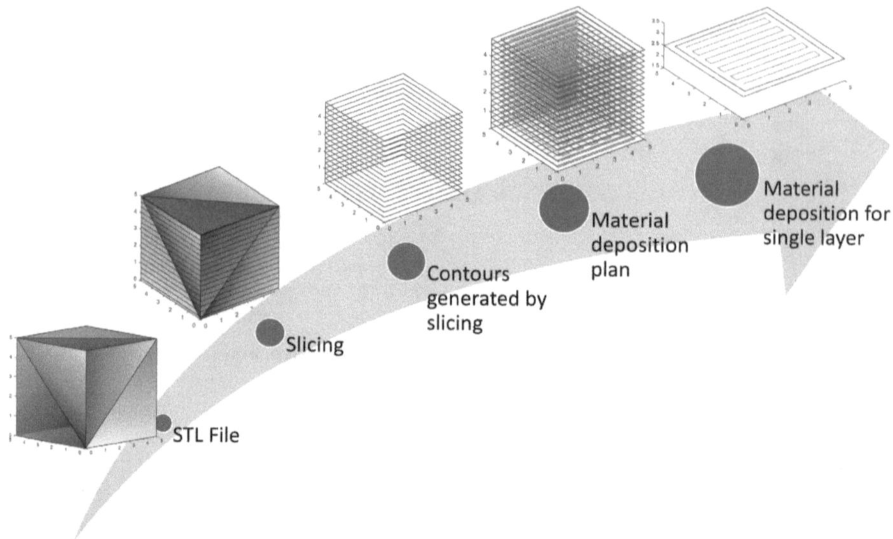

FIGURE 1.17
Common part processing scheme in CAPP software.

For overhanging parts, support structures must be defined. However, overhangs up to a certain angle may not require supports, depending on the process, material and print settings. Once the supports are defined, the part is sliced, converting the STL file into 2D contours. These contours establish the outer boundary for material deposition, as shown in Figure 1.17. The entire deposition scheme, along with machine parameters, is then translated into machine code for 3D printing.

Step 4 – Slicing

- Slicing is the process of splitting a 3D model into horizontal layers, allowing the computer to perform specific tasks.
- Once the slicing is finished, a new file format called G-code is created, with the extension .gcode.
- It transforms 3D models into instructions for the printer.
- Slicing concept will be dealt with in detail in Chapter 4, Section 4.3.

G-code provides the instructions for a 3D printer to construct an object layer by layer. This code can be transferred to the printer through various methods, such as a USB drive, memory card, LAN, USB cable, Wi-Fi or Bluetooth. Once received, the printer reads and executes these commands to build the object as designed.

Step 5 – Printing

The 3D printer reads the G-code instructions and begins the printing process. While many 3D printers can operate without constant supervision, it is recommended to regularly monitor low-cost, open-chamber FDM machines. This careful oversight helps avoid problems such as filament breaks, parts detaching from the bed, nozzle clogs and other potential issues.

Before initiating the print, it is important to properly prepare the printer. Even though many modern 3D printers perform automated checks before printing, manually verifying certain aspects is still a good practice. Some essential checks include the following:

1. Motion of the axes: confirm that the printer's *X*, *Y* and *Z* axes can move smoothly and accurately without any obstruction or misalignment.
2. Material availability: check that the printing material, often in the form of a filament or resin or powder, is loaded correctly and sufficiently for the printing job.
3. Support material: in some cases support structures may be necessary for overhanging parts of the object. Ensure that the support material is available and properly loaded into the printer.
4. Build platform adhesion: verify that the build platform is clean and level, ensuring proper adhesion of the printed object. Some printers may require use of adhesion aids like glue, hairspray or painter's tape to help the object stick to the build platform.

In the case of wire FFF, additional checks should be performed:

5. Filament feed: ensure that the filament is not entangled and is fed appropriately to the nozzle without any undue stress. This will help avoid issues like wire breakage or uneven cuts during the FDM process.

By performing these checks before starting the print job, one can minimize the risk of print failures and improve the overall quality of the final object.

Step 6 – Removal

The process of removing printed parts from the build platform is generally simpler with desktop 3D printers than with industrial-grade machines, which often require specialized training. For example, handling parts produced using SLM can be particularly critical due to the small metal particles present, which pose health risks if inhaled. Managing such equipment typically involves operating in controlled environments, necessitating additional expertise to ensure safety.

Part removal from industrial machines may involve interacting with machinery in a way that requires strict adherence to safety protocols. For instance, safety interlocks are essential to ensure that operating temperatures have cooled to safe levels or that no moving parts remain active during the removal process. These safety measures, when followed correctly, help to mitigate the risk of accidents and injuries during this critical stage.

Step 7 – Post-processing

- Depending on the technology and material utilized, the post-processing of 3D printed objects may differ. (This will be dealt in detail in Chapter 9.)
- For some machines, parts can be used directly without any post-processing, while other 3D printing processes necessitate post-processing (cleaning, finishing, grinding, painting, curing, surface treatment, etc.). Furthermore, the need for post-processing depends on the part's design, preprocessing and intended use.

A 3D-printed part often needs post-processing, such as removing supports and cleaning. In the FFF process, soluble materials can be used for support structures, which dissolve easily in water or a specific solvent without damaging the main part. On the other hand, parts produced through powder-based 3D printing may only require the removal of any excess powder from the surface.

Certain processes involve heat treatment or other specialized post-processing steps depending on the method of 3D printing. For metal 3D printing, specific surface treatments are necessary to handle residual stress, along with heat treatment to enhance the part's strength. Additionally, wire EDM is used to separate the part from the build platform in metal additive manufacturing. 3D printing follows a specific process chain that can vary slightly depending on the machine, software, material and parameters in use. Each of these factors significantly impacts the final product. Without careful planning, one risks wasting material, time and resources.

In summary, this chapter offers a thorough exploration of the 3D printing process, from CAD modelling through to post-processing. It has highlighted the critical importance of understanding how each phase interacts to achieve high-quality, and efficient results. As 3D printing technology evolves and becomes more prevalent across industries, mastering this process chain is essential for both professionals and hobbyists. By fine-tuning each step – whether it is CAD modelling, slicing, material choice or machine settings – 3D printing can reach its full potential and foster new manufacturing innovations.

This chapter lays the groundwork for comprehending the 3D printing process chain, encouraging readers to dive deeper into the finer details of this evolving technology and remain informed of emerging best practices.

Self-Check Exercises

1. Differentiate between subtractive and additive manufacturing processes with one example.
2. Prepare a list of items that are conventionally being prepared by subtractive manufacturing processes.
3. Compare material jetting and binder jetting additive manufacturing processes.
4. Compare binder jetting and selective laser sintering manufacturing processes.
5. Perform a web survey to identify parts that can be produced by each of the 3D printing processes.
6. Depending on the complexity of the object and the resources available, describe the method you would use to obtain the geometric data of the object, such as manual measurement or 3D scanning. Explain your choice and the potential challenges or limitations of the chosen method.
7. Choose a household object with moderate complexity (e.g., a kitchen utensil, a small tool or a decorative item) that you believe would benefit from reverse engineering and 3D printing. Describe the object and explain why you selected it for this assignment.
8. List the steps required in a typical 3D printing method.
9. List the form (wire, sheet, powder, liquid) of raw material and type of energy source required in each of the 3D printing process.
10. Differentiate between native and neutral file formats. Also give three examples of each format.

Discussion Topic

1. Discuss the use of 3D printing processes in the medical domain.

Multiple Choice Questions

1. Which of the following additive manufacturing processes is primarily used to repair or add additional material to existing components?
   (a) Material extrusion
   (b) Vat polymerization
   (c) Powder fusion
   (d) Directed energy deposition
2. In the context of 3D printing, what does FDM stand for?
   (a) Fused deposition modelling
   (b) Fused digital manufacturing
   (c) Flexible deposition modelling
   (d) Fast design manufacturing
3. Which additive manufacturing process uses a light source to cure a photopolymer resin in a vat selectively?
   (a) Material extrusion
   (b) Vat polymerization
   (c) Powder fusion
   (d) Directed energy deposition
4. What is the main difference between material jetting and binder jetting in 3D printing?
   (a) Material jetting deposits the build materials that make up the final part, while binder jetting deposits liquid adhesive onto another substrate.
   (b) Material jetting uses a laser to sinter powder, while binder jetting does not use heat.
   (c) Material jetting is used for industrial applications, while binder jetting is used for consumer products.
   (d) Material jetting is primarily used for metal printing, while binder jetting is used for plastic printing.
5. Which of the following materials is NOT commonly used in sheet lamination processes?
   (a) Paper
   (b) Polymer
   (c) Metal
   (d) Ceramic
6. In powder fusion 3D printing processes, what is the role of the energy source?
   (a) It selectively melts or sinters the material needed for each layer.
   (b) It heats the bed of powder to a consistent temperature.
   (c) It moves the print-head over the powder surface.
   (d) It spreads a thin layer of powder across a build surface.

7. Which of the following additive manufacturing processes has the least additive resolution but offers low costs and faster manufacturing time?
   (a) Material extrusion
   (b) Vat polymerization
   (c) Powder fusion
   (d) Sheet lamination
8. What is the primary use of UAM in the context of sheet lamination?
   (a) Bonding the layers together using heat
   (b) Bonding the layers together using sound
   (c) Cutting the layers into shape
   (d) Spreading the layers onto the build platform
9. Which of the following 3D printing technologies uses a focused energy source, such as a laser, electron beam or plasma, to melt the material?
   (a) Material extrusion
   (b) Vat polymerization
   (c) Powder fusion
   (d) Directed energy deposition
10. Which additive manufacturing process uses a recoater blade or roller to spread a thin layer of powder across a build surface?
    (a) Material extrusion
    (b) Vat polymerization
    (c) Powder fusion
    (d) Material jetting
11. How does the UAM process differ from SDL and CAM-LEM in sheet lamination 3D printing?
    (a) UAM uses sound to bond material layers together
    (b) UAM cuts the layers into shape before bonding them together
    (c) UAM requires a separate curing process after bonding the layers together
    (d) UAM uses heat to bond material layers together
12. Which file format is commonly used to convert a CAD file into a format that can be understood by an FDM 3D printer?
    (a) .PDF
    (b) .JPEG
    (c) .STL
    (d) .OBJ
13. Which one common post-processing step is required for parts printed using the SLS process?
    (a) Curing under UV light
    (b) Sintering in a furnace
    (c) Removal of excess powder
    (d) Application of a liquid binder

14. In SLS, what energy source is used to fuse powdered material?
    (a) UV light
    (b) Infrared (IR) light
    (c) Laser beam
    (d) Electron beam
15. Assertion (A): FDM is a popular 3D printing process for creating functional prototypes and end-use parts.
    Reason (R): FDM uses a wide range of thermoplastic materials that offer durability and strength.

Choose the correct option:

(a) Both A and R are true, and R is the correct explanation for A
(b) Both A and R are true, but R is not the correct explanation for A
(c) A is true, but R is false
(d) A is false, but R is true

Answers

1. (d) Directed energy deposition
2. (a) Fused deposition modelling
3. (b) Vat polymerization
4. (a) Material jetting deposits the build materials that make up the final part, while binder jetting deposits liquid adhesive onto another substrate
5. (d) Ceramic
6. (a) It selectively melts or sinters the material needed for each layer
7. (d) Sheet lamination
8. (b) Bonding layers together using sound
9. (d) Directed energy deposition
10. (c) Powder fusion
11. (a) UAM uses sound to bond material layers together
12. (c) .STL
13. (c) Removal of excess powder
14. (c) Laser beam
15. (a) Both A and R are true, and R is the correct explanation for A

2

Hardware and Software for 3D Printing

To cater to the diverse needs of users, various types of 3D printers are available in the market, differing in size, shape and methodology. These printers range from compact desktop devices, ideal for creating toys and prototypes, to sophisticated industrial machines designed for large-scale production. This chapter primarily focuses on various parts of a typical 3D printer, outlines generic specification of various types of 3D printers and finally illustrates the assembly of an open-source FDM 3D printer. Further, it discusses the various software required for 3D printing and factors that affect accuracy, precision and tolerance of 3D printed parts. The chapter comprises the following main sections:

2.1 Main Parts of a 3D Printer

2.2 Accuracy, Precision and Tolerance in 3D Printing

2.3 Software Required for 3D Printing

2.4 Generic Specification of Popular 3D Printers

After reading through the material, self-check exercises, objective questions and discussion topics presented in this chapter, the learner will be able to do the following:

(a) Identify various parts of a typical 3D printer.
(b) Assemble an open-source FDM 3D printer.
(c) Identify factors that affect accuracy, precision and tolerance of 3D printed parts.
(d) Identify the types of software required for 3D printing.
(e) Outline generic specifications of various types of 3D printers.

DOI: 10.1201/9781003562344-2

2.1 Main Parts of a 3D Printer

The increasing popularity of 3D printing has made it a versatile tool for creating a wide variety of models drawing in numerous enthusiasts globally. Apart from its ability to produce intricate designs, 3D printing technology is also capable of making functional components. To meet the diverse needs of users, the market offers various types of 3D printers, differing in size, shape and methodology.

The printers available vary from small desktop units, which are great for making prototypes and toys, to advanced industrial machines built for large-scale production. While there is a wide range of printers available, their fundamental components and functions are quite similar across different models. Generally, 3D printers can be classified as commercial/industrial or open source.

This chapter will mainly focus on three popular 3D printing technologies: fused deposition modelling (FDM), stereolithography (SLA) and selective laser sintering (SLS).

2.1.1 Construction Details of a Typical 3D Printer

A typical 3D printer comprises of following common components:

2.1.1.1 Motherboard or Controller Board

The primary component of a 3D printer is the motherboard, which is also referred to as the controller board. This essential part acts as the main connection point for all electronic parts found within the 3D printer. As the core of the machine, the controller board sends instructions to various components, such as motors, to ensure they operate together effectively.

A crucial role of the controller board is to control the temperature at which plastic filaments melt. This process allows the melted material to take the desired shape during the printing procedure. Additionally, the motherboard processes G-codes from the slicer software and communicates these commands to the printer components responsible for carrying out the printing task. In the following sections, we will outline the essential components of a 3D printer motherboard:

- The processor interprets the software code and generates a sequence of commands for the other components to carry out, leading to the production of a part. Processors on 3D printer motherboards come in two types: 8-bit and 32-bit.
- Connectors establish links between all the electronic components of a 3D printer and may consist of terminal blocks, DuPont connectors,

or USB connections. Controller boards can be either wired or wireless. A wired controller board has a USB port for external connections, while a wireless board utilizes Wi-Fi connectivity to communicate with other devices.

- Stepper drivers regulate the power sent to the individual coils within a stepper motor, enabling the motor shaft to rotate. A typical 3D printer contains four stepper drivers: one for each of the *x*-, *y*- and *z*-axes, and one for the filament extruder
- For printing a part, G-code instructions are essential. The required program from computer/laptop can be transmitted through various means such as Wi-Fi, a hardwired connection, or an SD card slot to the printer.
- The motherboard comes with preloaded firmware (software) that enables it to direct the printer in performing various tasks, such as initiating a print job or moving along an axis. Modern main-boards function as small computers equipped with the necessary software.

A motherboard usually takes the form of a flat, rectangular board with numerous metallic designs and wires connected to it. Important features of a motherboard include power and data connectors, capacitors, heat sinks and cooling fans. There may also be screw holes for attaching new components or securing the board within a device. Additionally, expansion slots may be available for future upgrades. Components that connect to the motherboard, either through wires or directly, are often referred to as motherboard components. These comprise of the following:

- Optical drives, such as DVD and CD-ROM
- Video cards and GPUs
- Sound cards
- Hard drives (SSD or HDD)
- Processors (CPU)
- Memory sticks (RAM)

2.1.1.2 Printer Software

The software can be categorized into three distinct types:

1. Software for editing and modifying the firmware that governs the electronics and motors of the 3D printer

   (This will be discussed in this section and Section 2.1.4)
2. Software for crafting solid models (CAD software)

   (This will be covered in detail in Chapter 3, Section 3.1)

3. Software for generating sliced models from solid models and creating tool paths (G-codes)
(This will be covered in Section 2.3 of this chapter and also in Chapter 4, Sections 4.2, 4.3 and 4.4)

It should be noted that certain 3D printer brands come with preinstalled firmware, such as MakerBot Desktop for MakerBot products and PreForm for Formlabs products. However, there are open-source options like Marlin available as well. Firmware is installed on the printer's motherboard and plays a vital role in managing various components such as the *X*-, *Y*- and *Z*-axes, as well as the motor that controls the speed of the filament entering the hot extruder head. It also controls the temperature of the extruder head and ensures proper alignment and temperature settings based on the provided code.

2.1.1.3 PSU or Power Supply Unit

The power supply unit (PSU) is responsible for providing power to all electronic and electrical components within the printer. It acts as a transformer and rectifier, converting AC power to DC and distributing the power to relevant components. The PSU can be mounted on the printer's frame or housed in a separate box as an independent unit.

When selecting a PSU, the output voltage is a crucial consideration. Generally, power supply units offer 12V or 24V DC output voltage. To choose the appropriate PSU for your printer, check the voltage requirements of the printer's hot end and select a corresponding power supply unit.

2.1.1.4 Frame

The frame of a 3D printer provides structural support for the electronic, electrical and mechanical components, as well as various 3D printer parts, such as stepper motors and the motherboard. It establishes the base for all the components to work together smoothly during the printing process. Common materials used to build frames include aluminium, plastic (acrylic sheets) and steel. A more rigid frame allows the printer to operate at higher speeds while maintaining precision and quality. Most 3D printers have three axes of rectilinear motion, with the extruder able to move horizontally along the *x*-axis and vertically along the *z*-axis. The printer bed enables the third axis of motion, moving horizontally along the *y*-axis.

2.1.1.5 User Interface/LCD Display

In order to operate a 3D printer, a user interface is essential. Most 3D printers are equipped with a basic screen (LCD display) along with buttons and dials. These components allow users to choose a model to print from an SD card, Pendrive and HDD of connected laptop/desktop and make simple adjustments to settings such as extruder and bed temperature. More advanced printers may feature multi-colour touch screens and user-friendly interfaces. As an alternative, a printer can be linked to a computer through a wired connection or Wi-Fi and can be managed using either OEM or third-party software to monitor the printing progress.

The LCD display controller facilitates 3D printing without the need to connect to a computer or utilize host software like Cura. It relies on an SD card to interpret G-code instructions and is well-suited for everyday printing, handling the majority of print jobs.

2.1.1.6 Connectivity

The connectivity of a 3D printer pertains to how users transmit G-code printing instructions to the controller. There are three main ways to connect to a 3D printer: using an SD card, a wired cable or Wi-Fi. Most printers offer SD card and cable connection options, while Wi-Fi connectivity is usually available in more sophisticated models.

2.1.1.7 Stepper Motors and Transmission Components

The device responsible for moving the printer hot end along the three axes needed for printing a part is the motion controller. Stepper motors assist in moving the printer head, printer bed, rods and screws. Stepper motors are chosen for their ability to maintain precise positioning, even though they produce rotary motion. This rotary motion is converted to linear motion using a power screw or a toothed belt. Typically, the *z*-axis is driven with a power screw, while the *x*- and *y*-axes can be driven with toothed belts. NEMA-17 stepper motors are commonly used in 3D printers and are available in three variations based on their torque rating: 20–25 (for low torque applications), 40–45 (standard choice motors – large torque applications) and 50–56 N-cm (the most powerful stepper motors).

The set of belts inside a 3D printer is the second motion component. Belts fit over the drive gears connected to the stepper motor, allowing smooth movement. There are two types of belts: a long strip that requires connectors to hold both ends, or a circular piece that requires no such connectors. Small teeth on the belt's surface enable synchronization with the stepper motors. The typical belt used is a rubber belt with a width of 6 mm, height of 1.38 mm, and tooth height of 0.75 mm. Belts enable movement in the *x*- and *y*-axes, while threaded rods cater to the *z*-axis.

2.1.1.8 End Stops

End stops serve as a reference point for the printer, helping it to establish its position and avoid exceeding its designated boundaries. They are essential for keeping the printer on track and preventing any off-track movements. This crucial 3D printing component enables the printer to secure an object along an axis, thus preventing it from reaching the end of the axis and potentially causing damage. While end stop switches are commonly used, especially in budget-friendly devices, there are other types of end stops available, including optical and electromagnetic varieties. These end stops cater to different accuracy requirements and printer costs.

2.1.1.9 Print Bed

The print bed serves as the platform on which the printed object is fabricated. Print bed dimensions can vary from 200 × 200 mm to 1 m × 1 m. They are typically mobile along the *y*-axis and may incorporate detachable panels to improve print adhesion and facilitate the removal of parts post-printing. Print beds can be constructed from glass or aluminium, with aluminium being the more prevalent choice due to its rapid heating properties.

Print bed levelling, achieved through integrated thumbwheels beneath the bed, ensures a seamless printing process and delivers a flawless final product. The print bed surface, being the uppermost layer, assists in the uncomplicated removal of the printed object and prevents adhesion. An effective print bed surface should facilitate proper adherence and minimize warping issues resulting from uneven cooling.

While aluminium print beds address warping, they are susceptible to expansion when heated, resulting in an uneven surface over time. Conversely, glass print beds do not face this concern but are less proficient at conducting heat, rendering them susceptible to warping. Furthermore, glass is fragile and can break easily if handled carelessly.

2.1.1.10 Print-Head

This is different for different printers based on the energy source and raw material they utilize. Further, printer head or equivalent setup in all the 3D printing methods/printers receives the coordinates of the part geometry through G-code file generated by printer software to finally create the object.

- ***FDM printers:*** the print-head is the component where the filament enters, melts and comes out to create the desired object. It has two main parts: the cold end and the hot end. The cold end, located at the top of the print-head, remains cool as its name suggests. The filament passes through the cold end and proceeds towards the hot end. A motor attached to the print-head enables the filament to move from the cold end to the hot end.

The hot end features a nozzle through which the melted filament flows. Prior to the nozzle, there is a heating chamber that turns the solid filament into liquid. The solid filament's forward motion pushes the liquid filament out of the nozzle based on the controller board's instructions. Nozzles come in various sizes, with the most common being the 0.4 mm brass nozzle. Other nozzle materials include stainless steel, ruby and tungsten carbide, each offering corrosion resistance and unique advantages. Popular print-heads are Zesty Nimble, Bondtech, Flexion, E3D Titan Aero, Slice Engineering Mosquito, DisTech Automation Prometheus V2 and E3D V6 etc.

- ***SLA and DLP printer:*** both SLA and DLP processes create solid objects by selectively exposing liquid resin to a light source. SLA uses a laser, while DLP employs a projector, to form extremely thin solid plastic layers that stack up to build a solid object. Unlike FDM printers, these printers do not have a print-head. Instead, a laser points at two mirror galvanometers that direct the light to the correct coordinates (based on G-code) on a series of mirrors. The light is then focused upwards through the tank's bottom, curing a layer of resin.

  The printing process begins when the build platform descends into a resin tank, leaving a space equal to the layer height between the build platform (or the last completed layer) and the tank's bottom. After curing, the layer detaches from the tank's bottom, and the build platform moves up, allowing fresh resin to flow beneath. This process repeats until the print is complete.
- ***SLS printer:*** unlike FDM printers, these printers do not have a traditional print-head. Instead, they utilize a laser that is directed by two mirror galvanometers to pinpoint the correct coordinates based on G-code. The laser light is then focused onto a series of mirrors and directed downward onto the powder layer, effectively fusing or melting it layer by layer.
- ***Material jetting printer:*** this type of 3D printer uses a print-head similar to inkjet printers and a UV light source for curing the jetted material. Unlike FDM machines, some printers, such as Stratasys, do not require post-curing. Instead of a continuous stream of material, droplets are dispensed only when necessary, released by a pressure change in the nozzle caused by thermal or piezoelectric actuators. Thermal actuators deposit droplets rapidly, using a thin film resistor to form the droplet. The piezoelectric method is often preferred as it allows for a wider range of materials to be used.
- ***Binder jetting printer:*** in contrast to the extruder of an FDM machine, this type of printer utilizes a print-head similar to inkjet printers. Binder jet 3D printing (BJ3DP) is a non-beam-based additive manufacturing (AM) method that involves jetting a liquid binder (typically a polymeric

liquid) onto layers of powdered materials, selectively joining them, and then undergoing a densification/curing process.

A carriage equipped with inkjet nozzles, akin to those found in desktop 2D printers, moves over the bed and selectively deposits droplets of a binding agent (glue) that bind the powder particles together. In full-colour binder jetting, coloured ink is also deposited during this step. Each drop is approximately 80 µm in diameter, ensuring high resolution. Binder jetting is a multi-step process that requires post-processing steps, such as curing and densification. Materials commonly used in binder jetting include metals, sand and ceramics, which are provided in powder form.

- ***Direct energy deposition printer:*** this particular type of printer features a print-head similar to FDM, but with a large nozzle replacing the extruder. The nozzle is equipped with a supply of raw material (powder or wire) and a heat source to melt the material at the tip. In some designs, the nozzle is mounted on a four- or five-axis robotic arm. The heat source can be a laser, electron beam or plasma arc, and the material feedstock can be metal powder or wire.

  Compared to metal wires, powders generally result in lower deposition efficiency, as only a portion of the total powder is melted and bonded to the substrate. While it is feasible to fabricate parts from scratch using this method, it is often used for industrial applications, such as repairing damaged turbine blades or propellers. This process, known as directed energy deposition (DED), enables the creation, repair or enhancement of parts with complex geometries and customized material properties.
- ***Sheet lamination printer:*** this printer type replaces the FDM print-head with motorized mirrors that guide a laser beam from the source onto the raw material sheet. The laser burns the material based on contour coordinates received as per G-codes, layer by layer.

2.1.1.11 Extruder and Feeder System (FDM Printer Only)

A critical component of 3D printers is the extruder. This part efficiently unspools the plastic filament and guides it through a heated chamber, allowing the material to reach the necessary temperature for printing. Subsequently, a specially designed nozzle deposits the material along a predefined path. To ensure precise control of the material flow through the nozzle, a gear-loaded stepper motor is used. The extruder assembly includes a gear attached to the stepper motor's shaft, which feeds the filament to the nozzle. There are three main types of extruders available for this purpose.

- Direct drive: extruders with direct drives are frequently positioned above the nozzle and hot end. With this kind of extruder, the plastic goes straight into the heated end and then is forced out of the nozzle.

- Bowden: in a Bowden extruder, the extruder is mounted on the printer frame. The extruder transfers the plastic to the hot end via a long, flexible tube. This configurationallows the printer to print faster since there is less weight near the nozzle.
- Dual extruders: Both the above types can be configured as dual extruders to print with multiple filaments simultaneously.

The feeder system comprises a network of tubing that enables filament materials to enter the extruder and exit the heated nozzle. An effective feeding mechanism facilitates multi-colour printing without the need for a dual extruder. The feeder system is a customizable technique that can be integrated into an existing printer.

2.1.1.12 Print Material (FDM Printer Only)

FDM 3D printers can use a variety of materials in the form of filament rolls, which are typically 1.75 mm or 3 mm in diameter. These printers can work with different types of plastics for printing:

- Thermoplastics: PLA, ABS and PETG are widely used thermoplastics known for their versatility and ease of use in 3D printing.
- Thermoplastic elastomers: elastomers, such as TPU, are flexible materials that resemble rubber after printing. However, they can be challenging to print, requiring a direct drive extruder, and are generally more expensive compared to other materials.
- Filled thermoplastics: this category encompasses materials that combine thermoplastics with fillers like metal powder, carbon fiber or sawdust. They are often used to create parts with unique appearances and improved strength.

2.1.2 Other Components and Deviations in SLA Type 3D Printer

In addition to the common components, a standard SLA 3D printer typically consists of four main parts:

- Resin vat: this container holds the liquid photopolymer, commonly referred to as "resin," which is a clear liquid plastic.
- Build platform: positioned above the vat, this platform can be raised or lowered according to the printing process.
- UV laser: this laser is responsible for curing the layers by solidifying the resin as it forms the desired object.
- Control interface: accessible on the printer, this interface manages the movements of both the build platform and the UV laser according to the 3D model's design and layer-by-layer instructions.

It is worth noting that numerous desktop SLA printers use an inverted configuration. In this setup, the laser is aimed upwards at the build platform, which starts off close to the bottom of the vat and gradually rises. This "bottom-up" approach is different from the "top-down" method and provides benefits such as lower resin usage and simpler removal of parts from the build platform.

2.1.3 Case Study on Assembly of an Open-Source FDM Type 3D Printer

Given the high cost and limited information on commercial 3D printers, this section provides instructions for assembling an open-source FDM 3D printer, making it accessible even for student projects. Although 3D printers can seem complex, they are easy to use with the right knowledge. Understanding the function of each component is essential. This guide encompasses the mechanical parts, open-source software, electronic controllers and extruders necessary for constructing an open-source FDM 3D printer. Figure 2.1 illustrates the typical structure of an open-source FDM printer.

2.1.3.1 Procedure in Brief

First, the components are carefully selected (Tables 2.1 and 2.2) and assembled, taking into account the basic framework and structure of the 3D printer. Once the assembly process is completed, the Marlin firmware (or similar software) is uploaded via Arduino software onto the Arduino ATMEGA 2560 AVR microcontroller, which controls the movement of actuators through the RAMPS 1.4 board shield and receives feedback through closed loop control

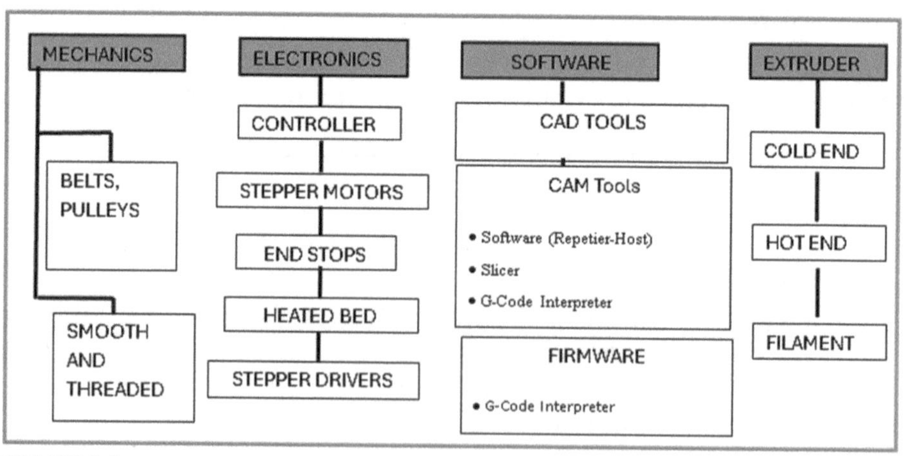

FIGURE 2.1
Structure of an open source 3D printer.

via sensors. Open-source software for computer-aided design (CAD) such as OpenSCAD (or similar) and software for printing like Repetier-Host (or other similar software) are installed on the computer to finalize the printing setup. The final step in the assembly process involves calibration, where the tip of the nozzle is adjusted to 0.4 mm (the diameter of the orifice) above the heated bed to ensure proper adhesion of the first layer to the bed. During the printing setup in Repetier-Host, the baud rate is adjusted according to the programming code in the Marlin firmware for communication between the printer and the computer.

A CAD model (.stl file) is downloaded from www.thingiverse.comor any other similar website with free-to-use CAD models and uploaded after connecting the computer with a printer via a USB port. The model is sliced to obtainan auto-generated report on the Slic3r program like estimated time to print, number of layers, length and volume of the filament used.

2.1.3.2 Selection of Components

In Table 2.1, you will find a list of different components along with their features and selection criteria. These components are essential for building a 3D printer. Table 2.2 outlines the available options for electronic controllers, software chains and extruders, along with brief explanations for the reasons behind each selection.

TABLE 2.1

List of Components and Selection Criteria for an Open-Source FDM-Based 3D Printer

S. No.	Component	Feature and Selection Criteria
1.	**SS RODS SMOOTH** 	**Diameter 8 mm, length 200 mm, 8 Nos** Two stainless steel rods for vertical Z-axis, two for X-axis and two for Y-axis. Stainless steel rods provide strength and do not wear out easily on continuous operation.
2.	**SS RODS THREADED** 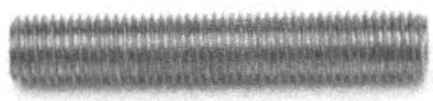	**Diameter 12 mm, length 300 mm, 1 Nos** **Diameter 8 mm, length 300 mm, 1 No** Two threaded stainless steel rods for the vertical Z-axis for quick and forceful movement. Two rods (12 mm) for the Y-axis to keep the bed sturdy during movement.

(*Continued*)

TABLE 2.1 (Continued)

List of Components and Selection Criteria for an Open-Source FDM-Based 3D Printer

S. No.	Component	Feature and Selection Criteria
3.	ACRYLIC SHEET BLACK  	**1200 mm × 900 mm × 5 mm** **600 mm × 50 mm × 10 mm** (Length × Width × Thickness) Quantity 1 each Cutting Charges – 500/- Provides supportive framework for brackets and SS rods. Easy to cut on a CNC machine and aesthetically good.
4.	GT 2 BELT 	**GT 2 6mm Timing Belt 5 metre** Readily available. Round tooth profile makes full flank contact with pulley. Synchronous.
5.	HEX NUT M10 	**16 Nos** I.D 8 mm are used to give linear motion on the *Z*-axis and fixed support to lower acrylic bed on the *Y*-axis.
6.	BALL BEARING 	**807 ZZ, 4 Nos and 624zz 1 No** Axis support for GT 2 belt on the *X*-axis and *Y*-axis and respective motion assistance.

TABLE 2.1 (Continued)

List of Components and Selection Criteria for an Open-Source FDM-Based 3D Printer

S. No.	Component	Feature and Selection Criteria
7.	HEAT BED 	**MK2 RED PCB Heat Bed 12/24 V** Size: 216 mm × 216 mm 35 µm Copper Etching **It remains the limiting size owing to its availability for the same.**
8.	COUPLER 	**Flexible coupling shaft coupler 8 mm to 10 mm 2 Nos** Connects the SS threaded rods (*z*-axis) and stepper motors. Owing to its spring action it helps in minimizing backlash during change of direction
9.	STEPPER MOTOR 	**NEMA 17 Bipolar Stepper Motors – 5 Quantity** Holding Torque – 4.4 Kg-cm Step Angle – 1.8° No. of phase – 2 Rated Voltage – 2.8V Rated Current – 1.68 A Shaft Length – 16mm Width – 43mm Weight – 300 gm Inductance per phase – 2.8mH Resistance per phase – 1.65 E
10.	NYLON TIE 	**8 Nos** Used to tie the belts with the brackets firmly.
11.	WASHER 	**32 Nos** Plain washer with I.D 8/12 mm Prevents wear and tear of acrylic sheet and distributes load on a large area, i.e., decreases stress concentration.
12.	GT 2 PULLEY 	**20 teeth, 5 mm bore GT 2 pulley for GT-2 belt of 6 mm** Smooth transition of *x*-axis carriage and *y*-axis heated bed.

(Continued)

TABLE 2.1 (Continued)

List of Components and Selection Criteria for an Open-Source FDM-Based 3D Printer

S. No.	Component	Feature and Selection Criteria
13.	**ABS BRACKETS** 	**Printed ABS Bracket** Downloaded as STL file from www.thingiverse.com Once downloaded the bracket assemblies are 3D printed from 3D printing service providers. (www.altem.com)
14.	**SCREWS** 	**Fasteners with nut** MS machine screws with pan head and combination drive and nut
15.	**J-HOT EXTRUDER** 	**SS Cylindrical body with brass nozzle of 0.4 mm with Fan** J-Head V2 extruder for PLA and ABS raw material. It has a groove mount for support on a 3D printed bracket.
16.	**ARDUINO ATMEGA 2560** 	Open-source electronic controller board with AVR CPU.
17.	**RAMPS 1.4** 	Electronic shield board to control stepper motors, heating elements via firmware installed on Arduino ATMEGA 2560

TABLE 2.1 (Continued)

List of Components and Selection Criteria for an Open-Source FDM-Based 3D Printer

S. No.	Component	Feature and Selection Criteria
18.	**LCD SMART CONTROLLER** 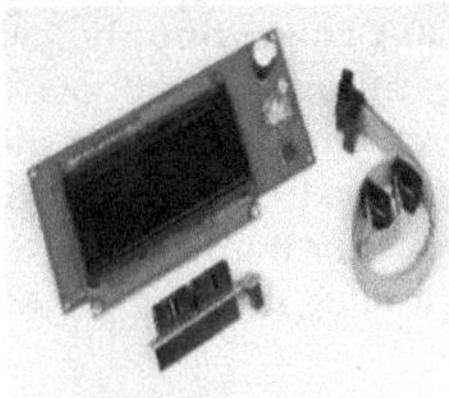	**LCD Blue Light Display Panel with 4 Lines and 20 Characters. Works on 5V** Gives live data during printing and option to control temperature and set motion parameters.
19.	**SD CARD** 	**SanDisk SD Card for storing STL files for printing.** Can be placed in Smart LCD Controller of RepRap to print directly without computer with pre-loaded STL files
20.	**GLASS** 	**216 mm × 216 mm× 4 mm (Length× Width× Thickness) glass mounted on the heat bed surface.** Used in order to give a uniform distance from the tip of the nozzle during calibration.
21.	**BULLDOG CLIPS** 	**Binder clips to hold glass with the heated bed.** Helps in temporary holding the glass with the heated bed.
22.	**USB Wire** 	**USB 2.0 A Male to B Male Printer Cable** Used for Controller-PC communication.

(Continued)

TABLE 2.1 (Continued)

List of Components and Selection Criteria for an Open-Source FDM-Based 3D Printer

S. No.	Component	Feature and Selection Criteria
23.	POWER SUPPLY 	**SMPS (Switched-mode power supply) 12V, 20A Power Supply, 200W Power** The specifications are above the normal requirement to assist any modification like extra extruder or change of controller or transforming 3D printer to CNC milling machine
24.	CONDUCTING WIRE 	For connecting actuators and sensors with the electronic shield board.
25.	THERMISTOR 	**100k ohm NTC 3950 Thermistor** **2 Nos** Thermistors (Thermocouple) are used to measure temperature of both heated bed and extruder. It also gives feedback to controller to assist in pre-heating.
26.	LED BULBS 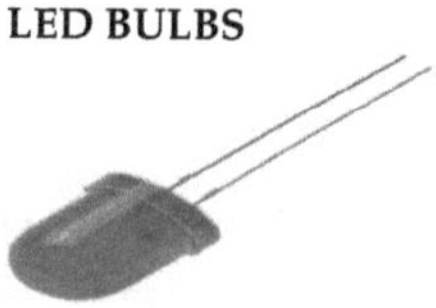	**10mm RED bulb** Can be used as an indication as to whether the printer is ON/OFF.
27.	LIMIT-SWITCH 	**End stops or limit switch. Marks the linear distance for -*x*, -*y* and -*z* direction.** Open-source based end stops to calibrate the end position of all the axis.
28.	LINEAR BEARING 	**LM8UU I.D 8 mm, length 24 mm Linear Ball Bearing** 5 pieces Specifically used as a guide surface (linear motion – LM), low friction with contact seal to ensure stability at high speed.
29.	STEPPER MOTOR DRIVERS 	**A4988 drivers controls the stepper motors.** Allegro Pololu A4988 stepper drivers for stepper motor control

TABLE 2.2

Selection Criteria for Controller, Extruder, Software and Raw Material for an Open-Source FDM-Based 3D Printer

S. No	Component/Software	Options Available	Selection With Reason
1.	**RAW MATERIAL**	• PLA (Poly-lactic acid) • ABS (Acronitrile butadiene styrene) • HIPS • HDPE • LDPE • PVA	**Poly-lactic acid** • Bio-degradable • Availability • Extensively research data available for comparison
2.	**EXTRUDER**	• 1730 Full-Metal Hot End • E3V-D6 • J-Head • Merlin • J-Head Lite V2 • V3 Hybrid • Pico • Prusa V3 • Budasch Nozzle • RRA Hot End	**J-Head Lite V2** • Bowden compatible • Can be used both in direct drive or Bowden configuration. • Circular groove mount for easy removability. • Brass nozzle
3.	**FIRMWARE**	• Marlin • Teacup • Smoothie • Sprinter	**Marlin** • Complete control of calibration • High cornering speed • PID temperature control • Proper arc support
4.	**HOST SOFTWARE**	• Repetier-Host • Pronterface • RepSnapper • Replicator G	**Repetier-Host** • Simple interface • Visual G-code interface • STL composer that allows placing multiple STL objects on one plate
5.	**G-CODE GENERATOR (SLICING SOFTWARE)**	• Slic3r • SFACT • Skeingorge	**Slic3r** • Fast G-code generator • Simple interface • Integrates with Repetier-Host
6.	**CONTROLLER**	• ARDUINO ATMEGA 2560 + RAMPS 1.4 • Raspberry Pi • REPLICAPE • Smoothieboard • DUET • ARDUINO + DICE • Melzi • Sanguinololu • ARDUINO + 3Drag Controller	**ARDUINO ATMEGA 2560 + RAMPS 1.4** • 56 i/o pin configuration • Automatic software reset • USB overcurrent protection

The choice of elements is a crucial aspect of the project and is based on the review of the existing literature, specifications and online information. The cost evaluation of the elements emphasizes the financial feasibility of constructing a 3D printer.

2.1.3.3 Uploading the Firmware to ARDUINO ATMEGA 2560

The next step is to load the firmware Marlin (or any similar software) onto the microcontroller illustrated in Figure 2.2. Before uploading the Marlin firmware, the ARDUINO controller (refer to Figure 2.3) needs to be interfaced with the PC. After downloading and installing the Arduino IDE software, connect the controller with the USB cable to the PC. Open the Arduino software, go to Tools > Board > Arduino ATMEGA 2560 as illustrated in Figure 2.4. This would automatically assign the port number to the controller every time it is connected to the PC. The port number is displayed on the same drop-down tab.

Remember to firmly connect the RAMPS 1.4 add-on shield to the controller. Ensure that all the male pins are connected to the power, analog and digital pins of the controller. Then, connect the device to the PC using the cable. Open the File menu, select "Open," and choose the Marlin firmware downloaded from the GitHub website. Verify the codes and upload them to the controller, following the instructions in Figure 2.5.

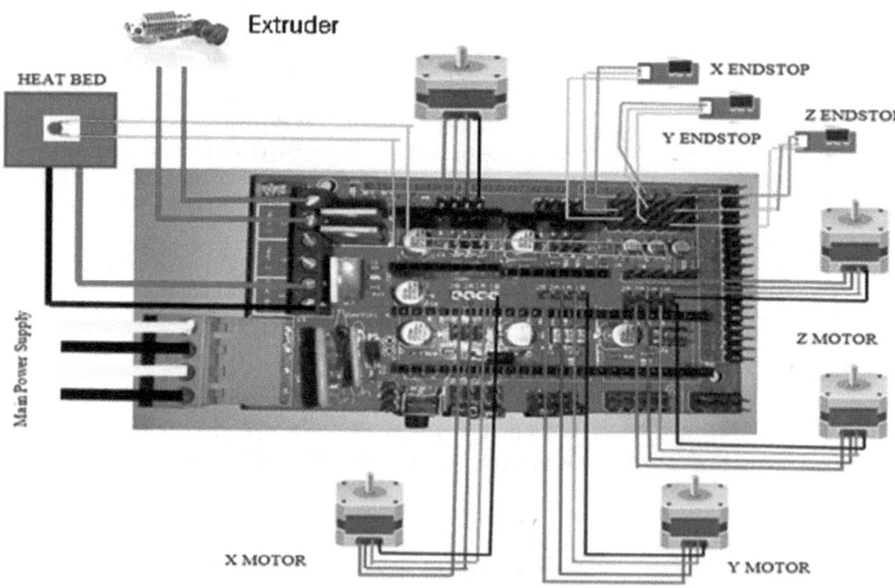

FIGURE 2.2
RAMPS 1.4 pin diagram.

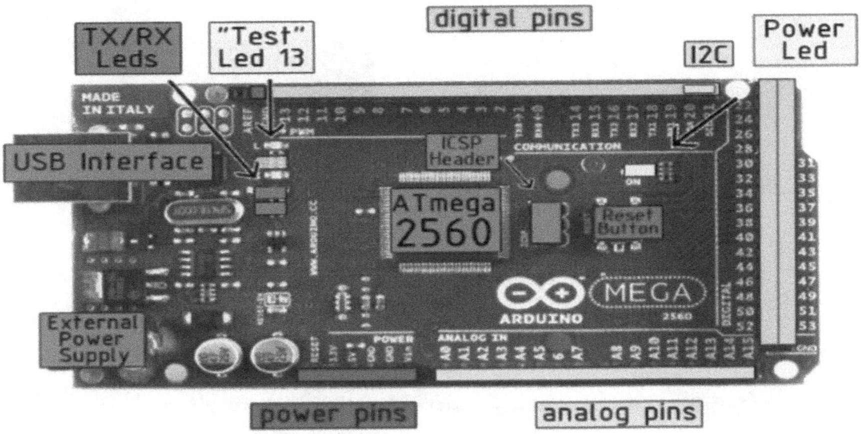

FIGURE 2.3
ARDUINO ATMEGA 2560 (controller).

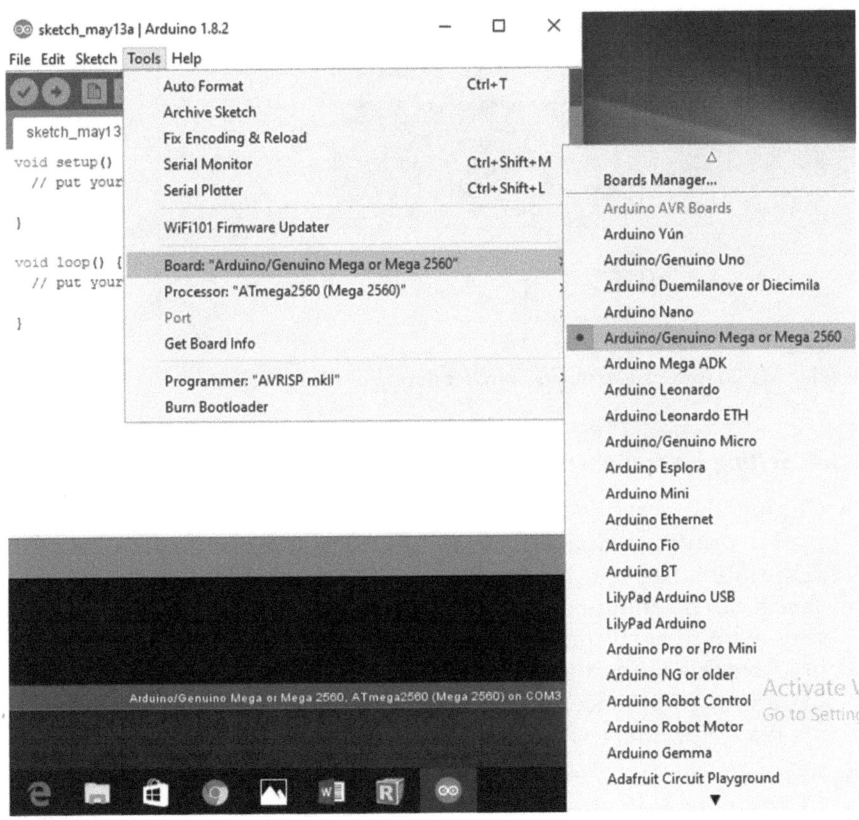

FIGURE 2.4
Selection of controller on Arduino IDE.

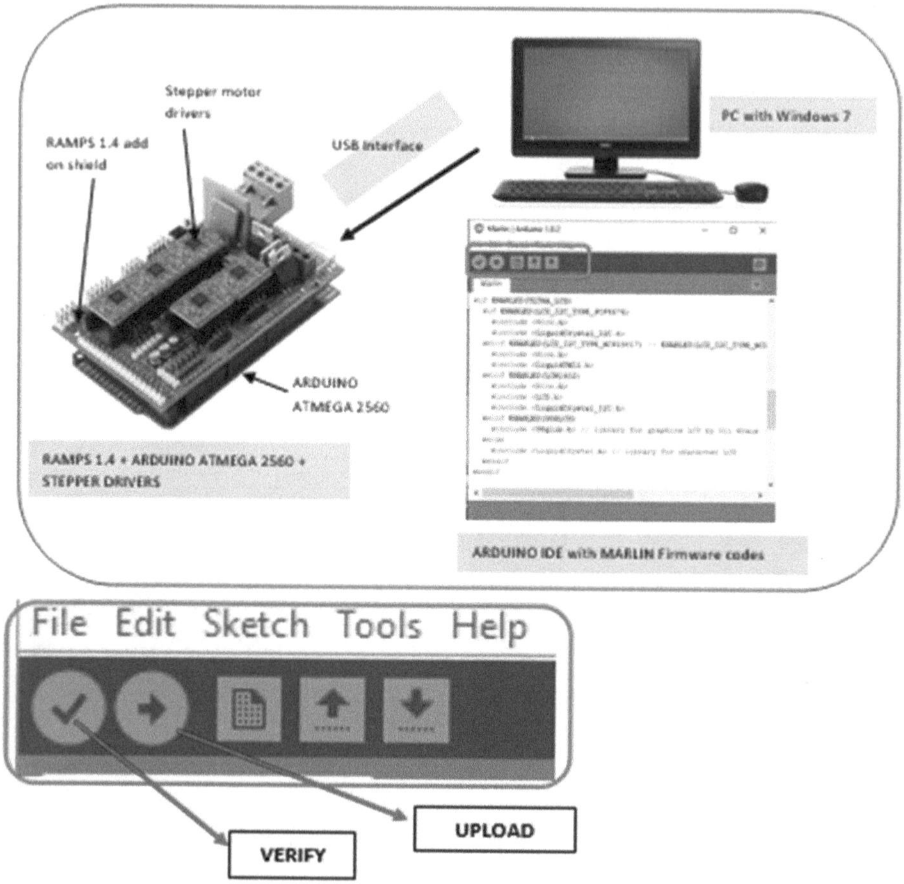

FIGURE 2.5
Uploading Marlin firmware to microcontroller through Arduino IDE.

2.1.3.4 Setting up Framework and Assembly

- Cutting the acrylic sheet and fixing brackets: the brackets designed for 3D printing are available as STL files and can be sent to various 3D printing services for production. The measurements for the acrylic sheets can be obtained from a local glasswork center. Acrylic sheets are chosen for their stiffness, resistance to corrosion, suitability for attaching brackets due to their similar material properties, and most importantly, their aesthetics. Once the acrylic sheet is cut, all the basic components are ready for final assembly.
- Assembly of components: the *X*-, *Y*- and *Z*-axes are represented by Cartesian coordinates, with the *X*-axis as the carriage axis, the *Z*-axis as the vertical axis, and the *Y*-axis as the heated bed axis in the default

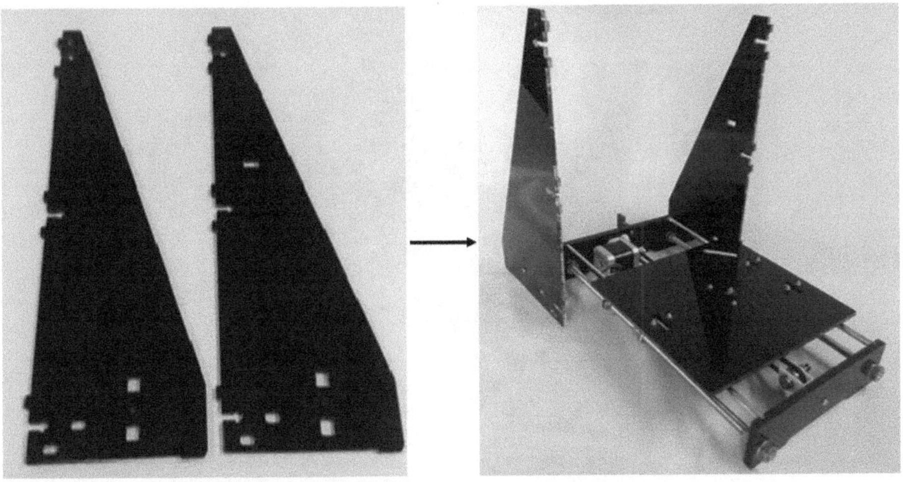

FIGURE 2.6
Mounting power supply and controller to the *Y*-axis.

design. Acrylic sheets of 5 mm and 10 mm thickness are utilized to provide an attractive appearance and sturdy support for mounting mechanical components. Two triangular acrylic sheets are attached to the *Y*-axis support and *X*-axis frame to accommodate the electronic controller and power supply. Figures 2.6, 2.7 and 2.8 depict the framework utilized. The components are assembled according to the basic framework of the Mendel-Prusa using the various tools and equipment listed in Table 2.3.

The *Y*-axis, or the build platform, is securely attached to the frame using 12 mm threaded rods, M10 hex nuts, and washers to evenly distribute the load. The smooth rods slide through linear bearings attached to the lower acrylic sheet platform.

After uploading the firmware to the ARDUINO board, the add-on shield is connected to the board using pins, and the stepper drivers are positioned on the shield. The controller is then attached to the left supporting acrylic sheet as shown in Figure 2.8, with pin hole markings for drilling on the sheet. The assembly of the 3D printer will require the following tools and equipment (Table 2.3). After uploading the firmware to the ARDUINO board, connect the add-on shield to the board using pins and position the stepper drivers on the shield. Next, mount the controller on the left acrylic sheet for support (see Figure 2.8). Mark the pin holes on the sheet for drilling. You will need the following tools and equipment (Table 2.3) to assemble the 3D printer. The assemble views with labeling are shown through Figures 2.9 to 2.11.

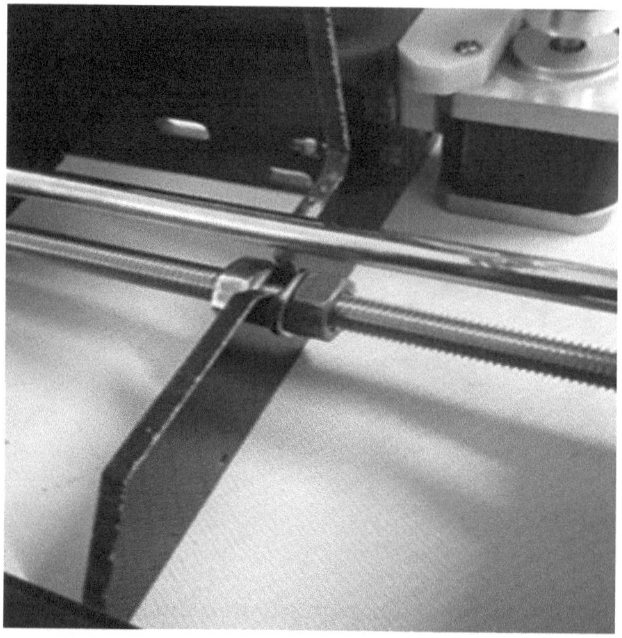

FIGURE 2.7
Fixing the *Y*-axis to the front frame.

2.1.3.5 Calibration

Once all the components are assembled the end stops (limit switch) and *Z*-axis calibration is performed to check the printing. Calibration of end stops defines the boundary for the *X*-, *Y*- and *Z*-axes (refer to Figure 2.12), which together make up the build volume of the 3D printer. One of the constraints here is the size of the heated bed 210 mm × 210 mm that constrains build volume of the parts. The *Z*-axis calibration is done to accurately position the orifice of the nozzle from the bed for uniform deposition of extruded thermoplastic material. To avoid any waviness in the part or during the printing process a 2.5 mm thick glass is fixed with the help of bulldog clips onto the heated bed. The glass being a flat and smooth surface helps in getting proper adhesion of the extruded material.

It is to be noted that the nozzle of the extruder needs to be positioned over the bed surface in such a manner that the distance apart is 0.3–0.4 mm.

2.1.3.6 Interfacing 3D Printer with Computer

The firmware has been uploaded to the microcontroller, so the calibration can be performed using the LCD controller. To connect the 3D printer to the PC, the Repetier-Host software has been installed on the same PC. The following steps are then taken to establish the connection:

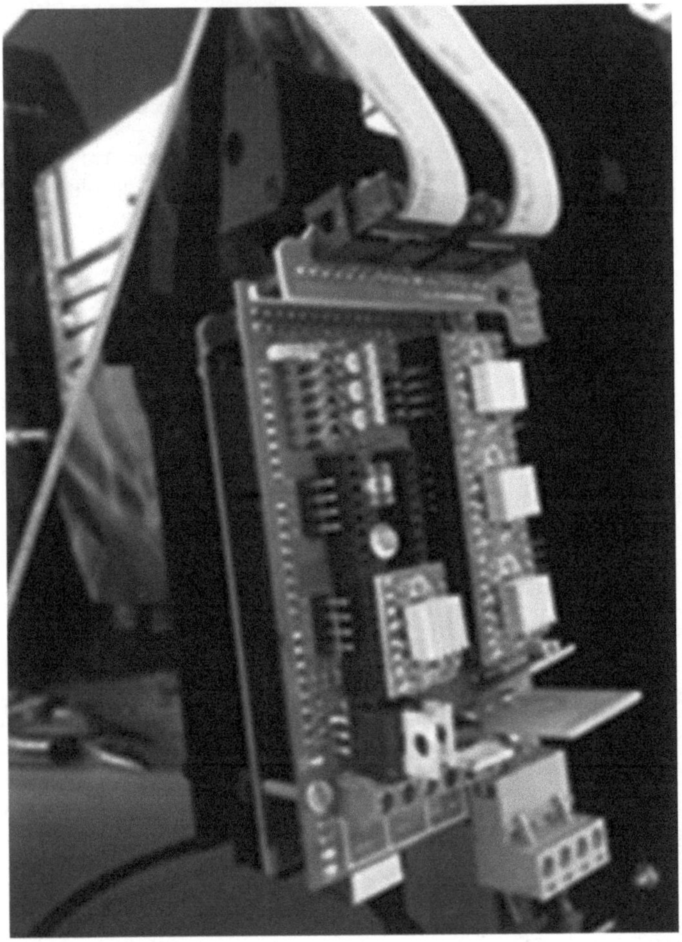

FIGURE 2.8
Mounting controller on left acrylic sheet.

(a) Software installation: Repetier-Host, Arduino and Slicer.
(b) Once the COM ports are found in the settings, the printer is connected to the computer via USB. In Arduino, the board ATMEGA is selected.
(c) The printer is powered on.
(d) The .stl file of the part to be printed is downloaded.
(e) The printer is linked to the software using the tab on the top left-hand side, which will turn green.
(f) The downloaded .stl file is loaded into the software.
(g) The print command is given after confirming the orientation on the heated bed.

TABLE 2.3

Tools and Equipment used for Assembly and Calibration of an FDM-Based 3D Printer

S. No.	Tool-Equipment	Specification
1.	**ALLEN KEYS** 	**1.5 mm, 2 mm and 2.5 mm** Used for couplers and extruder gear.
2.	**DOUBLE END OPEN SPANNER** 	**12×13, 14×15** Used to tighten M10 Hex Nuts with the partitioned acrylic sheet.
3.	**SCREW DRIVER SET** 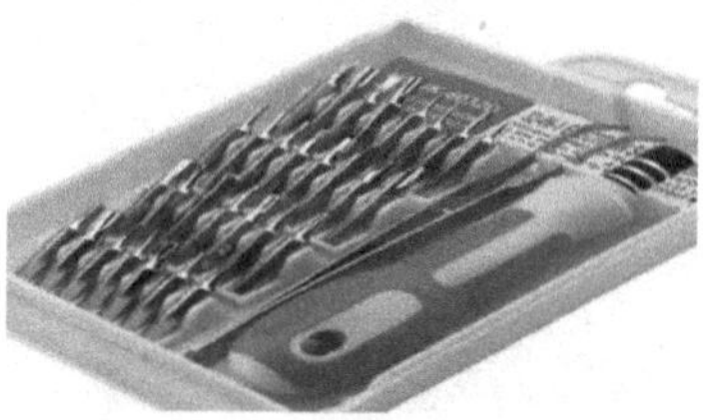	**32 pieces' screw driver set** Individual pieces to tighten screws and fasteners in couplers, joining brackets with acrylic sheet, stepper motors with brackets, heat bed with lower acrylic bed and controllers with acrylic sheet.
4.	**SOLDERING ROD cum WIRE** 	**240 V 60W Soldering rod** For soldering heated bed with wires and connecting to the controller. It can also be used to make DIY controllers.
5.	**SS RODS CUTTER** 	Available at local hardware shop. The cutter shears the rods as per the length required and later grinds the end points to avoid any cuts and bruises during assembly.

TABLE 2.3 (Continued)

Tools and Equipment used for Assembly and Calibration of an FDM-Based 3D Printer

S. No.	Tool-Equipment	Specification
6.	**EMERY PAPER** 	**Grit Size 80** It is used to polish the surface of the acrylic sheet and glass that is to be mounted on the heated bed.
7.	**VERNIER CALLIPER** 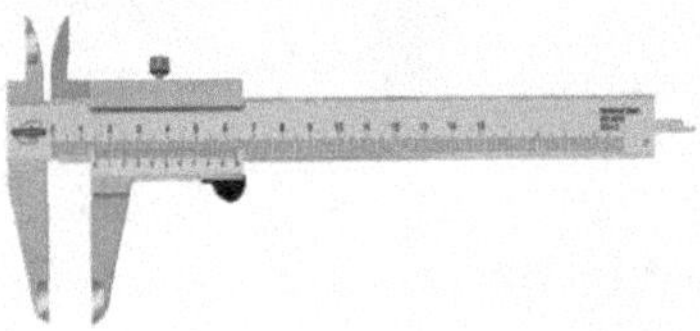	**Vernier Calliper** Used for calibration of the Z-axis.

2.2 Accuracy, Precision and Tolerance in 3D Printing

During its early development stages, 3D printing was mainly used to produce prototypes, leading to its original label as "rapid prototyping". However, as the technology progressed, it began to be used for manufacturing final products, resulting in its current name, "additive manufacturing". When discussing the manufacturing process, it is important to ensure that the end product can fulfill its intended function as designed. This necessitates the manufacturing of the product with precise dimensions and structure, especially when producing numerous parts, as precision is of utmost importance.

Additive manufacturing is also frequently used to create replacement parts for failed components in an assembly. Therefore, it is crucial for 3D printed parts to undergo accuracy and precision tests to guarantee their functionality. In certain cases, complete assemblies can be built using a 3D printer. In such instances, it is vital for each part to be manufactured within specific tolerance ranges to ensure interchangeability and compatibility with other components.

It is crucial to understand that a 3D printer's "high resolution" specification does not automatically ensure the accuracy or precision of the printed parts. Therefore, a comprehensive evaluation and testing of the quality and

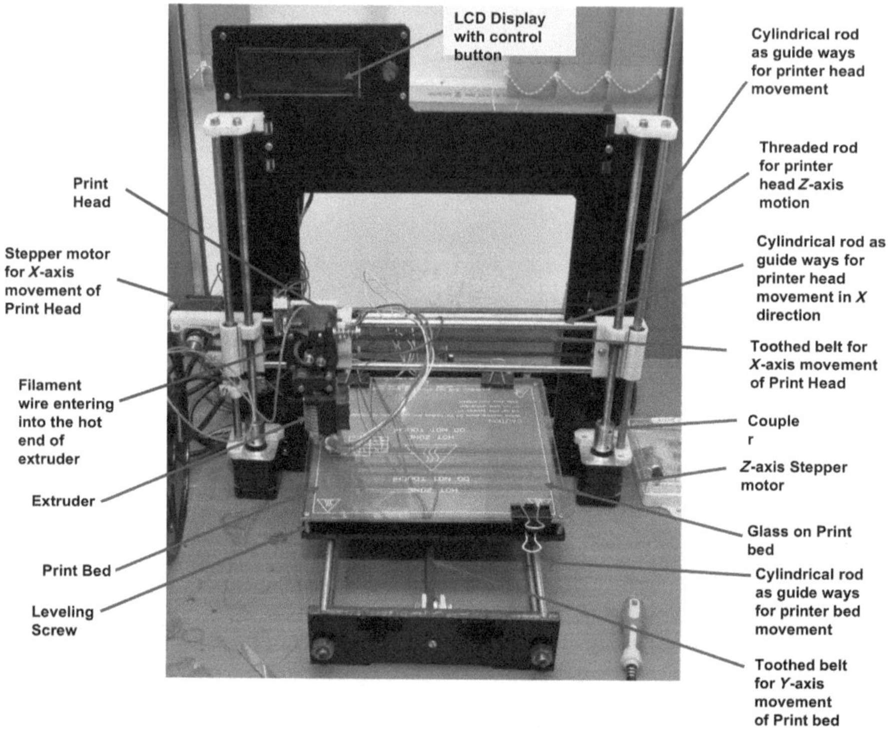

FIGURE 2.9
Assembled open-source FDM type 3D printer.

performance of each 3D printed component is essential to achieve optimal results in the manufacturing process.

In this section, we will discuss accuracy, precision and tolerance in 3D printing. We assume that the learners are well-informed about the basics of accuracy, precision and tolerance (for example, from an Engineering Metrology Course).

2.2.1 Definition

It is crucial to understand the concepts of accuracy, precision and tolerance (AP&T) in the domains of engineering, manufacturing and measurement. These concepts are defined as follows:

Accuracy: accuracy refers to the closeness of a measured value to the actual or true value of the quantity being measured. In simpler terms, accuracy indicates how correct a measurement is in comparison to the real-world value. High accuracy implies that the measured values are

FIGURE 2.10
Assembled open-source FDM type 3D printer (another view).

close to the true values, while low accuracy indicates a larger difference between the measured and actual values.

Precision: precision is the measure of consistency among a set of measurements of the same quantity. It illustrates the degree of variation among multiple measurements. Higher precision suggests that the measurements are closer to each other, while lower precision indicates a larger spread or inconsistency among the measurements. It is important to note that a precise set of measurements may not necessarily be accurate, meaning that the measurements can consistently deviate from the true value even if they are precise.

Tolerance: tolerance encompasses the acceptable range or limits of a physical dimension, property or value for a specific purpose or function. In the realm of manufacturing and engineering, tolerance denotes the permissible variation in dimensions, shape or other characteristics of an object or part. This allows manufactured parts to fit and function together properly, even in the presence of minor variations in their dimensions due to manufacturing errors or inconsistencies. Tighter

FIGURE 2.11
Closure view of print head with extruder and *X*-axis stepper motor; print bed with *Y*-axis stepper motor; print head with two *Z*-axis stepper motors on the assembled open-source FDM type 3D printer.

FIGURE 2.12
Position of limit switches for calibration on *x*- , *z*- and *y*-axis.

tolerances indicate a smaller allowable range of variation, leading to higher quality and more consistent parts. Conversely, looser tolerances permit a larger range of variation and may prove to be more cost-effective in certain scenarios.

Maintaining optimal tolerance is crucial as tighter tolerances can result in higher production costs.

Figure 2.13 demonstrates the concepts of accuracy, precision and tolerance in relation to the measured values. Within the plot, there are three distinct sets of data points, each representing a different scenario:

(a) Data points marked with red circles represent accurate and precise measurements. These measurements are both very close to the true value (high accuracy) and tightly clustered together (high precision). The red dashed line illustrates the average value of these measurements, which closely aligns with the true value.

(b) Data points marked with blue circles represent accurate but less precise measurements. These measurements are close to the true value (high accuracy) but exhibit a wider spread or variability among them (low precision). The blue dashed line depicts the mean of these measurements, which is still near the true value, but the individual measurements exhibit more variability.

(c) Data points marked with green circles represent inaccurate but precise measurements. These measurements consistently deviate from the true value (low accuracy) but are closely clustered together (high precision). The green dashed line shows the mean value of these measurements, which notably differs from the true value. However, the individual measurements show low variability.

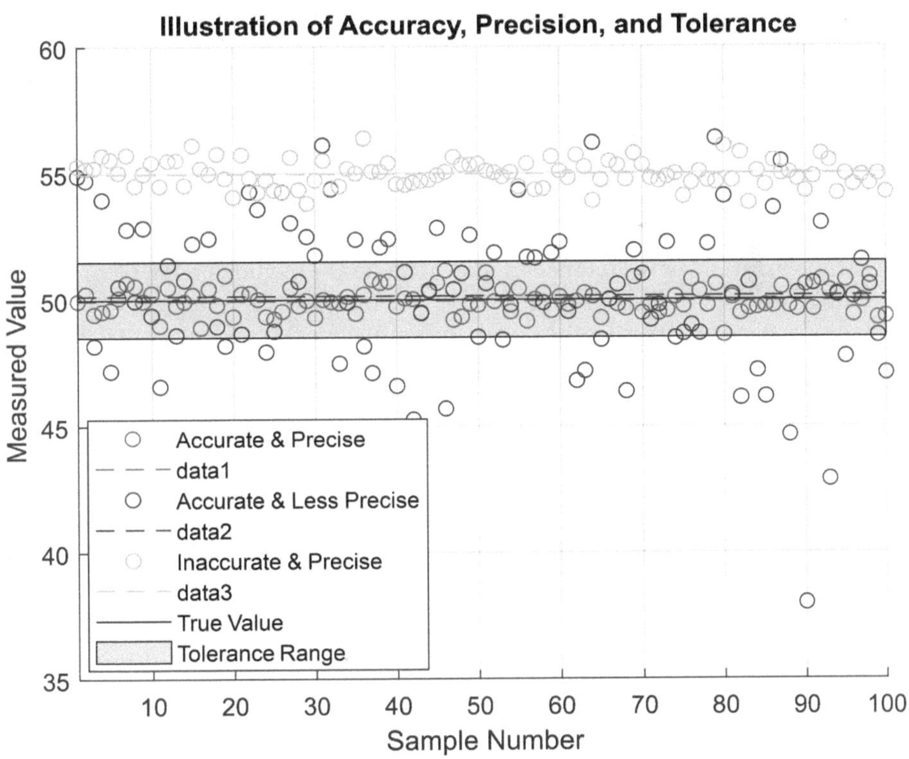

FIGURE 2.13
Illustration of accuracy, precision and tolerance.

The solid black line within the plot represents the true value, and the shaded region surrounding it signifies the tolerance range. This range indicates the allowable variation in the measured values for a specific purpose or function. In this instance, the tolerance is set at 1.5 units. If the measured values fall within this range, they are considered acceptable for the intended application.

Figure 2.13 provides a visual representation of the concepts of accuracy, precision and tolerance, thus contributing to a better understanding of their significance in fields such as engineering, manufacturing and measurement.

2.2.2 Factors that Affect Accuracy, Precision and Tolerance of 3D Printed Parts

Numerous factors impact the dimensional accuracy, precision and tolerance of 3D printed parts, collectively known as volumetric error. Volumetric error refers to the variation between the intended volume of the printed object as designed in the CAD model and the actual volume of the final printed part. This error is influenced by factors categorized into three groups based on the stages of the 3D printing process: preprocessing, processing and post-processing.

During the preprocessing stage, issues such as CAD model conversion, slicing and tool path planning can lead to dimensional changes in the part. These variations may arise from software limitations, slicing parameters or inaccuracies in tool path generation.

In the processing stage, factors like material behaviour (shrinkage or expansion) and the accuracy of the machine's axis movement, layer height and other printing parameters can affect the final product's AP&T.

Finally, the post-processing stage, encompassing support removal, surface treatment, stress relief and finishing processes, may also impact dimensions due to material removal, deformation or changes in material properties.

Let us now discuss these factors in greater detail, one by one.

CAD model conversion – when converting a CAD model from its native file format to a neutral file format like STL, errors can arise, particularly with complex, freeform shapes. Representing these shapes with triangular facets can result in some level of approximation.

Managing this conversion error is possible by increasing the resolution, which entails reducing the size of the triangles. For example, Figure 2.14 demonstrates this concept by depicting a sphere tessellated with different triangle sizes. Tessellating the sphere with larger triangles leads to more significant errors in its geometry compared to using smaller ones.

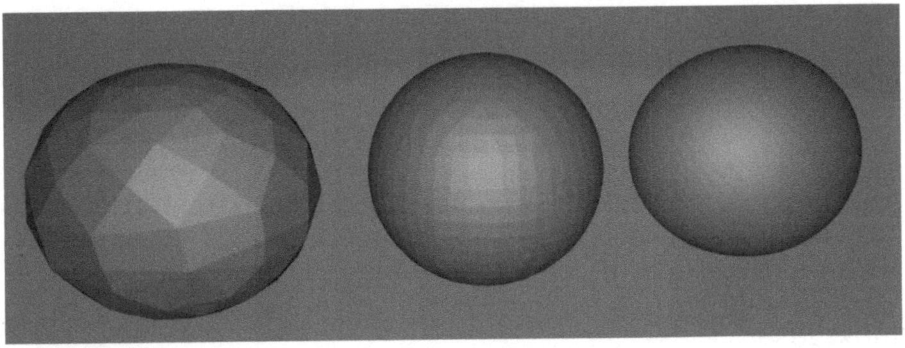

FIGURE 2.14
Spheres of the same dimension but tessellated with different parameters.

As mentioned earlier, SolidWorks software (and few others) offers options to adjust the resolution and other parameters during the CAD model conversion to STL. Modifying these settings can help uphold the dimensional accuracy of the CAD model, thereby minimizing the impact of conversion errors on the final design.

Slicing – when slicing the CAD model, there are volumetric errors that occur due to the slice height. Increasing the slice height leads to an increase in volumetric error. Since the entire object is being built layer by layer, the surface finish of the LM part is often subpar because complex surfaces are approximated into two-dimensional planar sections or layers. This approximation can result in volumetric errors in the final printed part, as there may be a difference between the volume of the deposited material and the volume specified in the CAD model.

Volumetric errors induced by slicing happen when the slicing process, which divides the CAD model into thin layers for printing, leads to inaccuracies in representing the object's geometry. These inaccuracies can be caused by factors like layer height, slicing settings and the chosen algorithm for generating tool paths. Consequently, the volumetric error can create differences in the final printed part compared to the original CAD model.

To reduce the impact of volumetric errors induced by slicing, it is crucial to optimize the slicing parameters, such as layer height and infill pattern, and to use advanced slicing algorithms that can more accurately approximate complex geometries. This can help minimize the discrepancy between the printed part and the CAD model, resulting in improved dimensional accuracy and better overall part quality.

As illustrated in Figure 2.15, it is important to observe that slice height affects the volumetric error and the dimensional accuracy.

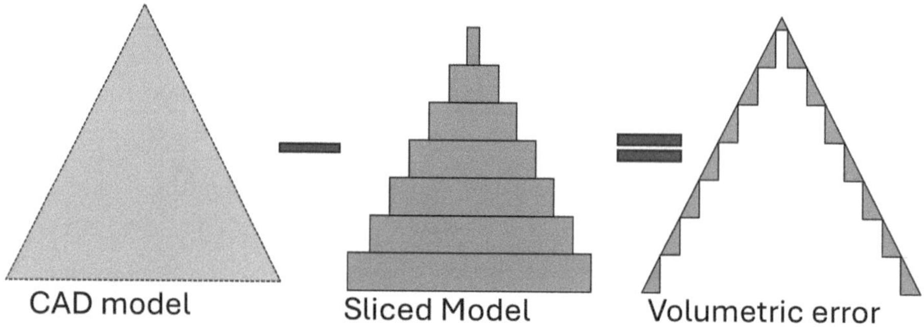

FIGURE 2.15
Volumetric error due to slicing.

Tool path – in 3D printing, the tool path essentially defines the course that the printing nozzle follows to craft an object. The instructions for the tool path encompass various parameters beyond mere coordinates and motion guidelines, significantly influencing the amount of material used, laser density and more. Any inaccuracies in the tool path can result in volumetric errors, leading to dimensional discrepancies in the final printed part. Some key factors contributing to these errors include the following:

(a) Inadequate tool path generation algorithms in certain slicing software may fail to generate precise tool paths for intricate surfaces, leading to errors in the printed part's shape.

(b) The correct infill pattern and density of a 3D printed part are contingent upon the tool path. Inaccurate calculations could cause volumetric errors due to over- or under-extrusion of material.

(c) Accurate toolpath compensation is vital to accommodate factors such as material shrinkage, printer calibration or nozzle size. If the slicing software does not adequately compensate for these factors, the printed part may deviate from the dimensions of the original CAD model.

(d) Inconsistent extrusion rates: the toolpath controls the material's extrusion rate during printing. Uneven toolpaths can lead to varying extrusion rates, causing irregularities in the printed part's volume.

(e) Shell and perimeter inaccuracies: the toolpath determines the number of shells and the thickness of the outer perimeters of a 3D printed part. If the slicing software fails to accurately calculate these parameters, it can result in volumetric errors due to deposition of either too much or too little material.

(f) To reduce volumetric errors caused by toolpath inaccuracies, it is crucial to utilize advanced slicing software that accurately generates toolpaths, optimizes slicing parameters specific to the printer and material, and regularly calibrates the 3D printer to ensure consistent performance.

(g) It is important to note that non-uniform scaling can distort the shape of an object as it alters the dimensions along each axis differently. This may result in an appearance that deviates from the original design. Therefore, it is crucial to carefully consider the desired outcome when selecting the appropriate scaling method for your object.

Material – in 3D printing, variations in temperature can lead to material shrinkage, expansion and warping, resulting in dimensional and geometric changes in the printed model. The accuracy of the final product relies heavily on the material properties.

In resin 3D printing, materials with a higher green modulus (modulus prior to post-curing) allow for the printing of delicate components with better precision and reduced risk of failure.

On the other hand, in FDM 3D printing, materials requiring higher extrusion temperatures are more prone to warping compared to those with lower extrusion temperatures. For example, ABS is notorious for its susceptibility to warping due to its higher shrinkage during the cooling process, unlike PLA.

2.2.3 3D Printing Process and their Tolerance

All 3D printers show differences in tolerance, which can vary between different 3D printing methods. Variations in precision among methods depend on their operational principles, but even within the same method, the precision of different machines may differ depending on the brand and components used in their manufacturing. The specifications provided by machine manufacturers act as a reference for what to anticipate. In this section, we will examine the specifications of some popular 3D printing methods.

Stereolithography (SLA): ± 0.2% (with a minimum limit of ± 0.2 mm)

SLA is one of the most accurate methods, alongside Polyjet. The laser diameter in SLA typically ranges from 100–150 µm, which is significantly smaller than in FDM processes. Tolerances mainly occur due to shrinkage. Thin, large surfaces can warp significantly in SLA, as with most processes.

Digital light processing (DLP): ± 0.1-0.2% (with a minimum limit of ± 0.1–0.2 mm).

DLP operates similarly to stereolithography but uses projectors instead of lasers. The tolerance depends on the resolution. Devices with small build volumes (≈ 10 cm side length) generally have tolerances of around 0.1 mm.

Selective laser sintering (SLS): ± 0.3% (with a minimum limit of ± 0.3 mm).

SLS is a widely used rapid prototyping technology, mainly due to its design flexibility (no support structures) and robust materials. However, the heat used in this process can cause shrinkage or warping.

HP multi-jet fusion: ± 0.2% (with a minimum limit of ± 0.2 mm).

HP multi-jet fusion, also known as high-speed sintering, primarily works with polyamide 12, similar to laser sintering. The factors influencing tolerance in this technology are comparable to those in laser sintering.

Fused deposition modelling (FDM/FFF): ± 0.5% (with a minimum limit of ± 0.5 mm).

FDM printers typically use ABS and PLA plastics. ABS shrinks by approximately 8% during cooling, while PLA shrinks by about 2%. These values may vary depending on the specific ABS or PLA material used. High-end FDM printers (e.g., Stratasys systems) deliver lower tolerances of up to 0.1 mm but are more expensive to operate.

Selective laser melting (SLM, direct metal 3D printing): ± 0.5% (with a minimum limit of ± 0.5 mm).

SLM is generally comparable to laser sintering, but the heat-conductive properties and temperatures involved in metal printing differ significantly from plastics. This technology faces similar challenges as laser sintering, with high temperatures causing warping. The printed parts are nearly pore-free, with surface roughness up to 20 µm. Critical surfaces may have an additional 1–2 mm of material, which is milled post-printing to achieve the desired tolerances and surface finish.

Polyjet: ± 0.1–0.2% (with a minimum limit of ± 0.1–0.2 mm).

Polyjet-printed parts are highly accurate due to the use of UV radiation instead of heat. Shrinkage or stress is usually minimal, even in thin-walled models.

Moreover, the calibration of the machine is another important factor that we need to consider before starting printing at any machine to fabricate the accurate part.

Post-processing – many 3D printing processes include a standard step called post-processing. For example, certain parts may require post-curing in SLA and DLP. Stress-relieving treatments could be necessary for preventing warping of thin components in SLM. Support removal, powder removal and surface finishing are typical steps for most 3D printed parts.

Some of these processes can affect the overall accuracy or cause geometric changes in the final part. For instance, high-temperature post-curing in resin-based 3D printing can lead to shrinkage. Additionally, various finishing techniques used to improve the surface quality of FDM parts can result in dimensional changes. Therefore, it is essential to consider the potential impacts of post-processing on the accuracy and dimensions of 3D printed parts.

We have examined the different factors that influence volumetric errors in 3D printing, which can impact the accuracy, precision and tolerance of the process. It is important to note that a machine's precision mainly depends on material and machine-specific parameters rather than file conversion parameters. Nonetheless, we have presented a comprehensive overview of the main factors that can influence the accuracy, precision and tolerance of 3D printed products. This provides valuable insights into optimizing the 3D printing process for the best possible results.

2.3 Software Required for 3D Printing

As discussed, the software required for 3D printing can be categorized into three distinct types:

1. Software for editing and modifying the firmware that governs the electronics and motors of the 3D printer

   (This we have already discussed in Section 2.1.4 of this chapter)
2. Software for crafting solid models (CAD software)

   (This will be covered in detail in Chapter 3, Section 3.1)
3. Software for generating sliced models from solid models and creating tool paths (G-codes)
   (This will be covered in Section 2.3 and also in Chapter 4, Sections 4.2, 4.3 and 4.4)

Throughout this section, we shall delve into the different software applications related to category 3 above employed for generating G-codes and machine codes for 3D printing.

2.3.1 Fusion 360

Autodesk's Fusion 360 is a powerful and flexible 3D design software that integrates computer-aided design (CAD), computer-aided engineering

(CAE), and computer-aided manufacturing (CAM) seamlessly into a single, easy-to-use platform. This advanced software is tailored to meet the requirements of both professionals and enthusiasts, providing a comprehensive solution for 3D modelling, simulation, rendering, and manufacturing.

Operating as a cloud-based platform, Fusion 360 enables real-time collaboration with team members and clients, ensuring effective communication and efficient project management. Equipped with a robust array of features, Fusion 360 empowers users to produce intricate assemblies, organic shapes, and complex parametric models, making it an indispensable tool for various industries, including automotive, aerospace and product design.

2.3.2 SOLIDWORKS

SOLIDWORKS, an advanced CAD software created by Dassault Systèmes, was first introduced in November 1995. Since then, it has become an essential tool for generating solid models, especially in the 3D printing field. However, SOLIDWORKS offers more than just computer-aided design, as it also includes computer-aided engineering, product data management and collaborative features, making it a comprehensive solution for design professionals.

At its essence, SOLIDWORKS uses a parametric, feature-based approach, which was originally developed by PTC's Pro/ENGINEER software. By leveraging the robust parasolid modelling kernel, SOLIDWORKS empowers users to create intricate 3D models and assemblies with unmatched accuracy and efficiency. Its user-friendly interface and extensive feature set make it an essential tool for engineers, designers and manufacturers in various industries.

Apart from its impressive modelling capabilities, SOLIDWORKS is recognized for its collaborative functions, enabling teams to work seamlessly and cohesively on projects. This feature promotes effective communication and streamlined project management, ensuring that design teams can efficiently bring their ideas to fruition.

In conclusion, SOLIDWORKS is a robust and adaptable CAD software that has transformed the way professionals create solid models for 3D printing and other purposes. With its advanced engineering, product data management and collaboration tools, SOLIDWORKS has become a standard in the industry for individuals looking to design and produce high-quality, innovative products.

2.3.3 Onshape

Onshape is an innovative CAD software that operates in the cloud. It was created by Onshape Inc., a company founded by former SOLIDWORKS executives and engineers. This cutting-edge platform offers a wide range of tools for 3D modelling, assembly creation and design collaboration, providing a flexible and convenient solution for design professionals and enthusiasts, as it can be accessed from any device with an internet connection.

Onshape is a robust tool for 3D printing. It provides an extensive set of modelling tools and is compatible with various 3D printing software and hardware. This makes it simple for users to create highly detailed models that can be tailored for different 3D printing methods, such as FDM, SLA and SLS.

The real-time collaboration feature offered by Onshape allows design teams to work together on the same project, streamlining the design process and enabling effective communication. This aspect is particularly beneficial in the context of 3D printing, as it enables teams to swiftly identify and address potential design flaws and necessary adjustments, thereby facilitating rapid prototyping and iteration.

Furthermore, Onshape supports a broad range of import and export options, ensuring compatibility with a variety of 3D printing file formats, including STL, OBJ and AMF. This adaptability simplifies the process of transferring designs to slicing software or directly to 3D printers, optimizing the entire 3D printing workflow.

In conclusion, Onshape is a powerful and user-friendly CAD software that is an essential tool for both 3D printing enthusiasts and professionals. Its strong modelling capabilities, real-time collaboration features and seamless integration with 3D printing software and hardware make it an indispensable resource in the fast-paced world of additive manufacturing.

2.3.4 TinkerCAD

TinkerCAD, a web-based 3D modelling software suitable for beginners, is created by Autodesk, the company known for Fusion 360. TinkerCAD is designed to be easy to use, making it perfect for newcomers, educators and those interested in 3D printing. Its simple interface and easy-to-use tools allow users to create, modify and export 3D models without requiring extensive CAD knowledge.

For those interested in 3D printing, TinkerCAD is a valuable resource, particularly for beginners. The software's user-friendly tools and features, such as drag-and-drop shapes and predesigned models, make it easy to create custom 3D designs. Additionally, TinkerCAD provides various 3D printing tutorials and chapters to help users understand the basics of designing for additive manufacturing.

One of TinkerCAD's key strengths is its support for 3D printing file formats like STL and OBJ. This enables users to easily export their designs directly to 3D printing software or services, streamlining the process from design to production. Furthermore, TinkerCAD interfaces with numerous 3D printing communities and online marketplaces, providing users with access to a wide range of resources and inspiration for their 3D printing projects.

TinkerCAD also encourages collaboration and sharing among users. Its cloud-based platform facilitates sharing designs with others, fostering a community focused on creativity and innovation. This collaborative environment is especially beneficial for educators, as it allows students to work together on projects and share their 3D printing creations with classmates.

In summary, TinkerCAD is a versatile and user-friendly 3D modelling software that serves as an excellent introduction to 3D printing. Its accessible interface, simplified design tools and compatibility with 3D printing file formats make it an ideal choice for beginners and educators. By creating a supportive and collaborative environment, TinkerCAD provides users with the tools they need to explore the world of 3D printing and bring their creative ideas to life.

2.3.5 UltiMaker Cura

The fast-paced realm of 3D printing consistently demands software that is efficient and user-friendly. UltiMaker Cura, a high-performing and adaptable slicing software, is specifically designed to meet the needs of professionals, hobbyists and 3D printing enthusiasts. Developed by UltiMaker, this software has become an essential tool for those looking to improve and streamline their 3D printing journey. Due to its impressive performance and extensive features, UltiMaker Cura has emerged as the preferred slicing software for millions of users worldwide.

UltiMaker Cura is available to everyone for use, modification and distribution as it is an open-source software. This open-source approach has built a community of users who collaborate, exchange ideas and contribute to the software's continuous development. As a result, Cura receives regular updates and enhancements, ensuring that it stays at the forefront of the rapidly evolving 3D printing industry.

A notable feature of UltiMaker Cura is its user-friendly interface, which is designed to be simple while offering advanced options for experienced users. The intuitive layout allows users to easily navigate through settings, import 3D models and start the slicing process. This user-friendliness has propelled UltiMaker Cura to become a highly sought-after choice for both beginners and experts.

In addition to its user-centric interface, UltiMaker Cura includes a wide range of features aimed at optimizing the 3D printing process. Users can customize print settings, such as layer height, print speed and infill density, to achieve optimal results. This level of customization ensures that prints maintain the highest quality, whether for personal enjoyment or professional applications.

UltiMaker Cura is known for its wide compatibility with a variety of 3D printers. The software not only works with UltiMaker's printers but also supports other top brands, making it a versatile choice. Its open-source nature allows the community to create and share profiles for an even broader range of 3D printers.

In addition to its compatibility, UltiMaker Cura offers advanced features that help users save time and resources. For example, the Adaptive Layers function adjusts the layer height based on the model's geometry, ensuring excellent print quality while reducing print duration. The Tree Support feature generates efficient support structures that are easy to remove, minimizing material wastage and post-processing efforts. These innovative features demonstrate UltiMaker Cura's commitment to delivering a cutting-edge 3D printing experience.

In summary, UltiMaker Cura has established itself as a leading slicing software in the 3D printing industry, thanks to its user-friendly interface, comprehensive features and extensive compatibility. With an open-source framework and active community, the software continues to evolve, ensuring its position at the forefront of 3D printing technology. Whether you are new to 3D printing or a seasoned professional, UltiMaker Cura offers the necessary tools and functionality to achieve outstanding results in all your 3D printing projects.

2.3.6 MeshLab

MeshLab is a powerful open-source software specialized in processing and modifying unstructured 3D triangular meshes, making it an essential tool in the field. Developed by the ISTI-CNR research center, MeshLab provides a wide range of features to meet the diverse needs of its users. From editing and refining 3D models to rendering and visualization, MeshLab offers impressive capabilities while ensuring ease of use.

As an open-source software, MeshLab ensures extensive accessibility for a wide range of users. Its source code is freely available, allowing developers and researchers to adapt and distribute the software to suit their specific requirements. This open-source approach has built a strong community of users who actively contribute to the software's continuous development, share insights and offer support to fellow MeshLab enthusiasts. Consequently, the software continually evolves, staying current with the latest advancements in 3D mesh processing technology.

MeshLab excels in tasks such as mesh cleaning, which involves removing duplicate vertices, filling holes and correcting mesh topology. It is particularly valuable for users dealing with 3D scans or models that require refinement before further processing or 3D printing. MeshLab also provides advanced algorithms for mesh simplification, enabling users to reduce the complexity of their models without sacrificing quality. This feature is essential for optimizing models for real-time rendering or improving computational efficiency.

In addition to mesh cleaning and simplification, MeshLab offers powerful tools for mesh editing and manipulation. Users can easily perform tasks such as cutting, merging and deforming meshes, as well as applying various filters to alter the geometry or appearance of their models. The software also supports a wide range of file formats, ensuring compatibility with other popular 3D modelling and processing tools. These features make MeshLab an ideal choice for users looking to streamline their 3D mesh processing workflow.

MeshLab's advanced rendering capabilities allow users to visualize and analyze their models in real-time. The software provides a broad selection of rendering options, including flat shading, wireframe and various shading techniques that simulate real-world lighting conditions. Users can also create and view ambient occlusion maps, curvature maps, and other visualization aids to better understand the geometry and structure of their 3D models. These visualization features empower users to make more informed decisions when editing and processing their meshes, ultimately leading to superior quality results.

In summary, MeshLab stands out as leading 3D mesh processing software due to its extensive features, user-friendly interface and open-source accessibility. Its potent tools for mesh cleaning, simplification, editing and rendering have made it an invaluable resource for professionals, researchers and hobbyists in the 3D modelling and computer graphics community. As MeshLab continues to adapt and evolve in response to the continuously changing 3D mesh processing technology landscape, it remains a reliable and versatile tool for users aiming to optimize their 3D models and achieve the best possible outcomes.

2.3.7 Simply 3D

In today's rapidly evolving world, the intersection of creativity and technology is more prominent than ever before. As industries advance and businesses grow, there is an increasing need for advanced tools to bring ideas to life. Enter SIMPLY 3D, a cutting-edge 3D design and modelling software that has transformed the world of 3D visualization and revolutionized how we conceptualize, create and interact with three-dimensional designs. SIMPLY 3D is a comprehensive software solution that meets the needs of both professionals and enthusiasts. It is a powerful tool that enables users to design, model and render stunning 3D visuals with exceptional ease and precision. With its user-friendly interface and extensive features, SIMPLY 3D has become the preferred software for architects, engineers, designers and artists looking to materialize their creative visions. One of the standout features of SIMPLY 3D is its intuitive user interface, which allows for effortless navigation and access to the wide range of tools necessary for creating extraordinary 3D models.

SIMPLY 3D's interface caters to users of all skill levels, making it an adaptable platform for anyone with a passion for 3D design.

Furthermore, SIMPLY 3D offers a wide range of modelling tools that cater to every aspect of 3D design. From simple shapes and objects to complex structures and environments, SIMPLY 3D's modelling features empower users to easily create detailed and elaborate models. Additionally, the software's parametric modelling capabilities enable users to establish connections between various model elements, simplifying the process of editing and adjusting designs as required.

SIMPLY 3D not only excels in modelling, but also offers advanced rendering capabilities that allow users to create photorealistic images and animations of their 3D models. With a wide range of rendering options, including real-time rendering and global illumination, SIMPLY 3D ensures that your designs look as realistic and engaging as possible. The software also supports various file formats, making it easy to import and export creations to other design tools and platforms.

At its core, SIMPLY 3D emphasizes collaboration, facilitated by integrated tools that enable teams to work together on projects, regardless of location. Through real-time editing and communication features, SIMPLY 3D promotes seamless collaboration among team members, ensuring effective contributions and a unified project vision.

As industries evolve and demand more complex and innovative designs, SIMPLY 3D's robust simulation tools allow users to test their creations under real-world conditions. From analyzing material behaviour under various stress factors to evaluating the impact of environmental elements on a design, SIMPLY 3D's simulation features provide valuable insights that help users optimize and refine their creations prior to implementation.

In summary, SIMPLY 3D is an innovative software solution that has transformed the world of 3D design and modelling. Its intuitive interface, wide array of modelling tools, advanced rendering capabilities and strong collaboration features make it a vital tool for both professionals and enthusiasts. By allowing users to produce visually impressive, lifelike 3D models and assess their designs in realistic scenarios, SIMPLY 3D enables individuals and teams to explore the boundaries of creativity and ingenuity.

2.3.8 Repetier-Host

Repetier-Host, extensively utilized open-source software in the 3D printing sector, is specifically designed for controlling and managing 3D printers. It is free and compatible with most printers, enabling users to monitor and control their printers from their computers.

The primary goal of Repetier-Host is to simplify the 3D printing process by providing a comprehensive set of features. One of its prominent capabilities is managing multiple printers, which saves time and enhances productivity by allowing users to oversee different printers from a single interface.

Another crucial function of Repetier-Host is its ability to preview 3D models before the printing process. This feature enables users to view the model from various angles and make necessary adjustments to ensure the accuracy and alignment of the final product with their expectations.

In addition, Repetier-Host offers various customization options. Users can tailor print settings such as layer height, infill density and print speed to achieve the desired print quality. Furthermore, they can customize the software's interface to meet their specific requirements.

Repetier-Host caters to experienced users with advanced features such as G-code editing, enabling modifications to the printer's code, and support for third-party plugins that extend the software's functionality.

Compatibility with a wide range of file formats, including STL, OBJ and G-code, means that users can import and export files from other software programs and seamlessly use them in Repetier-Host.

In conclusion, Repetier-Host is a crucial tool for anyone engaged in 3D printing. It streamlines the printing process, offers diverse customization options and provides support for multiple printers, making it valuable to both beginners and experienced users striving for optimal results from their 3D printers.

2.3.9 Slic3r

A critical tool in the world of 3D printing, slicing software acts as the link between digital models and physical objects, allowing users to convert 3D models into a set of instructions for 3D printers. Slic3r, created by Alessandro Ranellucci in 2011, has become widely recognized among hobbyists, professionals and enthusiasts, setting the standard in the 3D printing industry.

As an open-source software, Slic3r is available for free and can be modified, making it appealing to a wide range of users. Its source code is publicly accessible, allowing developers and researchers to customize and distribute the software to fit their specific needs, fostering a strong user community that shares knowledge and provides support.

With an intuitive and user-friendly interface, Slic3r is easily navigable for both novices and experienced users, offering a seamless workflow for importing 3D models, adjusting settings and creating G-code for 3D printers. Slic3r supports multiple file formats, ensuring compatibility with most 3D modelling tools and easy integration into existing workflows. Additionally, it is compatible with various 3D printers, enhancing its versatility and appeal to users. Slic3r offers a wide range of features aimed at optimizing the 3D printing process. Users can customize print settings, such as layer height, infill density and print speed, to achieve the best results for their specific requirements. The software also provides advanced features, including customizable support material generation and multi-material printing capabilities, addressing users with more complex needs. These features enable users to adjust their prints, ensuring high quality for personal and professional use.

Efficient slicing algorithms are another key feature of Slic3r. The software excels at processing large and intricate models, quickly and accurately creating G-code. Users can also utilize Slic3r's advanced features, such as adaptive slicing, which adjusts layer height based on the model's geometry to improve print quality and reduce print time. This combination of efficiency and accuracy establishes Slic3r as a reliable option for users looking to optimize their 3D printers' performance.

In conclusion, Slic3r is an innovative slicing software in the 3D printing industry. Its user-friendly interface, comprehensive features and open-source nature have made it a popular choice among professionals, hobbyists and enthusiasts. With its powerful G-code generation capabilities, customizable print settings and compatibility with a wide range of 3D printers, Slic3r remains a top choice for users. It continues to be a dependable and adaptable tool for those seeking to enhance their 3D printing experiences and achieve outstanding results.

2.4 Generic Specification of Popular 3D Printers

Following are the generic specifications (refer to Tables 2.4 to 2.11) of 3D printers based on different printing approaches:

TABLE 2.4

Generic Specification of the FDM 3D Printing Machine

3D printing machine	Stratasys FDM 3D printing machine
Maximum Build Envelope	(900 mm × 600 mm × 900 mm)
Layer Thickness	50 to 400 microns
Standard Accuracy	± 0.15% (lower limit ± 0.2 mm)
Minimum Wall Thickness	1 MM
Common Materials	Thermoplastics (PLA, ABS, PETG, PC, PEI etc.)
Input Data File Formats Supported	STL, OBJ, PLY, IGES, IGS, STEP, STP
Support Material	Water-soluble/Break-away
Computer System Requirements	Windows 10 (64-bit only)
Connectivity Options	USB, Ethernet, or Wi-Fi
Electrical Requirements for Printer	200-240 VAC, 1- Ph, 50/60Hz, 24A

TABLE 2.5

Generic Specification of the 750 SLA 3D Printing Machine

3D printing machine	750 SLA 3D printing machine
Technology	Stereo lithography apparatus (SLA)
Net Build Volume (*xyz*)	(750 mm×750 mm×550 mm)
Native Resolution	2000 dpi
Laser Wavelength	355 nm
Material	Thermo sets
Light Source Type	UV light source
Input Data File Formats Supported	STL, CTL, OBJ, PLY, ZPR, ZBD, AMF, WRL, 3DS, FBX, IGES, IGS, STEP, STP and X_T
Software	3D Sprint, 3D Connect
File Input to 3D Sprint	All 3D model file types
Operating Environment Temperature	18–28°C (64–82°F) 20–80%
Computer System Requirements	Windows 10 (64-bit only) 3D Sprint software (included with Printer)
Connectivity Options	USB, Ethernet, or Wi-Fi
Electrical Requirements for Printer	200-240 VAC, 1- Ph, 50/60Hz, 24A (SLA 750)

TABLE 2.6

Generic Specification of the DLP 3D Printing Machine

3D printing machine	Figure 4 stand alone
Technology	Digital light processing
Net Build Volume (*xyz*)	(124 mm × 70 mm × 195 mm)
Maximum Resolution	1920 × 1080 pixels
Layer Resolution	65 micron (390.77 effective DPI)
Wavelength	405 nm
Build Materials	UV curable plastics: Figure 4 TOUGH-GRY 10, Figure 4 TOUGH-GRY 15, Figure 4 ELAST-BLK 10, Figure 4 JCAST-GRN 10
Printing Speed	100 mm/hr
Light Source Type	UV light source
Accuracy	Six sigma repeatability
Input Data File Formats Supported	STL, CTL, OBJ, PLY, ZPR, ZBD, AMF, WRL, 3DS, FBX, IGES, IGS, STEP, STP and X_T
Software	3D Sprint, 3D Connect
Operating Environment Temperature	18–28°C (64–82°F) 20–80%
Connectivity Options	USB, Ethernet, or Wi-Fi
Electrical	100–240 VAC, 50/60 Hz, Single Phase, 4.0A

TABLE 2.7

Generic Specification of the METAL 3D Printing Machine

3D printing machine	SLS 300 3D METAL PRINTER
Software	Slicer
Import Format	STL
Typical Build Size	(300 mm × 300 mm × 300 mm)
Dimensional Accuracy	Dimensional accuracy: ± 0.3% (lower limit of ± 0.3 mm)
Layer Thickness	100–120 μm
Laser Power Type	50 W, CO_2
Power Consumption	400 W (printing) 2000 W (warm-up)
Printing Speed	12 mm (0.47 in) per hour /1 litre per hour
Printer Controls	13.3″ display with touch screen
Materials	Thermoplastics (usually nylon)
Material Options	Nylon 11: Tough, impact- and fatigue-resistant • Nylon 12: Strong and flame-retardant
System Requirements	Google Chrome 93 and up | WebGL 2.0 |4GB RAM (8GB recommended)
Electrical Requirements	1 × 230 V, AC 10 A, 50 Hz (EU) 1 × 115 V, AC 15 A, 60 Hz (US
File Types	.STL, .STEP, or .3MF

TABLE 2.8

Generic Specification of the METAL JETTING 3D Printing Machine

3D printing machine	ELEMX 3D METAL PRINTER
Software	Slicer
Import Format	STL
Net Build Volume (*xyz*)	(300 mm × 300 mm × 120 mm)
Max. Build Rate	0.5 pounds/hour / 84 ccc/hour
Dimensional Accuracy	XY: +/-0.6 mm; Z: +/-0.5 mm
Minimum Layer Thickness	0.24 mm
Density	>98.5%
Heat Treatment	As required
Surface Finish	Sand cast comparable
Chiller	78 cm × 44 cm × 66 cm
MATERIALS Input	356 (4008) Aluminium Alloy
Wire Diameter	.062 in / 1.6 mm
Powerrequirements	50 amp • 480V 3 phase • 4 wire

TABLE 2.9

Generic Specification of the BINDER JETTING 3D Printing Machine

3D printing machine	INNOVENTXBINDER JET 3D PRINTING MACHINE
Print Technology	Triple ACT (Advanced Compaction Technology)
Print Direction	Uni-directional (Bi-directional programmable)
Binder Jetting Module	Piezo-electric print-head (256 nozzles)
Max Build Rate	54 cc/hr (3 in3)
Print Resolution 80pL Print-Head	400 dpi
Layer Thickness	30 to 200 μm
Build Volume	0.676 L (41 in3)
Build Box Envelope (W × D × H)	160 mm × 65 mm × 65 mm
Electrical Requirements	120 V, 60 Hz, 1-phase 230 V, 50 Hz, 1-phase
Powders	Open platform, capable of printing metal, ceramic, sand and composite powders with a D50 of 3 to 100 μm
Binder Systems	AquaFuse™; FluidFuse™; PhenolFuse™; CleanFuse™

TABLE 2.10

Generic Specification of the DIRECT ENERGY DEPOSITION 3D Printing Machine

3D printing machine	Meltio M450
Print Technology	Wire-Laser Metal 3D Printing Technology
Process Control	Closed-loop, laser and wire modulation
Build Box Envelope (W × D × H)	145 × 168 × 390 mm
Enclosure	Laser safe, sealed, controlled atmosphere
Interface	USB, Ethernet, WiFi
Laser Type	6 × 200W direct diode lasers
Cooling	Active water-cooled chiller included
Laser Wavelength	976 nm
Wire Feedstock	Diameter: 0.8–1.2 mm, Spool Type: BS300
Total Laser Power	1200 W
Nozzle Size	65 × 82 × 194 mm
Nozzle Weight	1.14 kg
Nozzle Material	Stainless Steel 316L
Accessories	Laser Alignment System
Power Consumption	208/230 V single phase or 400 V three phase 2–5 kW peak depending on selected options.

TABLE 2.11

Generic Specification of the SHEET LAMINATION 3D Printing Machine

Technology	Sheet Lamination
Printing Process	Layer-by-layer bonding of material sheets using adhesives or heat and pressure
Compatible Materials	Paper, plastic sheets, metal foils, composite sheets
Material Thickness Range	0.05 mm to 1.0 mm
Build Volume	Range: 300 mm × 200 mm × 150 mm to 1000 mm × 800 mm × 600 mm
Bonding Method	Adhesive application, ultrasonic welding, or thermal lamination
Cutting Mechanism	Laser cutting
Layer Thickness	0.1 mm to 0.5 mm
Printing Speed	5 to 20 layers per minute (material-dependent)
Accuracy	± 0.1 mm to ± 0.3 mm
Software Compatibility	STL, OBJ, STEP file formats
Control System	Integration with CAD/CAM software
Power Requirements	110–240 V, 50/60 Hz, Single Phase
Machine Dimensions	Model ranges: 1500 mm × 1000 mm × 1200 mm to 2500 mm × 1500 mm × 1500 mm
Weight	200 kg to 1500 kg (based on size and model)
Environmental Conditions	Operating Temperature: 20°C to 30°C; Humidity: 20% to 70% RH
Safety Features	Emergency stop button, enclosed work area and dust extraction system
Applications	Prototype modelling, tooling, architectural models and metal preforms

Self-Check Exercises

1. Survey the web, select items, create support frame/structure and assemble an open-source **FDM** printer as outlined above.
2. Survey the web, select items, create support frame/structure and assemble an open-source **SLA** printer as outlined in the following web links:

   https://hackaday.io/project/21461-low-cost-open-source-lcd-based-sla-3d-printer

   www.hackster.io/news/lite3dp-gen-2-is-a-tiny-open-source-resin-3d-printer-562c645a9fcf

   https://youtu.be/IGPnx29km9U

   www.youtube.com/watch?v=UU6tWhV010M

   www.youtube.com/watch?v=9-QwKkldCM8

3. A 3D printed object of conical shape having a base diameter 20 and height 50 mm is printed with two different layer heights: 0.1 mm and 0.2 mm. Calculate the total volumetric error for the 3D printed model.
4. Compare and contrast the features, compatibility and user experience of two popular 3D printing software solutions (e.g., Repetier-Host and Slic3r). Provide a detailed analysis of their advantages and disadvantages, and discuss which software would be more suitable for different user scenarios (e.g., beginners, professionals, multi-printer management).
5. List the various components/parts of a typical FDM-based 3D printer.
6. List the various components/parts of a typical SLA-based 3D printer.
7. List the various components/parts of a typical SLS-based 3D printer.
8. Explain the function of the motherboard/controller board in a 3D printer.
9. List the essential components of a 3D printer motherboard/controller board.
10. List the different category of software required in 3D printing and their uses.
11. List the commercially available software under each category of software required in 3D printing.
12. List the specifications of the power supply unit used in FDM, SLA and SLS 3D printers.
13. Explain the function of stepper motors and various transmission components used in 3D printers.
14. Outline the specification of a stepper motor used in a 3D printer with justification.
15. Explain the use of end stops/limit switches in a 3D printer.
16. Write the features required in a printer bed and also the associated issues.
17. Compare the print-head of FDM, SLA, SLS 3D printers.
18. Describe the working of extruder and feeder system (FDM printer only).
19. Explain the calibration process in an FDM-based 3D printer.
20. Describe the interfacing procedure of a 3D printer with a computer/laptop.
21. Differentiate between accuracy, precision and tolerance with one example.
22. Describe the factors that affect accuracy, precision and tolerance of 3D printed parts.

23. Compare different 3D printers based on their accuracy to produce components.
24. Refer to different 3D printer manufacturer catalogues and prepare generic specifications of a SLA-based 3D printer to be purchased for an educational institute.

Discussion Topics

1. Discuss the use of 3D printing processes in the medical domain.
2. Discuss the components required to develop an open-source mini DLP printer.
3. Discuss factors that affect accuracy, precision and tolerance of 3D printed parts.
4. Discuss the open-source software available for three categories viz. CAD; slicing, printer setting tool-path generation; and firmware regarding 3D printing.

Multiple Choice Questions

1. What is the primary function of the processor in a 3D printer motherboard?
   (a) Regulating the temperature of the printer
   (b) Reading the G-code and sending instructions to other printer parts
   (c) Connecting all electronic components
   (d) Controlling the power applied to the stepper motor's individual coils
2. What is the purpose of stepper drivers in a 3D printer?
   (a) To read G-code commands
   (b) To regulate the amount of power applied to each of a stepper motor's individual coils
   (c) To connect electronic components
   (d) To control the temperature of the printer
3. Which of the following components is NOT considered a motherboard component?
   (a) Video cards and GPUs
   (b) Optical drives, such as DVD and CD-ROM

(c) Stepper drivers
(d) Hard drives (SSD or HDD)

4. Which of the following materials is NOT commonly used for constructing 3D printer frames?
(a) Aluminium
(b) Plastic (acrylic sheets)
(c) Steel
(d) Wood

5. What is the main advantage of using an aluminium print bed compared to a glass print bed?
(a) Aluminium beds are less fragile
(b) Aluminium beds heat up quickly
(c) Aluminium beds do not expand when heated
(d) Aluminium beds are better for levelling

6. Which of the following materials is NOT commonly used for making nozzles in FDM printers?
(a) Brass
(b) Stainless steel
(c) Ruby
(d) Aluminium

7. In SLA and DLP printers, what is the primary difference between the light sources used?
(a) SLA uses a laser, while DLP uses a projector
(b) SLA uses a projector, while DLP uses a laser
(c) SLA and DLP both use lasers, but with different wavelengths
(d) SLA and DLP both use projectors, but with different resolutions

8. Which type of 3D printer uses a print-head similar to inkjet printers to selectively deposit droplets of a binding agent onto layers of powdered materials?
(a) FDM printer
(b) SLA printer
(c) SLS printer
(d) Binder jetting printer

9. Which type of extruder configuration allows for faster printing due to reduced weight near the nozzle?
(a) Direct drive
(b) Bowden
(c) Dual extruders
(d) None of the above

10. Which of the following statements is NOT true about the calibration process of a 3D printer?
(a) Calibration of end stops defines the boundary for *X*-, *Y*- and *Z*-axes
(b) The size of the heated bed constrains the build volume of the parts
(c) The nozzle of the extruder should be positioned over the bed surface with a distance of 0.3–0.4 mm

(d) Calibration is performed only after interfacing the 3D printer with a computer

11. Why is it important to test the accuracy and precision of 3D printed replacement parts?
(a) To ensure interchangeability and compatibility with other components
(b) To guarantee high resolution in the final product
(c) To NOT reduce the cost of production
(d) To increase the speed of the manufacturing process

12. What does accuracy in measurements refer to?
(a) How consistent a set of measurements are to each other
(b) How close a measured value is to the true value
(c) The degree to which measurements can be reproduced
(d) The distribution of measurements around the mean value

13. Which statement best describes precision in measurements?
(a) It refers to how closely a set of measurements are grouped together
(b) It refers to how close a measured value is to the true value
(c) It refers to the permissible variation in dimensions or properties
(d) It refers to the quality of manufactured parts

14. In the context of tolerance, what does a tighter tolerance imply?
(a) A larger allowable range of variation
(b) Lower production costs
(c) A smaller allowable range of variation
(d) Less consistent parts

15. What is the relationship between precise measurements and accuracy?
(a) Precise measurements are always accurate
(b) Precise measurements are never accurate
(c) A precise set of measurements may not necessarily be accurate
(d) Accuracy and precision have no relationship

16. What is the volumetric error in 3D printing?
(a) The difference between the intended volume of the printed object as designed in the CAD model and the actual volume of the final printed part
(b) The accuracy of the machine's axis movement during printing
(c) The level of detail in the conversion of a CAD model to a neutral file format
(d) The discrepancy between the printed part and the CAD model due to slicing

17. Which of the following stages in the 3D printing process can impact dimensional accuracy, precision and tolerance?
(a) Preprocessing stage
(b) Processing stage
(c) Post-processing stage
(d) All of the above

18. What is one way to minimize the conversion error when converting a CAD model to a neutral file format like STL?
   (a) Decreasing the resolution
   (b) Increasing the slice height
   (c) Increasing the resolution
   (d) Decreasing the slice height

19. Which factor can contribute to volumetric errors due to toolpath inaccuracies?
   (a) Inadequate toolpath generation algorithms
   (b) Material shrinkage or expansion
   (c) Inaccurate CAD model conversion
   (d) Post-processing techniques

20. How can post-processing impact the accuracy and dimensions of 3D printed parts?
   (a) By introducing errors during CAD model conversion
   (b) By causing shrinkage or deformation during post-curing or finishing processes
   (c) By affecting the slicing parameters and layer height
   (d) By altering the toolpath and infill patterns

21. Which of the following software is cloud-based and developed by Autodesk?
   (a) Fusion 360
   (b) SOLIDWORKS
   (c) Onshape
   (d) TinkerCAD

22. Which software is known for its parametric, feature-based approach?
   (a) Fusion 360
   (b) SOLIDWORKS
   (c) Onshape
   (d) TinkerCAD

23. Which software is specifically designed for processing and editing unstructured 3D triangular meshes?
   (a) MeshLab
   (b) Fusion 360
   (c) Onshape
   (d) TinkerCAD

24. What is the primary function of the firmware in a 3D printer?
   (a) Designing 3D models
   (b) Slicing 3D models
   (c) Managing various components and settings
   (d) Printing 3D models

25. Which software is renowned for mesh cleaning and simplification features?
   (a) MeshLab
   (b) Fusion 360

(c) Onshape
(d) TinkerCAD

26. Which software was developed by the ISTI-CNR research center?
(a) MeshLab
(b) Fusion 360
(c) Onshape
(d) TinkerCAD

27. Assertion: Fusion 360 integrates CAD, CAE and CAM into a single platform.
Reason: Fusion 360 is developed by Autodesk to cater to the needs of professionals and enthusiasts alike.
(a) Both the Assertion and Reason are true, and the Reason is the correct explanation for the Assertion
(b) Both the Assertion and Reason are true, but the Reason is not the correct explanation for the Assertion
(c) The Assertion is true, but the Reason is false
(d) The Assertion is false, but the Reason is true

28. Match the following 3D design software with their developers:

A. Fusion 360	1. Onshape Inc.
B. SOLIDWORKS	2. Autodesk
C. Onshape	3. Dassault Systèmes

(a) A-2, B-3, C-1
(b) A-3, B-2, C-1
(c) A-1, B-2, C-3
(d) A-1, B-3, C-2

29. Which feature in UltiMaker Cura modifies the layer height based on the model's geometry?
(a) Adaptive layers
(b) Tree support
(c) Infill density
(d) Print speed

30. **Assertion:**TinkerCAD is ideal for novices, educators and those venturing into 3D printing.
Reason: TinkerCAD's user-friendly tools and pre-designed models facilitate the creation of custom 3D designs.
(a) Both the Assertion and Reason are true, and the Reason is the correct explanation for the Assertion
(b) Both the Assertion and Reason are true, but the Reason is not the correct explanation for the Assertion
(c) The Assertion is true, but the Reason is false
(d) The Assertion is false, but the Reason is true

Answers

1. (b) Reading the G-code and sending instructions to other printer parts
2. (b) To regulate the amount of power applied to each of a stepper motor's individual coils
3. (c) Stepper drivers
4. (d) Wood
5. (b) Aluminium beds heat up quickly
6. (d) Aluminium
7. (a) SLA uses a laser, while DLP uses a projector
8. (d) Binder jetting printer
9. (b) Bowden
10. (d) Calibration is performed only after interfacing the 3D printer with a computer
11. (a) To ensure interchangeability and compatibility with other components
12. (b) How close a measured value is to the true value
13. (a) It refers to how closely a set of measurements are grouped together
14. (c) A smaller allowable range of variation
15. (c) A precise set of measurements may not necessarily be accurate
16. (a) The difference between the intended volume of the printed object as designed in the CAD model and the actual volume of the final printed part
17. (d) All of the above
18. (c) Increasing the resolution
19. (a) Inadequate toolpath generation algorithms
20. (b) By causing shrinkage or deformation during post-curing or finishing processes
21. (a) Fusion 360
22. (b) SOLIDWORKS
23. (a) MeshLab
24. (c) Managing various components and settings
25. (a) MeshLab
26. (a) MeshLab

27. (a) Both the Assertion and Reason are true, and the Reason is the correct explanation for the Assertion

28. (a) A-2, B-3, C-1

29. (a) Adaptive layers

30. (a) Both the Assertion and Reason are true, and the Reason is the correct explanation for the Assertion

3

Preprocessing – The Digital Model

The first step in 3D printing is to have a digital model of the object to be 3D printed. Computer-aided design (CAD) is the process of creating precise and accurate representations of objects on a computer screen called digital models. This chapter enables the learner to identify the CAD file format based on the CAD software used and comprehend typical steps involved in 3D printing of an object from the digital model. It also presents case studies to obtain the digital model of a complex component through typical CAD software and also using a reverse engineering path. The chapter comprises the following main sections:

3.1 Computer-Aided Design Model and File Format
3.2 Reverse Engineering to Obtain a CAD Model

After reading through the material, self-check exercises, objective questions and discussion topics presented in this chapter, the learner will be able to do the following:

(a) Identify the CAD file format based on the CAD software used.
(b) List typical steps involved in 3D printing of an object from a digital model.
(c) Develop a digital model of a given object using CAD software.
(d) Obtain a digital model of a complex component through reverse engineering.

3.1 Computer-Aided Design Model and File Format

CAD involves generating precise and accurate digital representations of objects on a computer, eliminating errors and ambiguities. These digital representations, known as solid models, are crucial for inspecting and verifying designs before 3D printing, as well as for planning the fabrication process.

DOI: 10.1201/9781003562344-3

The initial step in 3D printing is obtaining a digital model of the object. This can be achieved in three ways:

- Selecting a pre-existing 3D model from an online source.
- Creating a new design using 3D modelling software such as CATIA, SolidWorks, UG-NX, Creo, TinkerCAD, Blender, or AutoCAD.
- Performing a 3D scan of an existing product or use CT scan data from a human body part.

Integrating CAD and CAM provides insights into converting CAD models into CAM files suitable for 3D printing. This seamless connection between design and manufacturing ensures efficient and accurate production of the desired object.

3.1.1 Solid Modelling

Solid modelling involves the use of a method to generate, modify and visualize virtual renditions of objects within software. These precise digital representations, created by combining different solid geometries, contain geometric details and are depicted using points, lines and surfaces. They encompass both topological and geometric data, with the former specifying whether a feature is located within or outside the space or geometry. When performing 3D printing, it is important for the modelled object to have volume and be a fully enclosed geometry, resembling a sealed body. Various computer programs, often known as CAD software, can be used to create models suitable for 3D printing.

CAD is the most widely used method for creating digital models. There are three methods that can be used:

- **Design:** you can create your own designs using 3D modelling programs like CATIA, Creo, UG NX, AutoCAD, SolidWorks, Blender, SketchUp, Cubify Invent, Maya, PhotoShop, TinkerCAD, or others. These software applications offer a variety of tools and features specifically designed to model objects for different industries, such as automotive, aerospace, consumer products and industrial equipment.
- **Download:** a multitude of files made by other users are available for you to find and access by browsing through some of the websites that include Thingiverse, YouMagine, GrabCAD, MyMiniFactory, and Shapeways. You can either buy these files or download them for free from these platforms, based on the terms and conditions that each creator has set. You can quickly locate and acquire a variety of digital 3D models customized to your specifications by going to these websites. It is available to download CAD files with neutral file format so that it can

be opened in any CAD software for further modifications. By visiting these websites, you can quickly find and download a variety of digital 3D models that suit your needs and preferences.

- **Scan:** 3D scanning, also known as reverse engineering, is an additional method for producing a 3D digital file. Generating solid models suitable for 3D printing is a good use for this technique. An existing physical object is digitized by first capturing its geometric properties and then recreating it in a digital format. Some well-known software applications, like MeshLab, Autodesk Fusion 360, and Geomagic Design X, facilitate reverse engineering. These programs offer many tools for processing data, creating solid models and fine-tuning meshes.

Moreover, specialized software for modelling within particular domains – like biomedical modelling – is available. This field frequently uses Materialise Mimics and 3D Slicer to design implants tailored to individual patients, anatomical models for surgical planning, and a variety of other medical applications. It is now feasible to create precise and detailed models for 3D printing in a variety of sectors and applications due to developments in CAD software and reverse engineering methods. Future 3D printing will be more effective, precise and widely available due to the availability of specialized software for particular fields, which also increases the possibility of innovation and tailored solutions.

3.1.2 Solid Representation Methods

There are four popular methods of representing a solid geometry.

1. Constructive solid geometry (CSG)
2. Boundary representation (B-rep)
3. Feature-based modelling (FBM)
4. Space decomposition (SD)

To represent a solid, it is effective to combine different methods. SolidWorks/Pro-E/CATIA/NX, popular software, employs feature-based modelling, enabling the direct representation of part geometry using higher-level entities linked to specific design functions or manufacturing features. The constructive solid geometry (CSG) technique utilizes primitive geometry and Boolean operations to produce intricate surfaces and objects. The CSG file format encompasses both topological and geometrical data relating to the object. It constructs volumes by Boolean operations of various 3D objects, including cones, cylinders, spheres, torus, blocks, closed spline surfaces and right angular wedges. Remarkably, these geometries can be succinctly represented using straightforward mathematical equations. Moreover, CSG modelling encompasses solids formed by sweeping any 3D geometry along

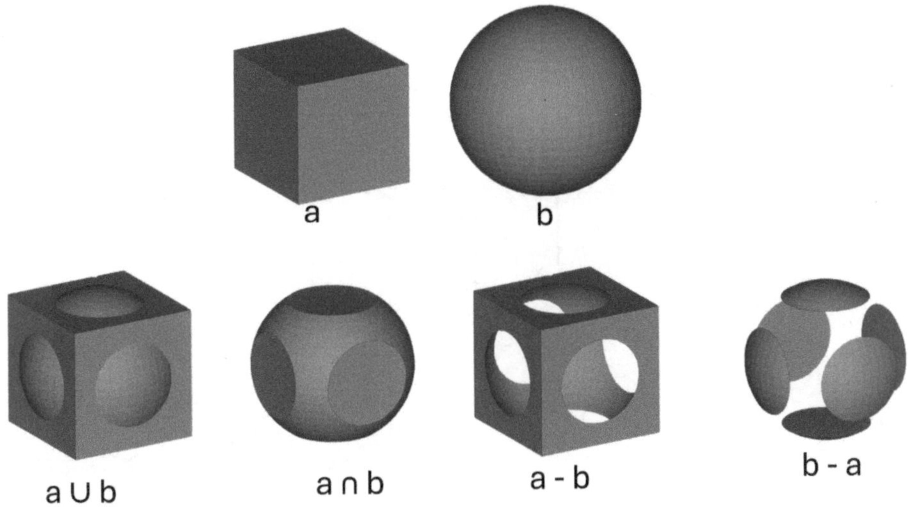

FIGURE 3.1
Constructive solid geometry-based representation.

a line or curve. Boolean operations, such as union, intersection and subtraction, are employed to craft the desired geometry, as depicted in Figure 3.1.

Boundary representation (B-rep) is a method of representing a solid model using its surfaces, as the name suggests. B-rep can be seen as an expansion of the wireframe model. Unlike CSG, B-rep stores faces, edges and vertices, and computes them when necessary. By definition, a solid is always enclosed by multiple surfaces, which include internal and external faces. Since faces can share vertices and have curved edges, B-rep improves the wireframe model by integrating face data and providing a more comprehensive understanding of the object's structure. This incorporation of additional information about the surfaces forming the boundaries enhances the understanding of the object's structure. In B-rep, there are two types of information: topological and geometric. Topological information outlines the relationships between vertices, edges and faces, similar to a wireframe model, and include edge and face orientation as well as connections. Geometric information is typically represented by equations for edges and faces.

The positioning of each face is crucial. A face is usually defined by a set of vertices, and these vertices must be arranged in such a way that the face's normal vector points towards the exterior of the solid, often achieved by ordering them counterclockwise using the right-handed rule. If the face is represented by an equation, the equation must be adjusted to ensure that the normal vector at each point on the face points away from the solid. Thus, understanding normal vectors enables the easy differentiation between the inside and outside of a solid in B-rep. This orientation process must be carried out for all faces. A six-faced object's boundary representation is shown in

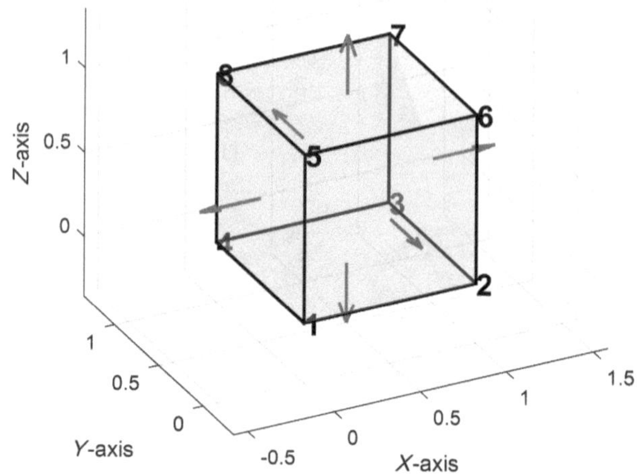

FIGURE 3.2
Boundary representation of a cube.

Figure 3.2. Every face has four vertices and is connected to other faces by an edge. In order for the face normal to point outward, the vertices used to represent the face must be placed in the correct order.

Feature-based modelling involves defining the geometry and topology of a solid model based on specific design features, such as extrusions, revolves, sweeps, fillets, sketches, patterns and mirrors. These features are added or subtracted to create a solid model, resulting in a more intuitive design process that closely mimics actual manufacturing steps. The following example demonstrates a solid model created (refer to Figure 3.3) using feature-based modelling techniques:

1. Use the sketch feature to create a 2 mm by 4 mm rectangle.
2. Extrude the sketch by 1.5 mm to form a rectangular prism.
3. Generate a cylinder on the top face of the prism by sketching circles and extruding them.
4. Extrude the circles to create cylindrical geometry on top of the rectangular prism.

In this instance, the design makes use of "sketch" and "extrude" features to create the solid model. The sketch feature permits the designer to outline fundamental 2D shapes, while the extrude feature converts these 2D sketches into 3D forms by stretching them along a specific direction. By utilizing feature-based modelling, designers can benefit from a more intuitive design process, the ability to easily modify the model, and a clearer understanding

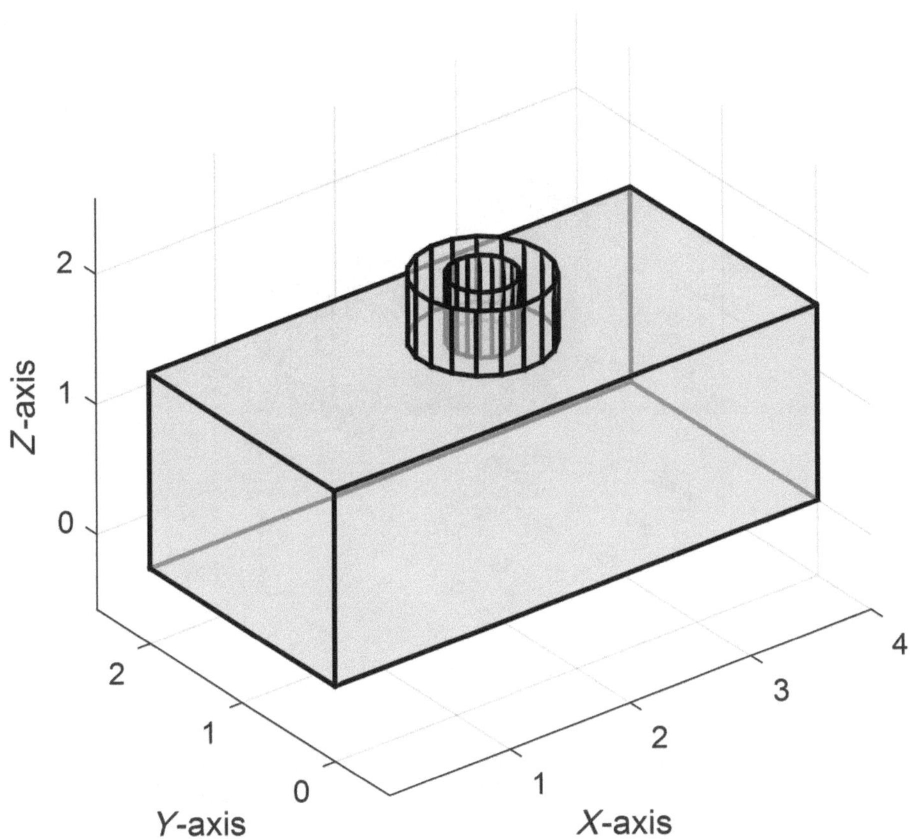

FIGURE 3.3
Feature-based modelling.

of the manufacturing steps required to create the final product. This approach enables designers to focus on the crucial aspects of the model and make necessary adjustments without having to overhaul the entire geometry.

Space decomposition is a voxel-based modelling technique used to represent 3D models. It involves dividing the 3D space into a regular grid, where each cell in the grid is represented by a small cube called a voxel (refer to Figure 3.4). These voxels can be manipulated individually to create or modify the 3D model. The use of voxel-based modelling is prevalent in various fields, including medical imaging, volumetric rendering and simulations. This approach offers a simple method for generating and modifying intricate 3D shapes. Voxel-based models are particularly adept at managing extensive data sets and complex geometries, which makes them well-suited for tasks demanding a high degree of precision.

Different CAD software packages utilize distinct CAD modelling methods. These methods can be employed to create various solid models. It is clear that

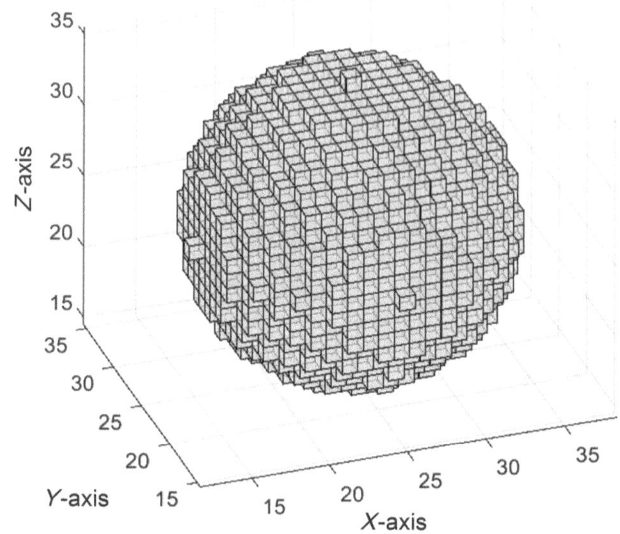

FIGURE 3.4
Voxel-based model of a sphere.

different CAD modelling methods result in diverse encodings of CAD output files. To link CAD with CAM, it is essential for the CAM software to be capable of interpreting all such files. However, the encoding methods are not confined to the types we have discussed here. Many CAD packages employ a hybrid modelling approach, so it is necessary to convert these diverse models into a uniform file format that follows a standard and can be easily decoded by the CAM software. Comparison of various CAD modelling approaches is described in Table 3.1

3.1.3 Case Study of Creating a CAD Model of a Component Using CATIA

Component to be modelled with dimensions is shown in Figures 3.5, 3.6 and 3.7.

Major steps to be followed:

1. Double click on the CATIA software launch icon
2. Open new part file
3. Set units
4. Draw feature-wise 2D sketches of the component
5. Covert these 2D sketches in the corresponding 3D features to create the final 3D CAD model

TABLE 3.1

Comparison of Various CAD Modelling Approaches

Aspect	Constructive Solid Geometry (CSG)	Boundary Representation (B-rep)	Feature-Based Modelling (FBM)	Space Decomposition (SD)
Core Principle	Combines simple geometric shapes using Boolean operations.	Represents shapes by their boundaries (faces, edges).	Focuses on defining and manipulating features (holes, bosses).	Divides space into simpler sub-volumes for representation.
Representation Method	Uses primitives (cubes, spheres) and Boolean operations.	Uses vertices, edges, and faces to describe objects.	Uses high-level features to describe the geometry.	Uses spatial subdivisions (e.g., voxels, octrees).
Modelling Focus	Focuses on the solid's volume and its composition.	Focuses on the surface and outer boundaries of objects.	Focuses on the functional aspects of features within the model.	Focuses on the spatial organization and partitioning of space.
Complexity Handling	Handles complex objects through combination of primitives.	Handles complex shapes through detailed surface descriptions.	Handles complexity by combining and modifying features.	Handles complexity by breaking it down into manageable regions.
Typical Applications	Simple geometric modelling, Boolean operations.	Detailed modelling of complex surfaces and shapes.	Engineering and design where features are critical (e.g., CAD).	Applications requiring efficient spatial queries and computations.
Editing Process	Editing involves modifying primitives or Boolean operations.	Editing involves manipulating surface elements.	Editing involves modifying or adding features.	Editing involves adjusting spatial partitions or regions.
Data Structure	Tree-based structure representing Boolean combinations.	Graph-based structure with vertices and edges.	Hierarchical structure with feature dependencies.	Grid-based or hierarchical spatial partitions.
Performance	Can be efficient for simple models but complex operations can be computationally intensive.	Can be efficient for detailed models but can become complex with highly intricate surfaces.	Typically efficient for design and manufacturing with feature-based operations.	Can be efficient for large-scale space management and simulations.

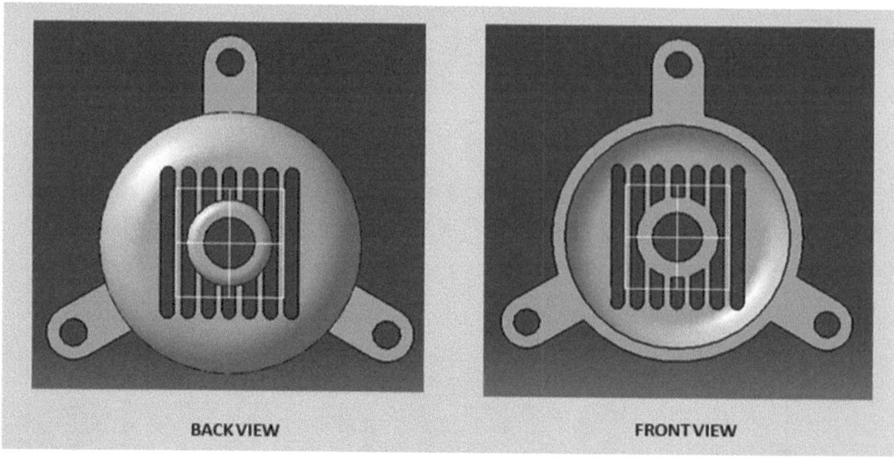

FIGURE 3.5
Front and back views of the component to be modelled.

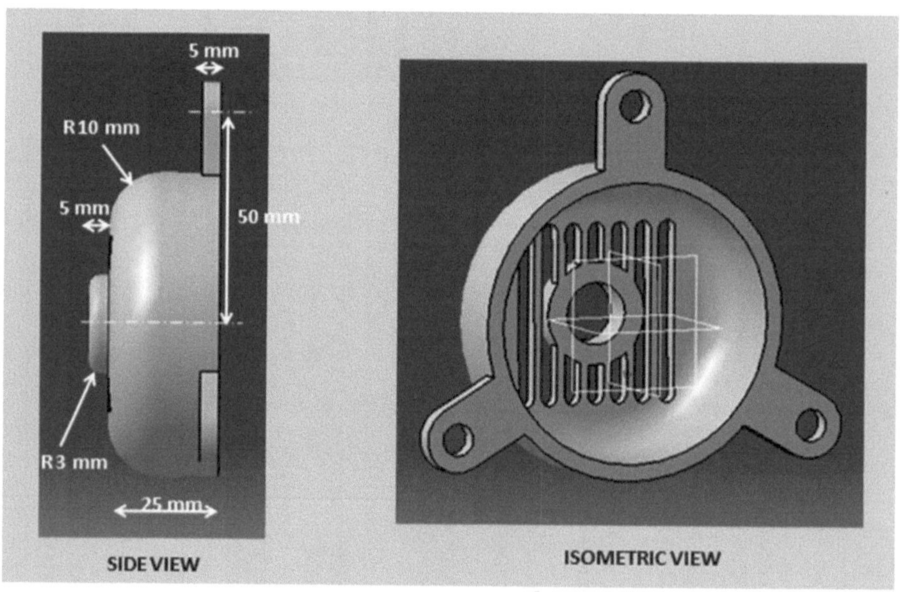

FIGURE 3.6
Side and isometric views and dimensions of the component to be modelled.

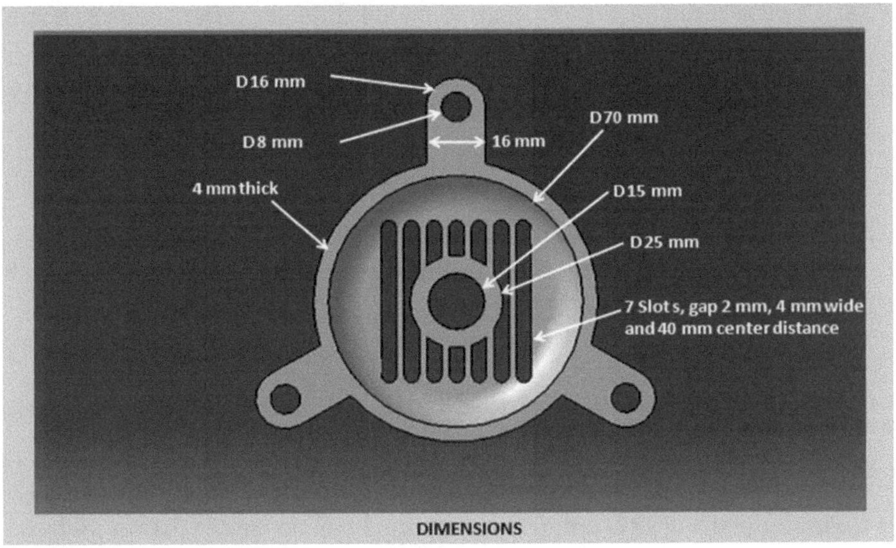

FIGURE 3.7
Front views and dimensions of the component to be modelled.

6. Save the file as CAT.prt native CATIA file format
7. Save this file as .stl neutral CAD file format
8. Open this .stl file in the 3D printer software

Micro steps to be followed:

1. Choose the *y-z* plane, create a circle with a diameter of 70 mm, and extrude it with a first limit of 25 mm (Feature-1).
2. Apply a 10 mm radius fillet to one of the circular edges of Feature-1 (Feature-2).
3. Shell Feature-1 with a 1 mm inside thickness and a 3 mm outside thickness, then remove one face (Feature-3).
4. On the back face of Feature-3, sketch a circular elongated slot according to the specified dimensions.
5. Cut a pocket into this sketch with a depth of 20 mm (Feature-4).
6. Utilize the rectangle pattern operation on Feature-4, creating four instances in the first direction with a 6 mm gap and one instance in the second direction with no gap. Repeat this process for the right side as well (Feature-5 and 6).
7. Sketch a 25 mm diameter circle and extrude it with a first limit of 5 mm and a second limit of 4 mm on the inner face of Feature-3 (Feature-7).

8. On the outer face of Feature-7, create a 15 mm diameter circle and remove material up to the next surface (Feature-8).
9. On the circular strip plane of Feature-3, sketch the projected shape as specified and extrude it by 5 mm (Feature-9).
10. Use the circular pattern operation on Feature-9, creating three instances with an angle of 120° (Feature-10).
11. Save the component. The resulting file will have a .CATprt extension.
12. Save the same file with a .stl extension using the "Save as" function to prepare it for 3D printer software.

3.2 Reverse Engineering to Obtain a CAD Model

Reverse engineering (RE) entails the creation of an accurate CAD model from an existing physical object, component, or body part. This process serves various purposes, including design modification, architectural understanding and functionality analysis. The advancement of 3D printing has significantly improved the speed and efficiency of replicating and fabricating part copies.

The first step in reverse engineering is obtaining the geometric data of the physical part to develop its CAD model. Simple objects with primitive geometry can have their data directly acquired using measuring tools. However, more complex parts require 3D scanning methods. In the medical field, human body parts' geometric data is collected from CT or magnetic resonance imaging (MRI) scans.

Reverse engineering has diverse applications across different domains, such as rapid product development, casting, CNC machining, medical imaging and demonstration, medical tooling and part inspection. This section aims to provide an understanding of the principles, techniques and applications of reverse engineering in today's fast-paced technological landscape.

3.2.1 Reverse Engineering Approaches

The process of creating a CAD model from existing parts made of basic shapes involves measuring the geometric details using different tools and then using CAD software to create a solid model. However, for parts with intricate geometry, like freeform surfaces or natural shapes that are hard to measure, 3D scanning or other methods are used to obtain the needed geometric data.

Point cloud data from a part's surface are obtained through a process called 3D scanning. Preprocessing operations, such as removing outliers and spikes, reducing point data and organizing it, are often necessary for this point data. When using a mesh-based surfacing technique, the well-organized point data is either meshed or segmented for a curve-based surface modelling approach to create a CAD model. Fortunately, due to sophisticated computational techniques and algorithms, pretreatment is now virtually unnecessary with the point cloud processing technologies available today.

Following the scanning process, the point cloud data is converted into a surface model or voxel model of the geometry. This surface model is then converted into an STL file or other formats suitable for 3D printing before being printed in 3D (refer to Figure 3.8).

Two types of devices, contact and non-contact, are utilized to collect surface data from a component. Additionally, medical imaging techniques are used to obtain point cloud data for layered or concealed objects, such as human bones, that cannot be scanned with 3D scanners.

(a) Coordinate measurement machines (CMMs) are frequently employed for inspection and tolerance purposes. These NC-driven devices can gather point data with high precision of many microns. However, they are inherently slow at collecting point data as each sampled point necessitates physical contact with the part's surface. Consequently, CMMs are inefficient for measuring freeform components and are mainly utilized for the automated inspection of simple fundamental shapes like slots, steps, holes and pockets.

(b) In contrast, non-contact devices do not necessitate physical contact with a part to gather point data, making them a more practical option for rapidly gathering large volumes of data. Laser scanners are the most commonly utilized non-contact measurement instruments as they can rapidly and accurately collect large volumes of point data. The accuracy of laser scanning equipment has improved, and its industrial application has significantly increased.

Computer tomography (CT) scanners are used to scan the material composition and interior geometry of components. They are also used to visualize the human body for medical purposes and to design surgical implants. High-power industrial CT scanners are used to gather information about the interior shape, porosity and density of industrial items. In the medical field, reverse engineering involves creating accurate 3D models of body parts or internal structures based on CT or MRI scans, typically available in digital imaging and communications in medicine (DICOM) format. This process is important for various applications such as surgical planning, implant design and education.

The process of reverse engineering medical parts begins (refer to Figure 3.9) with the acquisition and processing of CT or MRI scans containing specific anatomical details. These scans are then meticulously segmented by an experienced operator based on pixel values, which correspond to the x-ray attenuation coefficient of the body part. This crucial segmentation process

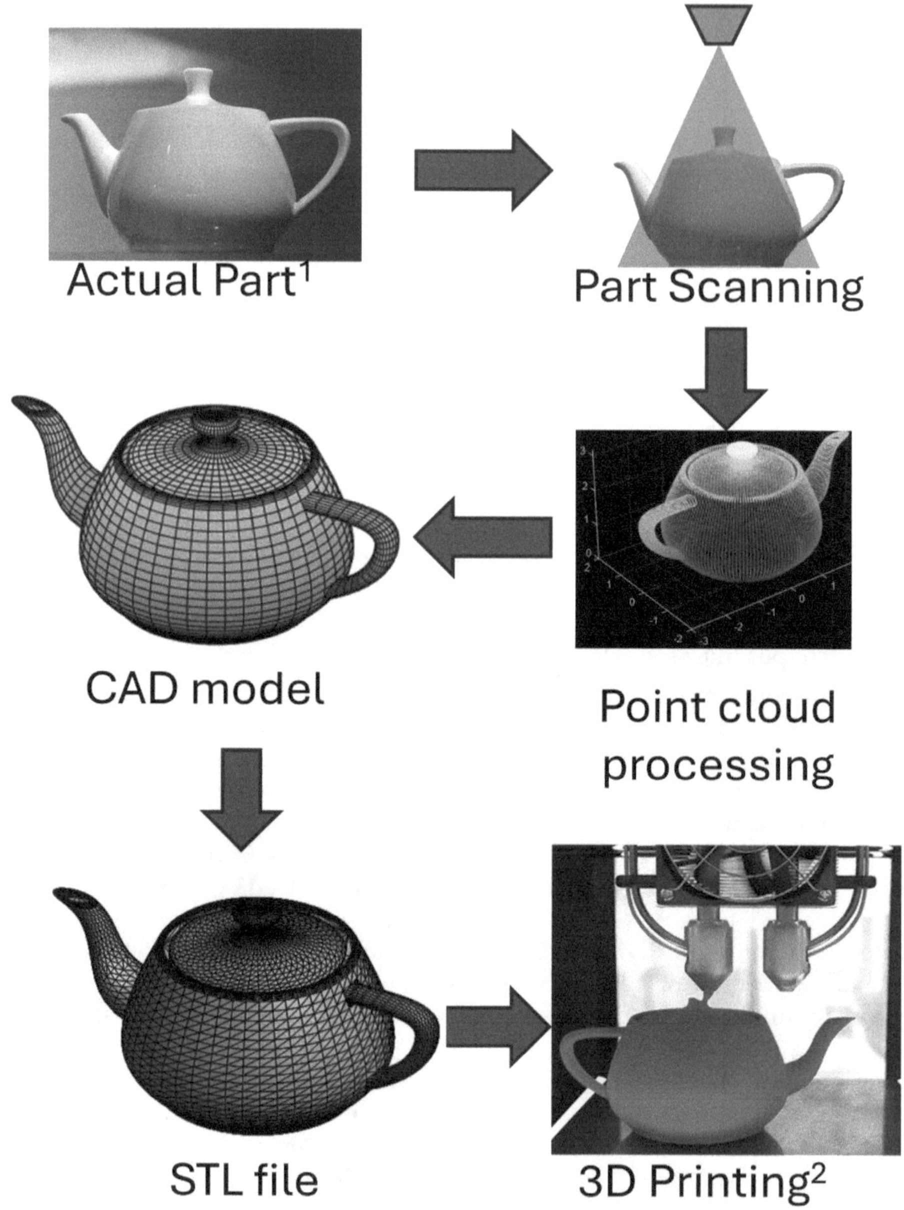

FIGURE 3.8
Reverse engineering steps (Utah teapot. (2024, February 14). In Wikipedia. https://en.wikipedia.org/wiki/Utah_teapot, https://inspirationtuts.com/the-important-utah-teapot/).

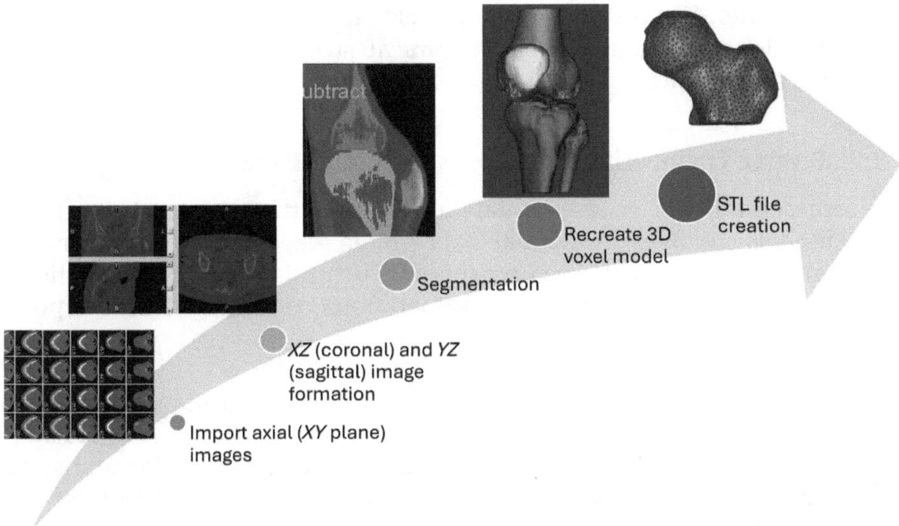

FIGURE 3.9
CT scan to 3D model conversion steps.

aims to precisely isolate the desired area of the CT scan file for further processing.

Following successful segmentation, the operator proceeds to create a mask that accurately represents the segmented part of the CT scan images' 3D array. This mask plays a crucial role in generating the point cloud data, which is subsequently transformed into a detailed voxel model. Voxel models serve as a comprehensive representation of 3D objects on a computer screen, achieved by dividing the object into a three-dimensional grid, with each cell in the grid termed as a voxel. This representation is invaluable for visualizing the intricate solid geometry of the scanned body part.

Once the voxel model is created, the point cloud data undergoes transformation into an STL (stereolithography) file format. STL files are widely utilized in 3D printing and additive manufacturing processes, as they succinctly describe the surface geometry of a 3D object without representing attributes such as colour or texture.

Finally, the STL file is utilized to 3D print the physical model of the medical part. These 3D printed models serve a myriad of purposes, from aiding surgeons in procedural planning and designing custom implants to educating patients about their anatomy or impending surgeries.

Reverse engineering of medical parts is an intricate and indispensable process that demands specialized software, expert knowledge and meticulous attention to detail. It encompasses tasks such as obtaining and processing medical data, segmenting images, creating a mask, converting point cloud data into a voxel model, generating an STL file, and ultimately 3D printing the final model. The resulting 3D models significantly augment medical professionals' understanding of anatomical structures and contribute to improved patient care.

3.2.2 Reverse Engineering Operations for Cloud Data

The representation of geometric data in 3D space through point cloud and mesh models involves individual data points, each with its coordinates. Sometimes, it is necessary to process this data by utilizing various operations. Here, we will delve into the general operations commonly used for point cloud data.

1. Noise reduction: point cloud data acquired from scanning devices or environmental factors often contains unwanted noise. To enhance the accuracy of the data, noise reduction algorithms like statistical outlier removal and moving least squares are utilized to filter out the noise.
2. Registration: when creating a 3D scan of a large or complex object, multiple scans may be necessary to capture the entire geometry. These individual scans are aligned and merged using registration algorithms such as iterative closest point (ICP) or global registration to form a complete and seamless point cloud.
3. Down-sampling: high-resolution point cloud data can be computationally demanding. Down-sampling algorithms like voxel grid filtering and random sampling reduce the number of points while preserving the overall shape and features of the point cloud.
4. Normal estimation: surface normal vectors, which are perpendicular to the surface at each point, are essential for tasks like mesh generation and surface reconstruction. Algorithms such as principal component analysis (PCA) and normal estimation using *k*-nearest neighbours compute the normal vectors for each point.
5. Feature extraction: point clouds obtained through LiDAR or photogrammetry contain 3D data sets. Identifying features such as edges, corners and planes is crucial for object recognition, segmentation and registration. Feature extraction algorithms like FPFH and SIFT detect and describe these features, with FPFH being effective in detecting local geometric features and SIFT being useful in detecting scale-invariant features.
6. Segmentation: point clouds may include multiple objects or parts that need to be segmented for further processing. Segmentation algorithms such as region growing, Euclidean clustering, and RANSAC (random sample consensus) partition the point cloud into significant segments based on geometric criteria or feature similarity.

When converting point cloud data into a mesh model, it is essential to create a continuous surface that accurately represents the object's geometry. This process is called surface reconstruction and involves using specialized algorithms to estimate the surface by connecting neighbouring points and forming a mesh of connected triangles. Surface reconstruction algorithms like Poisson surface reconstruction, ball-pivoting algorithm (BPA) and marching cubes are crucial for creating accurate 3D models from point cloud data.

To ensure accurate and high-quality 3D printing projects, it is important to have a solid understanding of basic operations and algorithms for processing and manipulating point cloud data. Fortunately, there are various software tools and libraries available, such as Point Cloud Library (PCL), MeshLab and CloudCompare, which offer a wide range of point cloud processing functions and algorithms for creating precise mesh models for 3D printing needs.

Reverse engineering for 3D printing has revolutionized industries such as product development, manufacturing and medical applications. Advanced 3D scanning methods, such as laser scanners and medical imaging techniques, provide precise geometric data to create accurate and high-quality CAD models of complex objects and internal structures. The subsequent conversion of this data into surface or voxel models and 3D printable formats has enabled rapid prototyping, customization and the development of innovative solutions across various domains. As technology advances, the potential applications and benefits of reverse engineering for 3D printing will expand, contributing to a more efficient, personalized and cutting-edge future in design and manufacturing.

Self-Check Exercises

1. Explain the four popular solid representation methods with the help of examples.
2. List the areas where reverse engineering is a preferred way to model the given object.
3. Differentiate between solid model and wire frame.
4. If you want to add or remove material along a normal or inclined vector to a 2D surface which commands will be useful in Pro-E, NX and CATIA CAD software?
5. If you want to add material or remove about an axis with angle of rotation to a 2D surface which commands will be useful in Pro-E, NX and CATIA CAD software?

6. If you want to add material or remove along a curved path to a 2D surface which commands will be useful in Pro-E, NX and CATIA CAD software?
7. Try to prepare a digital CAD model of any typical teapot/the following object in Figure 3.10.

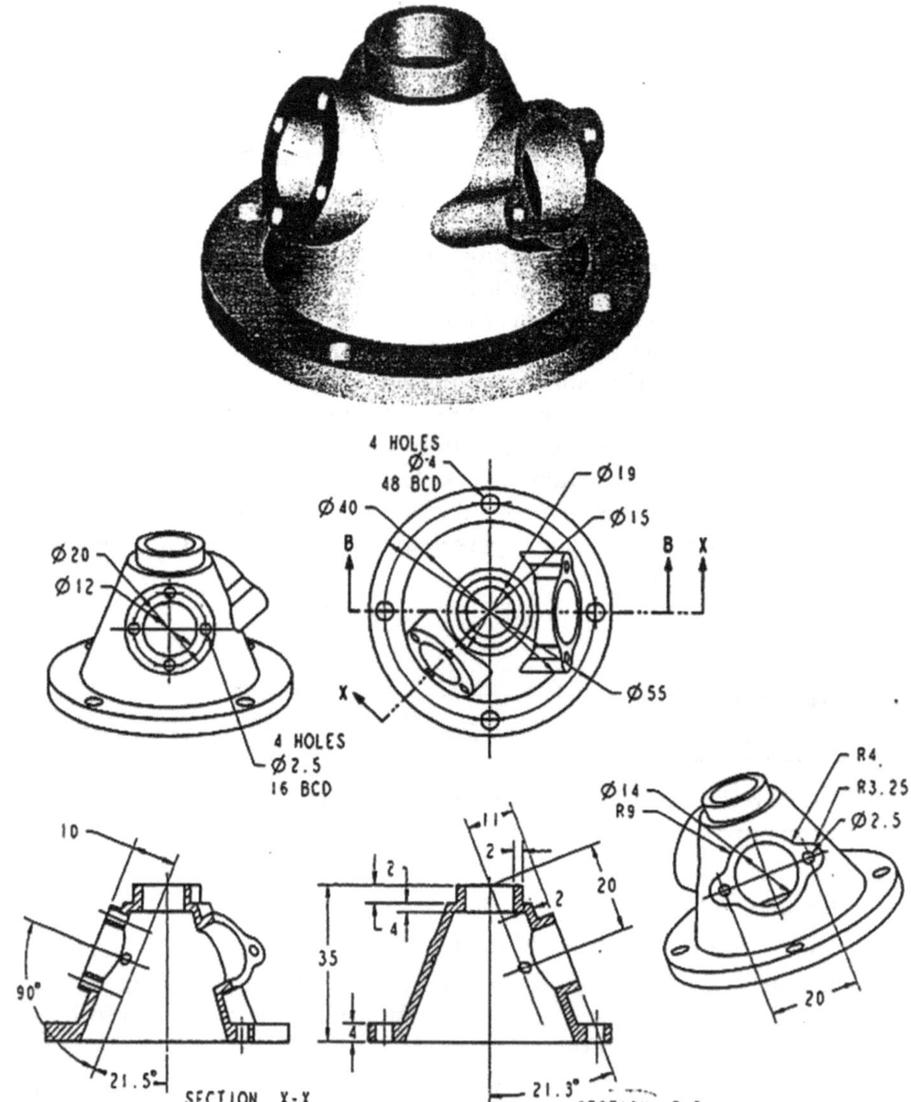

FIGURE 3.10
Self Check Exercise 7.

8. Depending on the complexity of the object and the resources available, describe the method you would use to obtain the geometric data of the object, such as manual measurement or 3D scanning. Explain your choice and the potential challenges or limitations of the chosen method.
9. Watch two videos on how to create a simple digital model using CATIA CAD Software and SOLIDWORKS CAD Software.
10. Choose a household object with moderate complexity (e.g., a kitchen utensil, a small tool or a decorative item) that you believe would benefit from reverse engineering and 3D printing. Describe the object and explain why you selected it for this assignment.

Discussion Topic

1. Discuss the use of CAD software vs scanning for creating digital models.

Multiple Choice Questions

1. What is the primary function of CAD software in the context of 3D printing?
   (a) To create virtual models of objects
   (b) To control the 3D printer's movement
   (c) To decide the material used for printing
   (d) To analyze the final printed object
2. Which of the following file formats is most suitable for 3D printing CAM packages?
   (a) IGES
   (b) STP
   (c) STL
   (d) DXF
3. Which solid representation method involves the use of primitive geometry and Boolean operations to generate complex surfaces and objects?
   (a) Constructive solid geometry (CSG)
   (b) Boundary representation (B-rep)
   (c) Feature-based modelling (FBM)
   (d) Space decomposition (SD)

4. In the context of 3D printing, what does CAM stand for?
   (a) Computer-aided manufacturing
   (b) Computer-aided modelling
   (c) Computer-aided movement
   (d) Computer-aided material
5. Why are triangles preferred for solid representation in 3D printing over other shapes like parallelograms or hexagons?
   (a) Triangles are more aesthetically pleasing
   (b) Triangles offer simplicity, speed and versatility in geometric computations
   (c) Triangles are easier to manufacture
   (d) Triangles require less material
6. What is the primary purpose of neutral file formats in CAD?
   (a) To enable full utilization of all features of a specific program
   (b) To facilitate easy file exchange between different CAD software
   (c) To store files specific to a particular CAD software
   (d) To provide a unique file format for each CAD software
7. Which of the following is an example of a native file format?
   (a) STEP (.stp, .step)
   (b) IGES (.igs, .iges)
   (c) AutoCAD (.dwg)
   (d) Parasolid (.x_t, .x_b)
8. In FBM, which feature is used to create a 3D object by extruding a 2D sketch along a specified path?
   (a) Sketch feature
   (b) Extrude feature
   (c) Revolve feature
   (d) Sweep feature
9. Which of the following is the primary purpose of the IGES file format?
   (a) To store design data specific to a particular CAD software
   (b) To facilitate the exchange of 3D model data between different CAD software
   (c) To represent a solid model using primitive geometry
   (d) To provide a unique file format for each CAD software
10. Which of the following is a key advantage of the space decomposition method of solid representation in comparison to other methods?
    (a) It provides a more compact representation of complex geometry
    (b) It allows for easy modification of individual features
    (c) It supports efficient spatial queries and proximity calculations
    (d) It simplifies Boolean operations on solid models

11. What is the primary purpose of reverse engineering in 3D printing?
    (a) To create a precise and accurate CAD model from an existing physical object or part
    (b) To improve the quality of 3D printed objects
    (c) To make the 3D printing process faster
    (d) To reduce the cost of 3D printing
12. Which of the following is NOT a typical application of reverse engineering?
    (a) Rapid product development
    (b) Casting
    (c) CNC machining
    (d) Creating 2D drawings from 3D models
13. What type of measuring device requires physical contact with the part's surface to obtain point data?
    (a) Laser scanner
    (b) Coordinate measuring machine
    (c) Computed tomography scanner
    (d) Magnetic resonance imaging scanner
14. Which type of scanner is used for capturing internal shapes and material density of parts?
    (a) Laser scanner
    (b) Coordinate measuring machine
    (c) Computed tomography scanner
    (d) Magnetic resonance imaging scanner
15. What is the process of isolating a specific area within a CT scan image called?
    (a) Registration
    (b) Segmentation
    (c) Surface reconstruction
    (d) Noise reduction
16. Which point cloud processing operation helps in reducing the number of points in the point cloud while retaining its overall shape and features?
    (a) Noise reduction
    (b) Registration
    (c) Down-sampling
    (d) Normal estimation
17. Which algorithm is used for partitioning the point cloud into significant segments based on geometric criteria or feature similarity?
    (a) Region growing
    (b) Euclidean clustering
    (c) RANSAC
    (d) All of the above

18. Which surface reconstruction algorithm connects neighbouring points and forms a mesh of connected triangles?
    (a) Poisson surface reconstruction
    (b) Ball-pivoting algorithm (BPA)
    (c) Marching cubes
    (d) All of the above
19. Which software tool is NOT commonly used for point cloud processing?
    (a) Point Cloud Library (PCL)
    (b) MeshLab
    (c) CloudCompare
    (d) Adobe Illustrator
20. In the context of 3D printing, which of the following steps is NOT a part of the reverse engineering process?
    (a) Capturing the physical object's geometry using a 3D scanner
    (b) Generating a point cloud or mesh model from the scanned data
    (c) Creating a CAD model based on the point cloud or mesh model
    (d) Slicing the CAD model to generate G-code instructions for the 3D printer

Answers

1. (a) To create virtual models of objects
2. (c) STL
3. (a) Constructive solid geometry (CSG)
4. (a) Computer-aided manufacturing
5. (b) Triangles offer simplicity, speed, and versatility in geometric computations
6. (b) To facilitate easy file exchange between different CAD software
7. (c) AutoCAD (.dwg)
8. (d) Sweep feature
9. (b) To facilitate the exchange of 3D model data between different CAD software
10. (c) It supports efficient spatial queries and proximity calculations
11. (a) To create a precise and accurate CAD model from an existing physical object or part
12. (d) Creating 2D drawings from 3D models
13. (b) Coordinate measuring machine
14. (c) Computed tomography scanner
15. (b) Segmentation

16. (c) Down-sampling
17. (d) All of the above
18. (d) All of the above
19. (d) Adobe Illustrator
20. (d) Slicing the CAD model to generate G-code instructions for the 3D printer

4

Preprocessing – Setting 3D Printing Parameters

Different CAD modelling software or reverse engineering tools are used to create a solid model (digital model) of the required part. The model in digital format needs to be supplied to any process planning software (3D printer software). The integration of CAD and CAM provides insights into how the CAD model of an object can be converted into a CAM file suitable for 3D printing. This seamless connection between design and manufacturing processes ensures efficient and accurate production of the desired object. This chapter gives information of .stl file formats and operation associated with it. It also provide information about the optimized part orientation in 3D printing process ensuring higher-quality prints and more efficient use of resources. Generation of supports for effective 3D printing of parts and use of effective method for slicing the CAD model to finally generate G-code file for printing the object are also covered in the chapter. The chapter comprises the following main sections:

4.1 Native and Neutral File Formats
4.2 STL File Format and Manipulations
4.3 Part Orientation and Support Generation
4.4 Slicing and Deposition Path Generation

After reading through the material, self-check exercises, objective questions and discussion topics presented in this chapter, the learner will be able to do the following:

(a) Differentiate native and neutral file formats.
(b) Create .stl file formats for 3D printing.
(c) Optimize part orientation in 3D printing process planning software, ensuring higher-quality prints and more efficient use of resources.
(d) Generate supports for effective 3D printing of parts.
(e) Use an effective method for slicing the CAD model.
(f) Generate a G-code file for printing the object.

DOI: 10.1201/9781003562344-4

4.1 Native and Neutral File Formats

CAD modelling is the first step in preparing for 3D printing. There are three different methods that can be used to create a solid model of the product intended for 3D printing (as discussed in Chapter 2). Various CAD modelling software or reverse engineering tools are employed to generate a digital model of the required part. This digital model needs to be imported into process planning software (3D printer software). To achieve this, the CAD file, typically in a native file format, must be converted into a compatible format.

There are two file formats used with CAD files:

Native file formats (refer to Table 4.1) are essentially dictated by the specific CAD (computer-aided design) software you have chosen to employ. By using these native formats, you are granted the opportunity to explore and take advantage of all the intricate features and functionalities the software has to offer, thereby maximizing your design capabilities and overall experience.

Neutral file formats (refer to Table 4.2) are intentionally designed to simplify the sharing of files among users of various CAD applications. This uncomplicated method facilitates smooth collaboration and compatibility across multiple platforms, fostering a culture of creativity and advancement in computer-aided design.

The IGES, STP and STL file formats are commonly used in CAD and 3D printing. These formats follow standardized protocols for encoding geometric data. Among them, the STL format is particularly well-suited for 3D printing as it represents solids using a triangular mesh, making it compatible and user-friendly for 3D printing computer-aided manufacturing (CAM) packages. To prepare a CAD model for 3D printing, the initial step in the

TABLE 4.1

Native CAD File Formats

Format	Extensions
AutoCAD	.dwg
CATIA V4	.model
CATIA V5, V6	CATPart, CATProduct, CGR
Creo	.prt, .asm
NX, Unigraphics	.prt, .jt, .j_t
Pro/E	.prt, .asm
Inventor	.ipt, .iam
MicroStation	.dgn
SolidWorks	.sldprt, .sldasm
Solid Edge	.par, .asm, .psm

TABLE 4.2

Neutral CAD File Formats

Format	Extensions
STEP	.stp, .step
STL	**.stl**
IGES	.igs, .iges
3D PDF	.pdf
Parasolid	.x_t, .x_b
VRML	.vrml
X3D	.x3d
COLLADA	.dae
DXF	.dxf

process chain involves converting the model into the STL file format. This conversion is essential for the successful execution of the 3D printing process, which will be further discussed in the upcoming chapter.

In comparison to the OBJ file format, which is another standard 3D file format, STL files for 3D printing are simpler and smaller, making them easier to share and publish. However, it is important to note that OBJ files contain information about colour and texture. On the other hand, PLY files are used for objects that have been 3D scanned.

Typically, CAD packages offer the option to save or export the CAD model in various formats, including STL.

4.2 STL File Format and Manipulations

STL is a widely used file format for 3D printing and CAD. It is identified by the .stl extension and the acronym stands for stereolithography, a popular 3D printing technology. It is also known as the standard triangle language or standard tessellation language. STL files are compatible with most 3D printers and CAD software, representing geometries through tiling the surface with millions of triangles. These files use a series of interconnected triangles (tessellation) to describe the surface geometry of a 3D model or object. The complexity of the design determines the number of triangles used and the resulting resolution.

4.2.1 Stl Interface Specification

STL files can store information in two different ways. These are

- Binary encoding and
- ASCII encoding

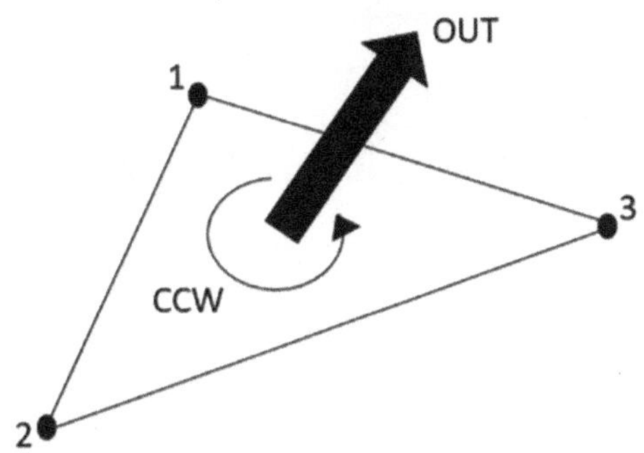

FIGURE 4.1
The orientation of the facet shall be determined by the direction of the normal unit and the order in which the vertices are indicated.

Binary files are smaller and easier to share while ASCII files are visually easy to read and check. Binary format is recommended to use for 3D printing. However, ASCII format is suggested for those who want to manually inspect the STL file for debugging.

The STL file is a list of facet (triangular) information. Each facet is uniquely identified by a normal unit vector (a line perpendicular to the triangle and a length of 1.0) and three vertices (corners). Normal and each vertex are described by three coordinates each, so a total of 12 numbers are stored for each facet.

Facet orientation: the facets are used to define the surface of a three-dimensional object. As such, each facet is part of the boundary between the inside and the outside of the object. The orientation of the facets (which path is "out" and which path is "in") is given redundantly in two ways that must be consistent. The direction of the normal is outward, first. Second, the vertices are seen in the anti-clockwise order when looking at the object from the outside (right-hand rule). These rules are shown in Figure 4.1.

Vertex-to-vertex rule: each triangle must share two vertices with each of its neighboring triangles. In other words, the vertex of one triangle cannot lie on the other side. This is shown in Figure 4.2. The object shown must be placed in the all-positive octant. In other words, all vertex coordinates must be positive-defined (non-negative and non-zero) numbers. The STL file does not contain any scale information and the coordinates are stored in arbitrary units.

4.2.1.1 STL ASCII Format

The ASCII format is primarily utilized for the development of new CAD interfaces. The large size of its files makes it infeasible for general use.

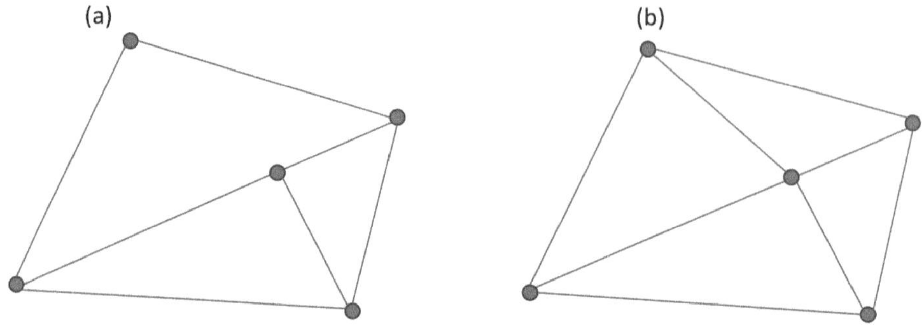

FIGURE 4.2
Vertex-to-vertex rule. (a) shows a violation of rule; (b) shows the correct configuration.

The syntax of the ASCII STL file is as follows:

```
solid name
{facet normal n_j n_j n_k                 }+
        outer loop
                vertex v1_x v1_y v1_z
                vertex v2_x v2_y v2_z
                vertex v3_x v3_y v3_z
        endloop
    endfacet
endsolid name
```

Boldface indicates the keyword, it must appear in lowercase letters. Note that there is space in "facet normal" and "outer loop," while there is no space in any of the keywords starting with "end." The indentation must be with spaces; the tabs are not allowed. The notation"{...} +" implies that the contents of the brace brackets can be repeated one or more times. Italic symbols are variables to be replaced by user-specific values. The numerical data for facet normal and vertex lines are single precision floats, e.g. 1,23456E+789. A facet normal coordinate may have a leading minus sign and it may not have a vertex coordinate.

4.2.1.2 STL Binary Format

A binary STL file has the following structure:

- An 80 byte header that can be used as a title.
- A 4 byte unsigned long integer, the number of facets.

- For each facet, a facet record of 50 bytes (the result is that each facet is represented by 50 bytes, 12 for the normal, 36 for the three vertices, and two for the spacer.)

The facet record has the form:

- The normal vector, three floating values of 4 bytes each;
- Vertex 1 *XYZ* coordinates, three floating values of 4 bytes each;
- Vertex 2 *XYZ* coordinates, three floating values of 4 bytes each;
- Vertex 3 *XYZ* coordinates, three floating values of 4 bytes each;
- An unsigned integer, of 2 bytes, that should be zero;

The binary format uses IEEE integer and floating point numeric representation.

The syntax for the STL binary file is as follows:

Bytes	Data type	Description
80	ASCII	Header; No data significance
4		Number of facets in file
4	float	*i* for normal
4	float	*j*
4	float	*k*
4	float	*x* for vertex 1
4	float	*y*
4	float	*z*
4	float	*x* for vertex 2
4	float	*y*
4	float	*z*
4	float	*x* for vertex 3
4	float	*y*
4	float	*z*
2	Unsigned integer	Attribute byte count

+

The syntax for attributes is not explicitly documented in the formal specification, though it is stated that the attribute byte count must always be set to zero.

A common question arises regarding why triangular meshes are preferred for solid representation in 3D printing instead of other shapes like parallelograms or hexagons. The answer lies in the inherent advantages of triangles in geometric calculations. Due to their rigid structure, triangles are ideal for computational purposes. They allow for simpler and faster intersection calculations compared to more complex shapes. A triangle is uniquely defined by three non-collinear points, whereas other shapes, like hexagons, would require more points to represent them accurately.

Additionally, triangles are the simplest planar figures that belong exclusively to one plane, eliminating the possibility of multiple solutions in calculations. This characteristic makes them easier and more efficient to work with across various applications. Another key advantage is their ability to approximate complex surfaces. Any surface, no matter how intricate, can be broken down into triangles, making them highly versatile for modelling. Furthermore, triangles can only be sub-divided into smaller triangles, ensuring consistency in geometric representation.

In conclusion, triangles offer numerous advantages in geometric computations, including simplicity, speed and adaptability. Their rigid structure, ease of use and capacity to model complex surfaces make them a preferred choice for many applications. This is why STL format is widely used as an intermediary between CAD and CAM processes in 3D printing.

Example of ASCII file: the following file shows a simple ASCII STL file describing the right-angled, triangular pyramid structure as shown in Figure 4.3.

```
Solid triangular_pyramid
 Facet normal 0.0 -1.0 0.0
   Outer loop
       Vertex 0.0 0.0 0.0
       Vertex 1.0 0.0 0.0                    Triangle face 1
       Vertex 0.0 0.0 1.0
   end loop
 end facet

 Facet normal 0.0 0.0 -1.0
   Outer loop
       Vertex 0.0 0.0 0.0
       Vertex 0.0 1.0 0.0                    Triangle face 2
       Vertex 1.0 0.0 0.0
   end loop
 end facet
 Facet normal 0.0 0.0 -1.0
   Outer loop
       Vertex 0.0 0.0 0.0
       Vertex 0.0 0.0 1.0
       Vertex 0.0 1.0 0.0
   end loop
```

```
end facet
  Facet normal 0.577 0.577 0.577
  Outer loop
        Vertex 1.0 0.0 0.0
        Vertex 0.0 1.0 0.0
        Vertex 0.0 0.0 1.0
end loop
end facet
 end solid
```

Another example of ASCII file: the following Figure 4.4 depicts an STL file of a cube that is composed of 12 triangular facets. Each facet is defined by three vertices and one unit normal. The red arrow represents the normal of each facet. Each facet must be represented in CCW sequences of the vertices as shown in Table 4.3.

The data shown in Table 4.1 is used to write an STL file. The ASCII format of the STL file can be read and written by humans. Here is an example of **two** facets of a part of an STL file that is in ASCII format:

```
solid MATLAB_1

facet normal 0 0 -1
  outer loop
  vertex 0 0 0
  vertex 1 1 0
  vertex 1 0 0
  endloop
endfacet

facet normal 0 0 -1
 outer loop
 vertex 0 0 0
 vertex 0 1 0
 vertex 1 1 0
 endloop
endfacet
```

```
Self-Check Exercise:

Identify to which triangular
 face of Figure 4.4 this coding
 is related to.
```

```
Self-Check Exercise:

Identify to which triangular
 face of Figure 4.4 this coding
 is related to.
```

The ASCII STL file representing the normal and vertices of the facet. The normal points the direction while vertices show *x*-*y*-*z* direction coordinates.

Another example:
In Figure 4.5, the solid model of a teapot is converted into an STL format. The STL model represents the object's boundaries using triangular facets. These triangular facets can be observed on the surface of the teapot. Each facet in the STL file is defined by three vertices and a unit normal vector that points

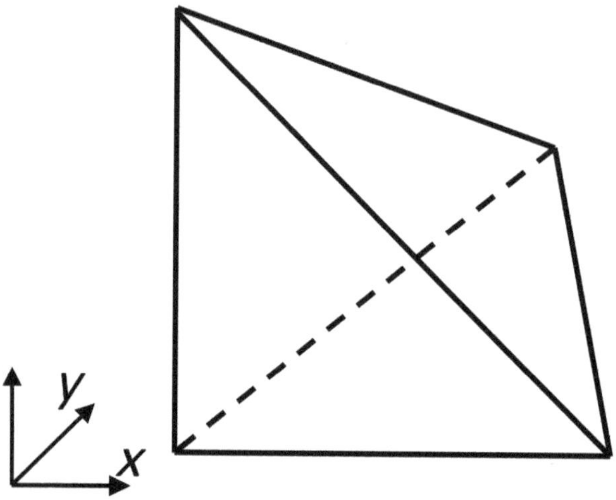

FIGURE 4.3
A right-angled triangular pyramid as defined in the STL sample file.

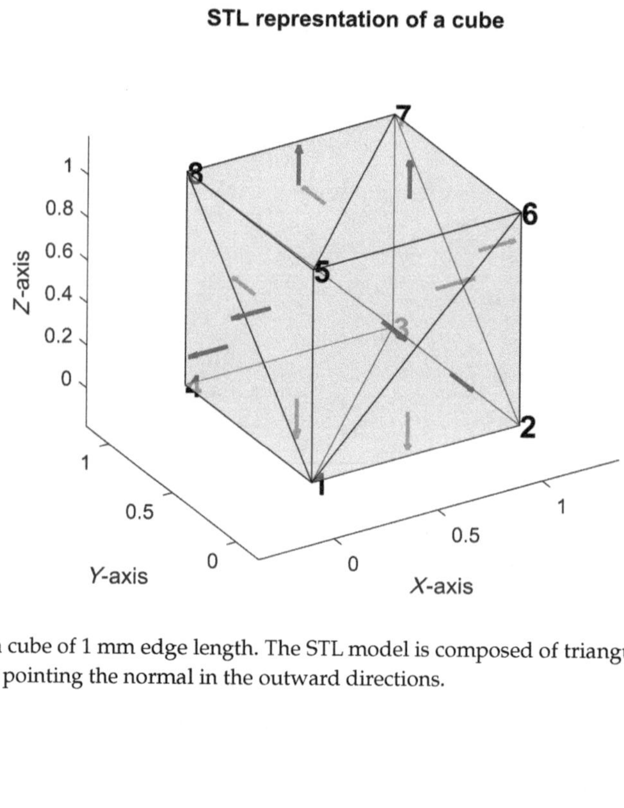

FIGURE 4.4
STL model of a cube of 1 mm edge length. The STL model is composed of triangular facets and the red arrows pointing the normal in the outward directions.

TABLE 4.3

Vertices Sequence and Normal Vector of 12 Triangular Facets that Represent a Cube Together

Facet	Vertices	Normal
1	1 3 2	0 0 -1
2	1 4 3	0 0 -1
3	5 6 7	0 0 1
4	5 7 8	0 0 1
5	1 2 6	0 -1 0
6	1 6 5	0 -1 0
7	3 4 8	0 1 0
8	3 8 7	0 1 0
9	1 8 4	-1 0 0
10	1 5 8	-1 0 0
11	2 3 7	1 0 0
12	2 7 6	1 0 0

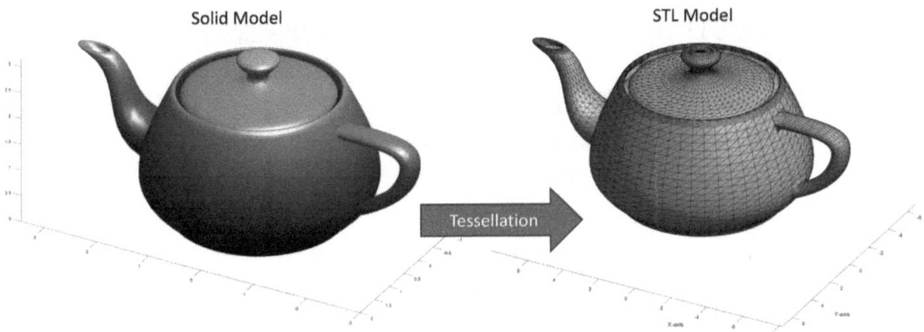

FIGURE 4.5
Solid model and STL format model of a teapot.

outward from the geometry. This indicates that there is no material extending in the direction of the normal vector for any given facet. Additionally, the sequence of vertices in the STL file adheres to the right-hand rule, as previously discussed in the earlier section.

4.2.2 Advantages and Disadvantages of Stl Files

4.2.2.1 Advantages of STL Files

- STL files are universally supported by almost all 3D printers, making them a widely recognized and accessible format.

- They are highly versatile and can be used to create a wide range of objects with precision.

Due to the absence of colour and texture data, STL files are typically smaller in size and process more quickly compared to other formats. This makes them an excellent choice for printing objects in a single colour and material.

4.2.2.2 Disadvantages of STL Files

- STL files have limited capacity for incorporating colour and texture, which is why they are primarily used for prototyping rather than producing final products.
- The format only recognizes the external surface and shape of a 3D model, without any information about the interior structure.
- Metadata, such as scale, units, copyright, authorship, material properties, colour or texture, cannot be stored in an STL file. It exclusively contains geometric data.
- Modifying STL files can be challenging due to the lack of editable design elements.
- STL files are best suited for models with flat surfaces. Curved surfaces are approximated by triangular facets, and even at high resolution, the representation of these curves may appear coarse.

4.2.3 Opening and Creating Stl Files

4.2.3.1 How to Open an STL File

- Mac: several online tools can be used to view STL files in a web browser or an STL viewer app can be downloaded to preview the design across devices.
- Windows: install the 3D Viewer app. Once installed, right-click on your file and select "Open with," then choose the 3D Viewer. The 3D model image will appear.

4.2.3.2 How to Create STL File

- First draw the model on any CAD software program, then save and export the completed design to your computer as an STL file.
- Exporting STL files may be different in each CAD software. As a quick reference, here are some of the steps for exporting STL files with different software.
    (a) *TinkerCAD*

Design > Download for 3D printing > .stl

(b) ***Autodesk Inventor***

1. Select IPro > Print > 3D Print Preview
2. Select Options then choose desired resolution > Click Ok
3. In the preview window, Select Save Copy As or Send to 3D Print Service
4. Save As to STL file (*.stl)

(c) ***SketchUp***

In SketchUp, creating STL files directly within the program may not be possible. An extension for .stl must be downloaded first in order to export STL files.

1. Download and install DXF or STL plugin
2. Select Tools > Export to DXF or STL > select the units (in millimetres) for the model

(d) ***SolidWorks***

1. File > Save As
2. Set Save As file type to STL
3. Options > Resolution > Fine or Custom > Ok
4. File > Save As
5. STL > Options. For a smoother STL file, change the Resolution to Custom
6. Change the deviation to 0.01 mm
7. Change the angle to 5 (smaller deviations and angles will produce a smoother file, but the file size will get larger)

(e) ***Solid Edge***

1. Application Button > Save As (opens dialog box)
2. From the Save As drop down menu, select STL documents (*.stl)
3. Select the Options button from the Save As dialog box
4. Adjust Conversion Tolerance and Tolerance Units (millimetres recommended). The lower the conversation tolerance, the finer the tessellation
5. Adjust the surface plane angle (dependent on your desired surface smoothness). The lower the surface plane angle, the greater the accuracy (noticeable in small details). As a rule, the finer the tessellation and the greater the accuracy, the larger the size of the STL file, and the longer it takes to generate it
6. Under Output File as: Check Binary. Binary STL files are much smaller than STL files saved in ASCII format
7. Click OK > Save

(f) ***CATIA***

1. Select STL command (we recommend setting maximum segmentation to 0.015 mm)

2. Select the model > Yes > Export
   Note: CATIA V5 is capable of creating STL files from CATPart files, but not from assemblies (CATProduct files) or geometrical representations (car files). Therefore, source files, including those saved in a neutral format (i.e., STEP or IGES), must be saved as CATParts. If the source design was saved as an assembly, it is imported to CATIA as a CATProduct.

(g) ***Fusion 360***
1. File > 3D Print
2. Under Refinement / Refinement options, select export settings.
3. Output > untick "Send to 3D Print Utility" > OK > Save

(h) ***IronCAD***
1. Right-click on the part
2. Click Part Properties > Rendering
3. Set Facet Surface Smoothing to 150
4. File > Export
5. Select .STL

(i) ***PTC Creo***
1. 3D printing in PolyJet
2. File > Print > 3D Print
3. Define Material
4. Define STL resolution

(j) ***Pro/ENGINEER***
1. File > Export > Model
2. Set type to STL
3. Set chord height to 0. The field will be replaced by minimum acceptable value
4. Set Angle Control to 1
5. Click OK

(k) ***NX (Formerly UG NX)***
NX software, allow to save not only entire parts as STL files, but also selected surfaces of a part. In addition, assembly output enables to save several components as a single unit while maintaining each component as a separate volume (shell).
1. File > Export > STL
2. Rapid Prototyping dialogue box will appear
3. Output Type: Binary
4. Binary STL files are much smaller than STL files saved in ASCII format
5. Triangle Tolerance: 0.015 mm. This is the maximum distance allowed between the surface of the original design and

the tessellated surface of the STL triangle, and affects the smoothness of the model surface

6. Adjacency Tolerance: 0.015 mm. If the distance between the two surfaces is less than this setting, they are considered as attached. This setting must be less than the printing resolution. For example, when printing models at a resolution of 30 micrometres (microns), the setting must be no more than 0.03 mm.
7. Auto Normal Gen: Check box
8. Normal Display: Check box
9. Triangle Display: Check box
10. Click OK
11. Export Rapid Prototyping dialogue box will appear
12. Name your file and click OK

- Below are some of the most important settings when exporting STL files. These settings will result in good quality prints and file size.

  Angle control: To improve the print resolution, it is recommended to have gaps between triangles and change its angle deviation between adjacent triangles. In this way, it will be easier to adjust how close objects are layered or tiled together.

  Chordal deviation: Chord height is the maximum distance between the surface of the original design and STL mesh. Choosing the right tolerance will give you smoother and not pixelated 3D part. The chord tolerance should be set between 0.01 to 0.001 mm. This setting usually results in good 3D prints.

4.2.4 Repairing an STL File

For 3D printing, surfaces are converted to a mesh format that describes geometry as a cloud of connected triangular faces and vertices. If mesh conversion is done badly, the result is a model with all kinds of edges, holes and floating parts, as well as areas with many intersecting triangles. If it is done properly, the 3D print is a manifold mesh without holes, indistinguishable from the original design.

A typical STL file repair workflow contains the following steps:

1. Auto-repair– the STL repair software's wizard will attempt to fix all major errors, including holes, separate shells and intersections.
2. Separating shells – a mesh consists of collections of connected triangles. It can contain multiple continuous surfaces that ideally are joined together while obsolete ones need removal.
3. Closing holes, bridging gaps – some comprehensive STL repair programs allow different ways of hole filling, such as planar, tangent, ruled or freeform.

4. Resolving overlaps and intersections – this typically requires recalculating entire portions of the mesh.
5. Filtering out double faces, double vertices, inverted normals, and sharp, narrow triangles.
6. Stitching open edges and remaining holes.
7. Manual repair by deleting and creating triangles.
8. Remeshing to optimize triangle count.
9. Exporting to the chosen mesh format.

Common software tools for repairing STL files:

- Meshmixer
- Meshlab
- Magics
- Blender
- Netfabb

Figure 4.6 is showing the basic repair utility available with the Meshlab tool software.

4.2.5 STL File Manipulation

Once a part is converted to an STL format, there are limited operations that can be performed. This limitation arises because the triangular structure of the file does not support drastic modifications. There are two key reasons for this.

First, the relationship between triangles is purely defined by shared points and vertices. You can move a point or vertex, but this only affects the immediate surrounding triangles. Making broader changes, such as adjusting a large set of points, becomes much more challenging and less precise.

Second, small adjustments to the geometry can cause disproportionately large changes to the triangle distribution. For example, when working with simple shapes like the cut cylinder shown in Figure 4.7(a), even a minor alteration to a measurement can result in a significant shift in how the triangles are distributed across the surface. This makes fine-tuning the model difficult and unpredictable.

While it is possible to modify an STL model by reducing the number of triangles, achieving smooth or precise boundaries in most models remains difficult. For instance, adding a fillet, as shown in Figure 4.7(b), results in a substantial alteration to the STL file. Similarly, attempting to move the oval shape representing the cut surface would disrupt the constant radius curve of the triangles forming the fillet.

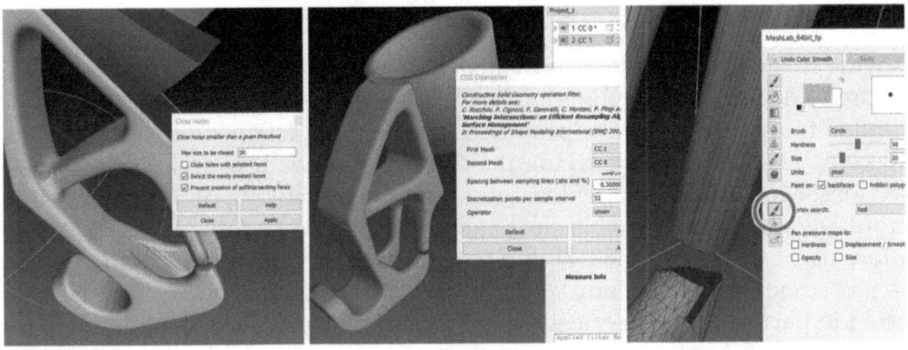

FIGURE 4.6
Basic repair utilities in Meshlab: Close holes, Boolean, and brush selection.

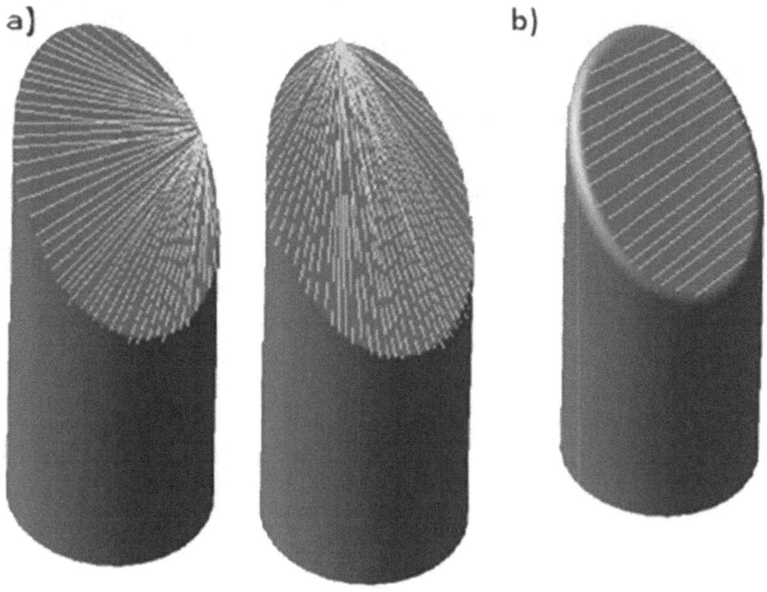

FIGURE 4.7
The STL files of a cut cylinder. (Notice that while the two models in (a) are very similar, the position of the triangles is somewhat different. Adding a simple filet to (b) shows an even greater change in the STL file.)

In many cases, models for additive manufacturing (AM) are created by individuals or teams from different departments or companies than the original designers. Often, those building the model may not have access to the original CAD data. As a result, it may become necessary to make modifications to the STL file before the part can be produced. The following sections will outline commonly used tools for editing STL files.

4.2.5.1 Viewers

A wide range of STL viewers is available, many of which can be downloaded for free. One example is the Marcam STL viewer (see Figure 4.8). Like other similar tools, this software provides limited interaction with the STL file, allowing users to view the triangular mesh, apply shading and display cross-sections. Most free tools are designed for passive viewing of STL data, while more advanced tools enable data modification. These advanced tools may allow users to rewrite the STL file or add supplementary information, such as measurement data. Many of these viewers are offered by companies that provide part-building services, serving as an incentive to use their services while also helping to reduce errors that can occur during data transfer, such as incomplete STL conversions or misinterpretation of design intent.

Other STL file manipulation tools offer additional features, such as the following:

- Verifying the integrity of STL files based on the issues mentioned earlier.
- Adding support structures, including tapered features that simplify removal.
- Optimizing the use of AM systems by ensuring efficient packing of components and minimizing the number of support structures.
- Embedding serial numbers or markings for easy identification and assembly.

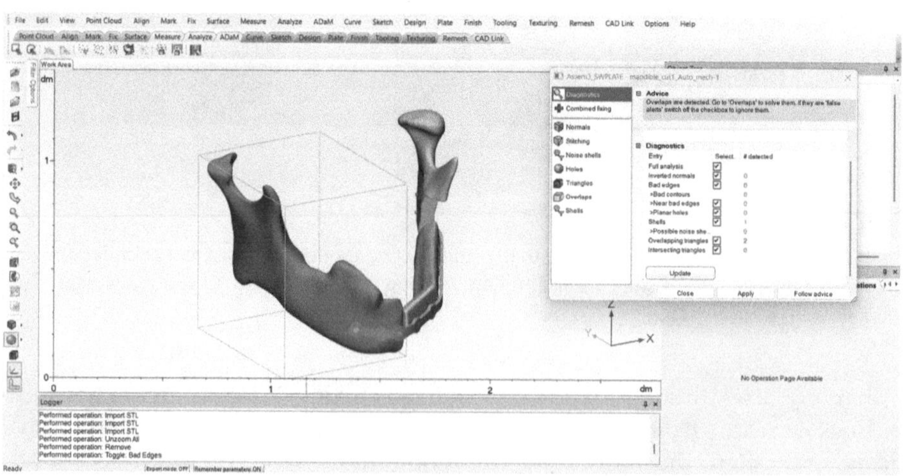

FIGURE 4.8
Visualization of STL file in MIMICS from Materialise.

- Remeshing STL files that were generated through reverse engineering or other non-CAD-based programs. Such files are often excessively large, but their size can be reduced without sacrificing accuracy.
- Segmenting large models or merging multiple STL files into a single data set.
- Performing Boolean operations, such as subtracting model data from a blank template to create a mold.

4.3 Part Orientation and Support Generation

Part orientation refers to the way an object is positioned on the build platform during the 3D printing process. It is important to recognize that the orientation of a part significantly influences its overall build quality, as well as the efficiency of processing and post-processing. Additionally, the orientation determines the amount of support material needed, as different angles and positions may require varying levels of support. Proper part orientation can help minimize the need for support structures, reducing both material usage and post-processing time.

When producing multiple copies of the same part or different parts simultaneously, effective placement and orientation are key. This ensures maximum utilization of the build volume, enabling efficient production of multiple parts in a single print. Additionally, optimal orientation and placement can simplify the slicing process, allowing the software to generate more accurate toolpaths for the parts.

In this section, we will examine the importance of part orientation and its effects on 3D printing outcomes, including support requirements for overhangs and the slicing process for complex models. By understanding the interaction between part orientation, support generation and slicing, you will be better equipped to prepare 3D models for printing, ultimately improving both the quality and efficiency of the final product.

4.3.1 Part Orientation

In 3D printing process planning software, various methods are available to assist in determining the optimal part orientation. These methods fall into two categories: manual methods and automatic methods. As the names imply, manual methods involve using transformation operations such as move, rotate and copy to manually adjust and place parts on the build platform. In contrast, automatic placement methods use algorithms to position parts on the build platform based on specific parameters. These parameters may include minimizing support structures, optimizing strength in a particular direction, reducing the bounding box, decreasing surface area or minimizing dimensions along the x or y axis.

By understanding and applying both manual and automatic methods, users can effectively optimize part orientation in 3D printing process planning software, leading to higher-quality prints and more efficient resource use.

4.3.1.1 Manual Placement Methods

Manual placement involves adjusting a part using fundamental transformation operations like translation, rotation, mirroring and scaling. In process planning software, these transformations are applied by multiplying the transformation matrix by the part's coordinates. For example, Figure 4.9 shows the menu options in Cura software, where the graphical user interface (GUI) displays icons for move, scale, rotate and mirror commands.

By using these manual placement tools, users can efficiently modify the orientation and position of their 3D models within the software, ensuring optimal results during printing.

FIGURE 4.9
Menu option in Cura software showing Move (Translate), Scale, Rotate, and Mirror options (top to bottom).

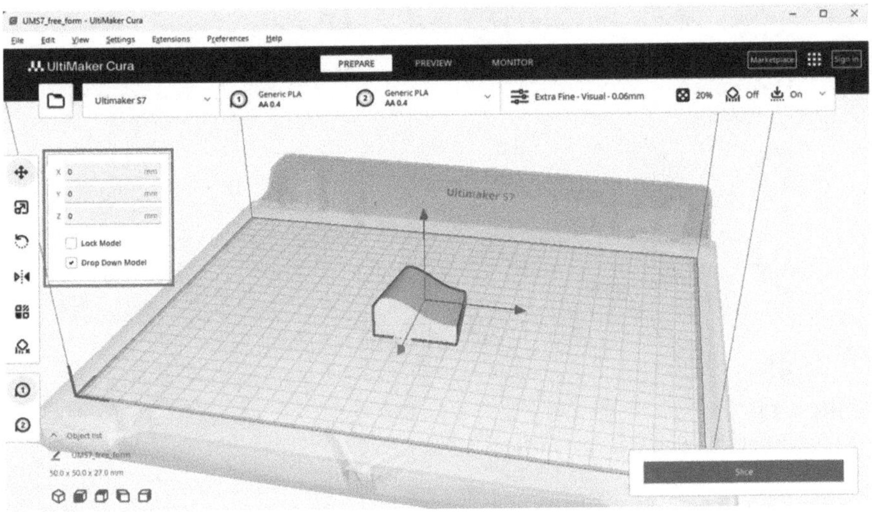

FIGURE 4.10
Cura screenshot showing translation parameters in red box.

Move–move is used to translate the part from one place to another on the build plate form. It used a transformation matrix:

$$T = \begin{bmatrix} 1 & 0 & 0 & T_x \\ 0 & 1 & 0 & T_y \\ 0 & 0 & 1 & T_z \\ 0 & 0 & 0 & 1 \end{bmatrix}$$

In this matrix T_x, T_y and T_z represent the translation distances along the x, y and z axes, respectively. To apply the translation to a 3D point (x, y, z), you can multiply this transformation matrix with the point's coordinates in the form of a 4×1 column vector.

So in software, you need to define T_x, T_y and T_z as shown in the following Figure 4.10.

Scale – scale is used to increase and decrease the size of the object on the build platform by defining the scale dimensions or scale factor. The transformation matrix for scale is as follows:

$$S = \begin{bmatrix} S_x & 0 & 0 & 0 \\ 0 & S_y & 0 & 0 \\ 0 & 0 & S_z & 0 \\ 0 & 0 & 0 & 1 \end{bmatrix}$$

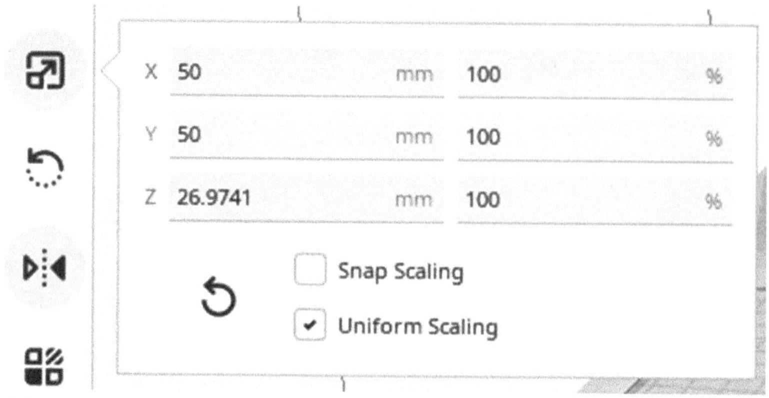

FIGURE 4.11
Cura screenshot showing scale parameters.

In this matrix, S_x, S_y and S_z represent the scaling factors along the x, y and z axes, respectively. To apply the scaling to a 3D point (x, y, z), you can multiply this transformation matrix with the point's coordinates in the form of a 4×1 column vector.

In Cura software you can define the scaling parameters along x, y and z direction as shown in Figure 4.11.

As shown in Figure 4.11, there are two types of scaling that can be applied along the axes of an object: uniform and non-uniform scaling. When uniform scaling is selected, it adjusts the parameters along the x, y and z axes proportionally, preserving the object's original proportions. In contrast, when non-uniform scaling is chosen, distinct parameters must be defined for each axis independently.

It is important to recognize that non-uniform scaling can distort the object's shape, as it changes the dimensions along each axis differently. This may result in an appearance that diverges from the original design. Consequently, careful consideration is necessary when selecting the appropriate scaling method to achieve the desired outcome for your object.

Rotation – rotation is used to change the orientation of an object on the build platform by defining the rotation angles. The transformation matrices for rotation around the x, y and z axes are as follows:

Rotation around x axis (angle θ_x):

$$R_x = \begin{bmatrix} 1 & 0 & 0 \\ 0 & \cos(\theta_x) & -\sin(\theta_x) \\ 0 & \sin(\theta_x) & \cos(\theta_x) \end{bmatrix}$$

Rotation around y axis (angle θ_y):

$$R_y = \begin{bmatrix} \cos(\theta_y) & 0 & \sin(\theta_y) \\ 0 & 1 & 0 \\ -\sin(\theta_y) & 0 & \cos(\theta_y) \end{bmatrix}$$

Rotation around z axis (angle θ_z):

$$R_z = \begin{bmatrix} \cos(\theta_z) & -\sin(\theta_z) & 0 \\ \sin(\theta_z) & \cos(\theta_z) & 0 \\ 0 & 0 & 1 \end{bmatrix}$$

In these matrices, θ_x, θ_y and θ_z represent the rotation angles around the x, y and z axes, respectively. To apply rotation to a 3D point (x, y, z), you can multiply the corresponding transformation matrix (or a combination of these matrices) with the point's coordinates in the form of a 4×1 column vector.

The rotation process enables you to precisely control the object's alignment and orientation within the 3D space. By adjusting the rotation angles around each axis, you can achieve the desired position for your object on the build platform. This is particularly useful for optimizing the object's placement for manufacturing or assembly purposes, ensuring a seamless fit with other components or maximizing the use of available build space.

InSlicer software, the object can be rotated interactively by dragging the rotation circles that are indicated in red, green and blue colours, as shown in Figure 4.12.

It can also be decided in the software which face you want to lay on the bed. By pressing the () button it places a face of the part on the bed automatically. Additionally, the next icon () is used to place a particular face on the platform.

Mirror – mirroring is used to create a reflection of an object on the build platform by defining the mirror planes. The transformation matrices for mirroring along the x, y and z planes are as follows:

Mirror along the x axis:

$$M_x = \begin{bmatrix} -1 & 0 & 0 \\ 0 & 1 & 0 \\ 0 & 0 & 1 \end{bmatrix}$$

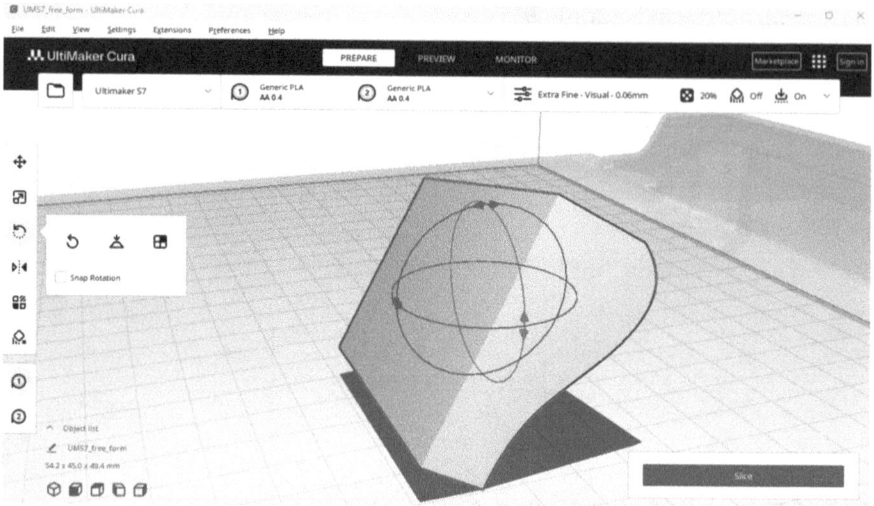

FIGURE 4.12
Rotate menu of Cura software, red, green, and blue circles showing interactive rotation path.

Mirror along the y axis:

$$M_y = \begin{bmatrix} 1 & 0 & 0 \\ 0 & -1 & 0 \\ 0 & 0 & 1 \end{bmatrix}$$

Mirror along the z axis:

$$M_z = \begin{bmatrix} 1 & 0 & 0 \\ 0 & 1 & 0 \\ 0 & 0 & -1 \end{bmatrix}$$

In these matrices, the values along the diagonals represent the mirror planes along the x, y and z axes, respectively. To apply mirroring to a 3D point (x, y, z), you can multiply the corresponding transformation matrix (or a combination of these matrices) with the point's coordinates in the form of a 3×1 column vector.

The mirroring process enables you to create a symmetric copy of the object within the 3D space. By reflecting the object along each axis, you can achieve the desired reflection for your object on the build platform. This is particularly useful for creating symmetrical designs, generating opposite-handed

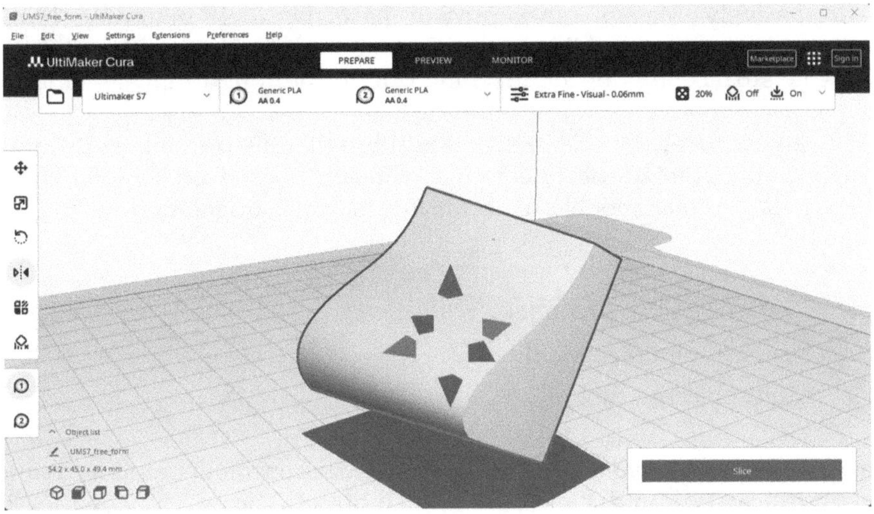

FIGURE 4.13
Cura screenshot showing interactive arrows to create mirror object.

parts for assembly, or optimizing the object's placement for manufacturing purposes.

In Cura software, there are interactive arrows that mirror the object about principal planes as shown in Figure 4.13.

4.3.1.2 Automatic Placement Methods

In addition to manual operations, some software programs, such as Materialise Magics, offer built-in automatic operations for placing parts on the build platform. With the automatic placement feature, parts can be efficiently nested on the platform. Two nesting options are available:

1. Geometry-based nesting
2. Bounding box-based nesting

Remember that when importing multiple parts, you can use the automatic placement algorithm to instantly position them on the platform. Even if the platform is not large enough to accommodate all parts, the software will search for a solution beyond the platform borders, arranging the parts for optimal viewing.

Bounding box-based nesting can lead to wasted capacity if a part has a large bounding box but a small projected area. In contrast, Magics uses the actual form of the parts for geometry-based nesting, resulting in more efficient use of space.

Alternatively, Magics can represent parts by their bounding boxes, leading to faster nesting. However, this method may not fully utilize the available surface area of the machine due to the rough representation of the parts.

The automatic part orientation algorithm in 3D printing software is designed to optimize the orientation of objects based on various objectives. Three popular models for achieving optimal orientation include the following:

1. Cost minimization model
2. Support minimization model
3. Multi-objective model

Cost minimization model – The cost minimization model aims to reduce the overall cost of producing a 3D printed object. The cost function for 3D printing can be expressed as follows:

Total Cost = Preprocessing Cost + Machine Utilization Cost for Building the Part + Material Cost + Post-processing Cost

Preprocessing Cost = (Model Positioning Time + Support Generation Time + Model Slicing Time + Tool Path Generation Time) × (Staff Cost + Facility Cost) + Machine Setup Time × (Staff Cost + Machine Utilization Rate)

Build Cost = Build Time × Machine Utilization Rate

Material Cost = Cost of Modeling Material + Cost of Support Material

Post-processing Cost = Post-processing Time × Staff Cost + Cost of Other Materials and Facilities (e.g., water, heat source, etc.)

In the cost optimization model, the goal is to minimize the total cost of printing the part. The objective function aims to minimize the overall cost, subject to the variables of orientation along the x, y and z directions, and the angular position of the object along the principal axis.

By utilizing these models, 3D printing software can automatically determine the most efficient and cost-effective orientation for objects, ensuring optimal use of resources and minimizing production costs.

Support minimization model – we know that the volume of support depends on the projected area of the overhanging part and it can be determined from the following equation:

$$V_e = \sum_{i=1}^{n} V_i = \sum_{i=1}^{n} A_i \times h_i$$

In the context of an STL file, let us consider a scenario where "n" represents the total count of downward-facing facets that necessitate support structures.

For each individual facet *i*, we can denote its projected area on the horizontal slicing plane as A_i. Additionally, we must take into account the average support height (h_i), which is determined by evaluating the corresponding facet in relation to either the platform or the next intersecting facet, depending on which of the two exhibits a lesser height. By comprehensively analyzing these parameters, we can gain a more in-depth understanding of the various facets and their structural requirements within the STL file.

Here the objective function will be minimization of the V_e by optimizing the variables related to part orientation.

Multi-objective model – in a multi-objective m odel, the orientation of a part is optimized by considering multiple objectives and applying suitable weighting criteria. These criteria can include the following:

- Maximizing the number of perpendicular surfaces.
- Maximizing the number of upward-facing horizontal surfaces.
- Optimizing the alignment of hole axes in the build direction.
- Maximizing the number of curved cross-sections illustrated in the horizontal plane.
- Maximizing the area of the base surface.
- Minimizing the slope of the base surface.
- Minimizing the overall build volume.

By incorporating these criteria, the model ensures an optimized orientation of the part that accounts for various design factors and objectives.

Materialize Magics is a professional software that have many features and inbuilt algorithms to automatically align the part. We are discussing some important commands that used to set the orientation of the model(s) on the build platform.

Minimizing bounding box – the orientation of the part will be adjusted to minimize its bounding box while maintaining the original scale of the part. This approach aids in streamlining the quoting process and enhancing the efficiency of the build preparation. Refer to Figure 4.14.

The **shape sorter** is a specialized tool of Materialise Magics engineered to organize components with comparable geometries. To initiate the sorting process, a primary or "master" part is designated as a reference. Subsequently, the remaining selected parts are systematically arranged around this master part, taking its orientation into account as showing in Figure 4.15.

Once the parts are organized according to the master part's orientation, an option to group these components becomes available. This grouping feature facilitates efficient nesting of the parts on the build platform. By implementing this nesting strategy, users can optimize the arrangement and placement of

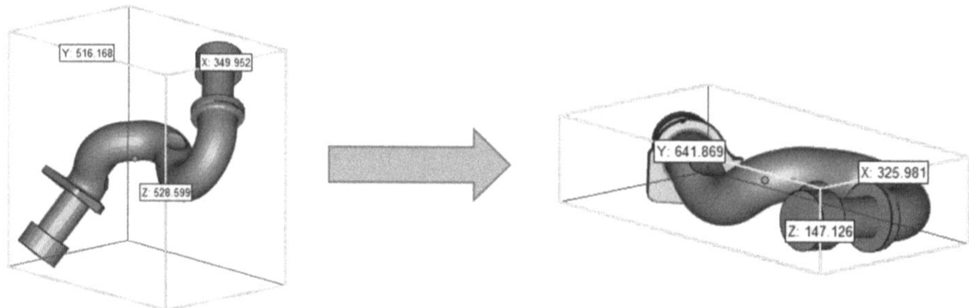

FIGURE 4.14
Minimizing bounding box (image taken from Materialise Magics user manual).

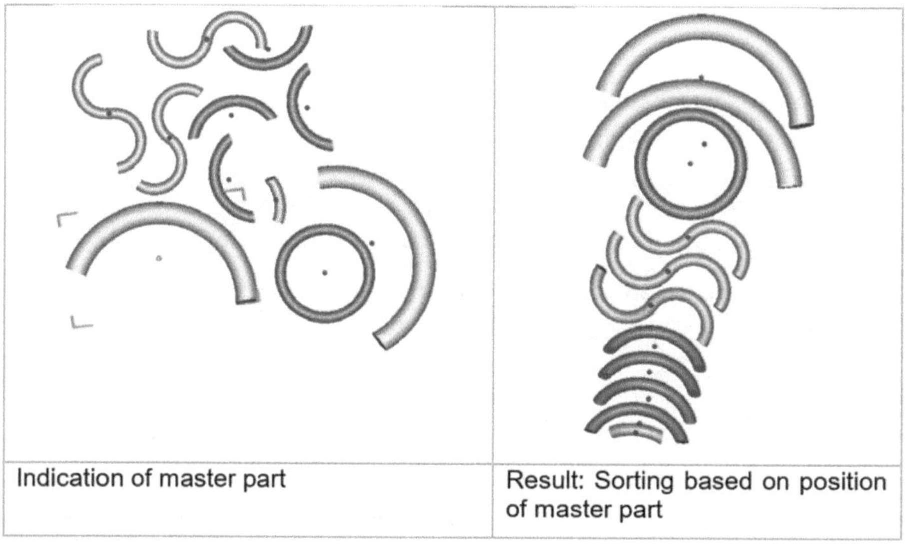

FIGURE 4.15
Shape sorter (image taken from Materialise Magics user manual).

parts, ultimately enhancing the overall utilization of the platform's available space.

In essence, the shape sorter simplifies the organization of parts with similar geometries by using a master part as a reference point, then enabling users to group and nest the components effectively on the build platform for maximum efficiency.

3D Nester – The Magics 3D Nester software guarantees an optimal loading configuration for your sintering machines. Taking into account the geometry of each part, the software intelligently nests your components to maximize the quantity of parts within the build envelope and/or minimize the overall build time. Simultaneously, the software ensures that none of the parts come into contact with one another or the container, effectively preventing any potential collisions.

4.3.2 Support Generation

Support structures are essential for creating overhanging features in additive manufacturing (AM). As mentioned in earlier sections, some AM processes can operate without explicit support, often achieved by orienting the part to eliminate overhangs.

For any STL part and build direction, support structures can be easily calculated using the facet normals of the triangles in the STL file in relation to the build direction. If the build direction is vertical, triangles with normals pointing upward typically do not require support, while those with downward-facing normals may need a support structure.

Different additive manufacturing processes utilize various materials, each with differing capacities to handle overhangs. Consequently, the need for support structures may depend on both the material and process parameters, such as layer height. In some cases, certain materials and settings may eliminate the requirement for support for overhanging features.

In Cura software, two types of support structures are available: column-type support and tree support (see Figure 4.16). When generating these structures, users must define several parameters, including the support material extruder, support angle, support pattern, support density, support horizontal expansion, support infill layer thickness, gradual support infill steps, and gradual support infill height. Users can also

FIGURE 4.16
Normal (left) and tree (right) supports generated by Cura software.

choose to enable or disable options such as support interface, support roof, and support floor by checking or unchecking the corresponding boxes in the software.

When generating support structures for a 3D model, users can select from available techniques. The "Normal" support method creates a support structure directly beneath the overhanging areas, extending vertically downwards. In contrast, the "Tree" support method forms branching structures that reach toward the overhangs, providing support at the tips of the branches. This approach allows the branches to navigate around the model, maximizing support from the build plate when feasible.

The support placement setting determines where support structures are positioned. Users can select either "Touching Build Plate" or "Everywhere" for placement options. When set to "Everywhere," support structures will also print on the model itself, offering additional support where needed.

The support angle parameter indicates the minimum inclination angle of a facet that requires support. The support density and support pattern parameters can be adjusted to change the infill characteristics of the support structures. For instance, altering the support pattern modifies the infill pattern, while adjusting the density affects the infill density of the support structure.

Support generation is crucial in additive manufacturing processes, such as stereolithography and metal sintering. The ability to quickly and efficiently create support structures is equally important in rapid prototyping. The final quality of a part relies heavily on the accurate verification and adjustment of the generated support structures. Materialise Magics software offers a variety of support types, enabling users to combine different support structures on a single surface to optimize the manufacturing process and achieve the highest quality for the final part.

In this chapter, we have explored the vital roles that part orientation and support generation play in 3D printing. We have learned that part orientation significantly influences various aspects of 3D printed objects, including build quality, processing and post-processing, as well as the determination of necessary support structures.

By properly orienting parts, it is possible to minimize material usage and reduce post-processing efforts by decreasing the need for support structures. In cases where multiple parts or different components are produced simultaneously, effective placement and orientation are essential for maximizing the use of build volume and improving the slicing process, leading to more accurate toolpaths for composite models.

In summary, a thorough understanding of the interplay between part orientation, support generation and slicing is essential for making informed decisions when preparing 3D models for printing. By considering these factors, you can enhance the quality and efficiency of your final products, ensuring the success and cost-effectiveness of your 3D printing projects.

4.4 Slicing and Deposition Path Generation

In 3D printing, slicing is a critical process that converts three-dimensional geometry into a series of two-dimensional contours. These contours act as boundaries for material deposition, ensuring that the material is accurately placed within designated areas to create the desired shape. Typically, slicing occurs along the deposition axis, which is generally oriented in the $z+$ direction, using a plane that is parallel to the x-y plane to slice the CAD model. To ensure dimensional accuracy, the slice height must correspond to the layer thickness specified in the material deposition plan.

After the slicing process is complete, the deposition path is generated. There are usually two types of deposition paths: (1) primary material deposition, which constructs the main object, and (2) support material deposition, which provides the necessary support structures during printing. The primary material is deposited within the contour boundaries, while the support material is placed outside these contours as required.

In this part of chapter, we will delve into the principles of slicing and deposition path generation in 3D printing, highlighting their significance in producing accurate and well-structured prints, as well as discussing various strategies for optimizing the process.

4.4.1 Slicing

The slicing process can be categorized in two ways:

1. Direct slicing
2. Indirect slicing

Direct slicing refers to the technique of slicing a part directly from any CAD format without requiring an intermediate file format. The model is typically represented using analytical curves and mathematical definitions of geometries. Because the original model is utilized for slicing, all part features remain intact, ensuring that there is no data loss in this process.

In contrast, indirect slicing entails converting the CAD model into an intermediate file format, such as STL. This conversion can lead to the loss of certain features due to data transformation and tessellation associated with STL files. Minute geometric details may be compromised during this process. However, this loss can be mitigated by increasing the resolution of the tessellation.

Figure 4.17 displays two slice contours generated by slicing both the root CAD model and the STL model of a cylinder with identical radius dimensions. The tessellated model of the cylinder transforms its curved surface into a triangulated representation, resulting in straight lines when

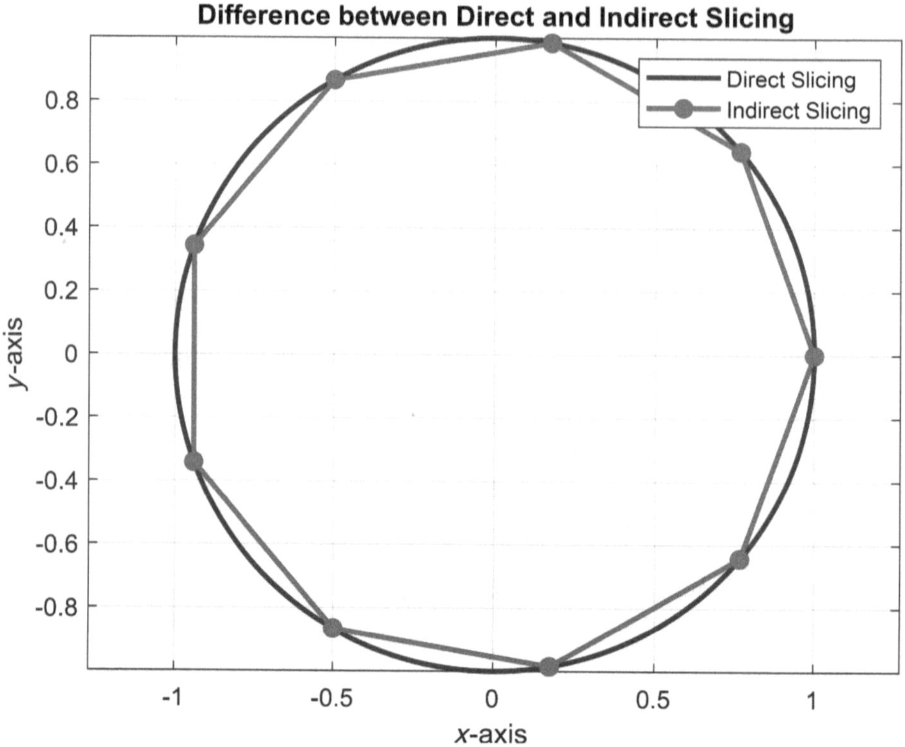

FIGURE 4.17
Direct and indirect slices generated by slicing a cylinder's root CAD model.

sliced. In contrast, direct slicing of the actual model yields a perfect circular contour. This comparison underscores the distinction between direct and indirect slicing, with direct slicing offering a more accurate depiction of the original geometry.

Additionally, based on slice height, slicing can be classified into two categories:

1. Uniform slicing
2. Adaptive slicing

As indicated by its name, uniform slicing involves slicing the CAD model with a consistent layer thickness (slice height). In contrast, adaptive slicing employs varying layer thicknesses throughout the build direction to achieve specific objectives. Adaptive slicing helps preserve certain features and minimizes cusp heights, and it may also eliminate the need for support structures.

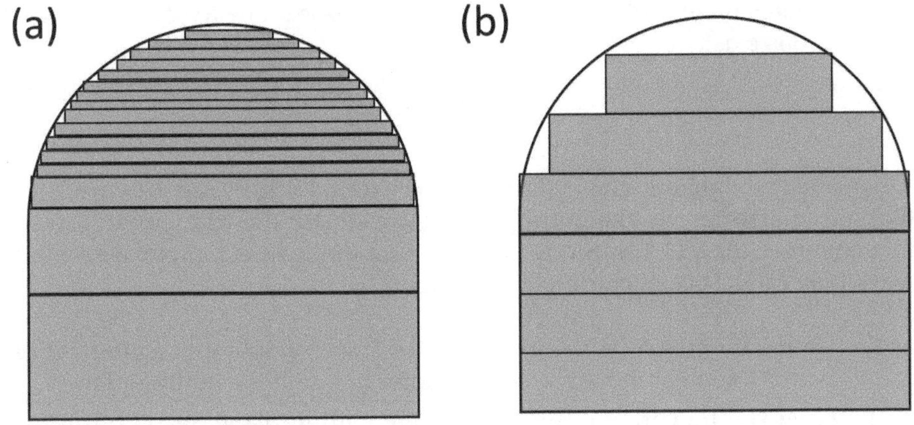

FIGURE 4.18
(a) Adaptive slicing; (b) uniform slicing.

The increasing interest in adaptive slicing techniques is due to their effectiveness in reducing the staircase effect, which often occurs with traditional slicing methods, thereby improving the overall surface finish. By adjusting slice thickness based on the geometric characteristics of the object, adaptive slicing targets areas with significant curvature and steep sidewall angles. This approach effectively reduces dimensional discrepancies associated with the staircase effect, as illustrated in Figure 4.18.

4.4.1.1 Algorithm to Slice Objects Uniformly

During the slicing process, a CAD model is segmented into several layers by employing a cutting plane that aligns parallel to the build plane. The distance between each layer, referred to as the layer height, is a user-defined parameter that dictates how the CAD model is sliced. The cutting plane intersects the facets of the CAD model in various configurations, generating intersection points. Line segments are created by using two intersection points from identical facets and are then stored randomly in a file. Subsequently, these line segments are organized to construct a contour. The intricacies of the slicing procedure are outlined below:

The algorithm employed during the slicing process is capable of accessing the data of all facets, which encompasses their coordinates and normal vectors. It analyzes the Z coordinates of every facet, compares them, and identifies the Z coordinates of the bottom and top facets (Z_{min} and Z_{max}). The Z coordinates of the cutting plane can be computed using a specific formula, such as $Z_n = Z_{n-1}$ + slice height, where $n = 1$ to k, $k = (Z_{max} - Z_{min})$/slice height. It should be noted that the k will be rounded off to the nearest lower integer (bottom round-off). The algorithm assesses the Z coordinates of each facet and compares them with the cutting plane's Z coordinate to ascertain if any intersections exist.

$$Z_{min} \leq Z_n \leq Z_{max}$$

Z_{max} and Z_{min} values for a single facet can be used to easily identify the facets that are intersected by the cutting plane. Facets are the triangular surfaces that compose the STL file. Since there are numerous facets, there are several ways they can be cut, including as follows:

1. **Case I:** the cutting plane intersects the facets, yielding two intersection points. None of the three vertices intersect with the cutting plane.
2. **Case II:** one vertex is situated on the cutting plane, resulting in one intersection point.
3. **Case III:** the cutting plane intersects the facets, with one point lying on the cutting plane and producing two intersection points.
4. **Case IV:** two vertices are on the cutting plane, and one vertex is on either side of the cutting plane.
5. **Case V:** all three vertices of the facet are positioned on the cutting plane.

Intersection points are acquired, and line segments are created using the intersection points of the same facets. The line segment is established using a straight-line equation involving the coordinates of the intersection points (X_1, Y_1, Z_1) and (X_2, Y_2, Z_2). The algorithm examines all the facets, and the line segments are stored in a file in random order. In the subsequent step, these line segments are arranged to form a closed contour.

The construction of the contour involves connecting the line segments obtained during slicing to create the outer surface or boundary of the model to be printed. Each line segment has a head and a tail. The head of one line segment connects to the tail of the adjacent line segment, continuously joining the lines in this manner until the head of the final line segment connects to the tail of the first line segment, creating a closed contour. The first line segment from the slicing file is transferred to the contour file. The algorithm then copies one line segment from the file and checks if it is a neighboring line segment to the previous line using the coordinates of the line segments. If the coordinates of one end of a line segment match the coordinates of the end of another line segment, they are considered neighboring lines. The line segment is transferred to the contour file if it is a neighboring line. Otherwise, another line segment is chosen, and the same process is followed until a neighboring line is transferred. This procedure continues until each line segment is transferred, and a closed contour is formed.

Figure 4.19 depicts a cube and its contours generated with respect to its cross-section.

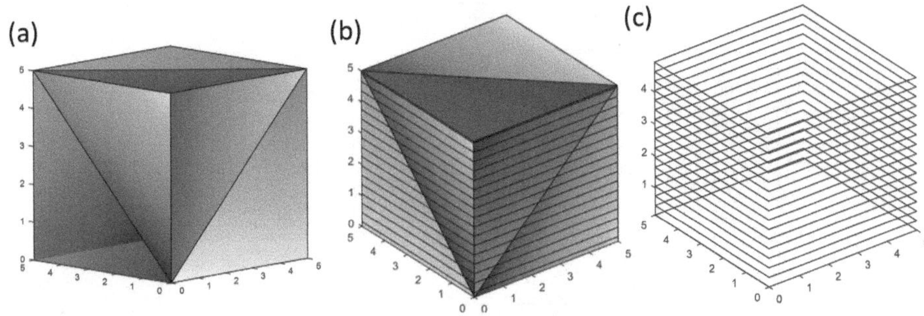

FIGURE 4.19
(a) The STL file of a cube, (b) sliced STL, (c) generated contours after slicing.

Contour organization – upon slicing, multiple contours may be generated at various Z heights, with some of them occurring at the same Z height. For hollow parts or parts with holes, two or more contours at the same Z height represent the boundaries of the external body, as well as any holes or other features such as pockets and slots. It is crucial to determine which contours represent the exterior or interior, where the exterior signifies the absence of material outside the contour. Organizing the contours helps to identify hole-like features, where material filling should be avoided.

To automatically identify the contours a relationship matrix is created according to the following conditions:

$R(i,j) = +1$, if contour i contains contour j

$R(i,j) = -1$, if contour i contained in contour j

$R(i,j) = 0$, if two contours are separated

$R(i,j) = +2$, if $i = j$

Consider a case of a contour as depicted in Figure 4.20, where a total of nine contours are present. We must organize these contours in a manner that enables us to identify the exterior and interior contours based on the rules mentioned in the earlier paragraph. By applying these rules, we can populate the contour relationship matrix. For instance, $R(1,1) = +2$ since $i = j$, similarly $R(1,2) = +1$, and so on.

Since there are no intersecting contours, an inpolygon test can be utilized to identify the parent–child contour relationships. Several algorithms can be employed to perform the inpolygon test. A few algorithms are as follows:

1. **Ray Casting Algorithm:** this algorithm involves casting a ray from the test point in any direction and counting the number of times it intersects the polygon's edges. If the number of intersections is odd,

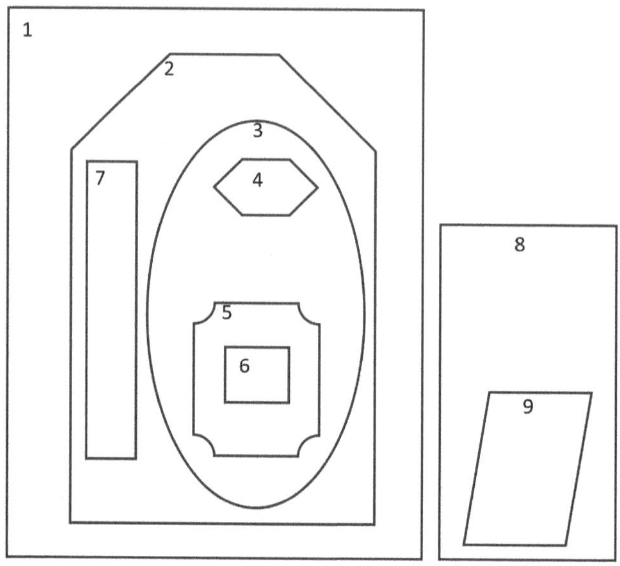

FIGURE 4.20
A group of contours showing a slice of a random part.

the point lies inside the polygon; if it is even, the point lies outside. The algorithm works for both convex and concave polygons.

2. **Winding Number Algorithm:** this method calculates the winding number for a test point concerning a polygon. The winding number is the number of times the polygon "winds" around the point. If the winding number is non-zero, the point lies inside the polygon. This algorithm works for both simple and complex polygons and is more robust than the ray casting algorithm in handling degenerate cases.
3. **Point-In-Polygon (PIP) Algorithm:** this algorithm is based on the angle summation approach. It calculates the sum of angles made by the test point with all pairs of adjacent vertices of the polygon. If the sum is equal to 360° (2π radians), the point lies inside the polygon. This method works for both convex and concave polygons but may have issues with degenerate cases.
4. **Crossing Number Algorithm:** similar to the ray casting algorithm, this method counts the number of times a line extending from the test point to infinity (in any direction) crosses the polygon's edges. If the crossing number is odd, the point lies inside the polygon, and if it is even, the point lies outside. This algorithm works for simple polygons but may fail for complex polygons with self-intersections.

Each of these algorithms has its advantages and limitations, and the choice of the most suitable one depends on the specific application and polygon complexity. Using any suitable algorithm the contours are identified and the matrix is generated.

So for a total of nine contours, shown in Figure 4.20, the contour relationship matrix will be as follows:

i/j	*1*	*2*	*3*	*4*	*5*	*6*	*7*	*8*	*9*
1	2	1	1	1	1	1	1	0	0
2	–1	2	1	1	1	1	1	0	0
3	–1	–1	2	1	1	1	0	0	0
4	–1	–1	–1	2	0	0	0	0	0
5	–1	–1	–1	0	2	1	0	0	0
6	–1	–1	–1	0	–1	2	0	0	0
7	–1	–1	0	0	0	0	2	0	0
8	0	0	0	0	0	0	0	2	1
9	0	0	0	0	0	0	0	–1	2

This matrix is the father–son relationship of various contours. This is a way to generate the data, but there could be other ways for the same objective. This matrix we can utilize to extract the information regarding its position, i.e., exterior or interior. There are two conditions:

1. For one contour i, if $j \in \{1,2\ldots m\}$, $R(i,j) \neq -1$ then it is separate exterior contour.
2. For one contour i, if $j \in \{1,2\ldots m\}$, $R(i,j) = -1$ then it is contour contained by contour j. If the number of time –1 occur it is odd then it is an interior contour else it is an exterior contour.

It is important to note that Contour 1 (located in row 2) does not include any element labelled –1, indicating that it is an external contour. Likewise, Contour 8 also lacks an element –1, which allows us to conclude that both Contours 1 and 8 are external and distinct contours.

In contrast, when a cell contains an element –1, it signifies that the *j*th contour is encompassed within the specific contour. For instance, Contours 2 through 7 each have a –1 value in the first cell, indicating that these contours are enclosed within Contour 1. The relationship matrix can also be used to distinguish between interior and exterior contours. For example, Contour 2 has a single occurrence of –1, which is odd, indicating it is an interior contour. Conversely, Contour 3 features two occurrences of –1, which is even, suggesting that it is an exterior contour.

Thus, this method enables us to identify the contours where material needs to be filled, as illustrated in Figure 4.21, which highlights the area designated for material filling.

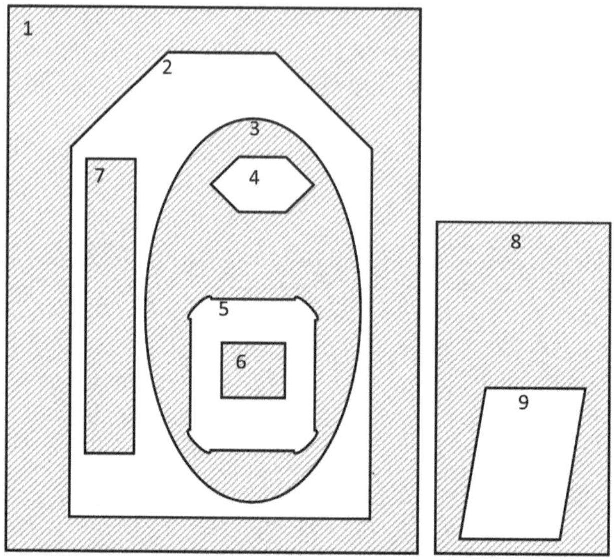

FIGURE 4.21
Contour identification for material filling.

Once the exterior and interior contours have been identified, we assign properties to the contours by sorting the edges of the exterior contour in a counterclockwise direction, while the edges of the interior contour are sorted in a clockwise direction.

The next step in the process is slicing the STL file. During this stage, the STL file is divided, preparing the triangles for production. Each manufacturing device utilizes its own specialized software for slicing files. To slice an STL file, the user must first open the slicing software and import the desired file. After the import, various slicing options become available, including layer thickness, print quality, extrusion temperature, material type, infill percentage and options for support material or print rafts. These selections are based on specific design requirements and can typically be found under the "settings" tab in the software of the manufacturing machine. An example of this is illustrated in Figure 4.22, which shows the Cura software (for a 3D printer) displaying options such as layer thickness (extra fine, fine, normal, fast, and extra fast), infill density, infill pattern, and material type.

After selecting the appropriate slicing options, the STL file is automatically segmented and prepared for fabrication (refer to Figure 4.23). As the complexity of the object increases, the number of triangles required also grows, resulting in a larger file size for the CAD model and necessitating a suitable support structure for the accurate production of the part.

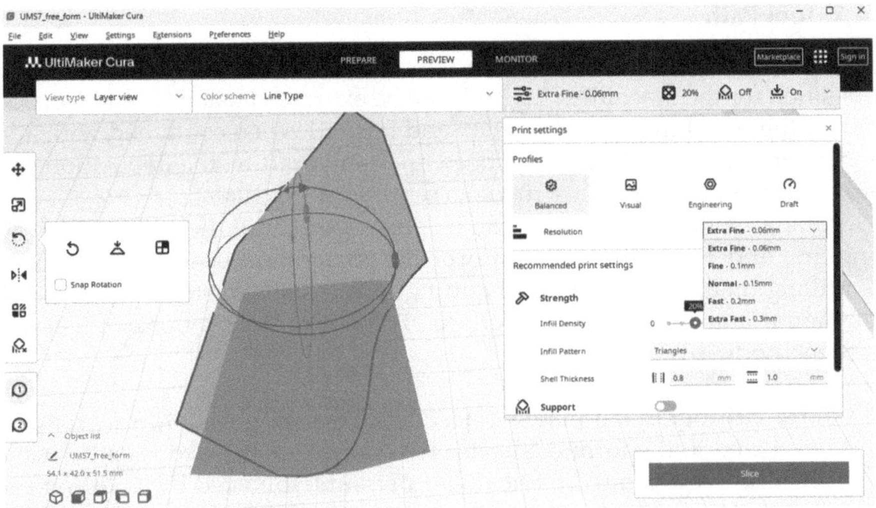

FIGURE 4.22
Cura screenshot showing slicing options.

FIGURE 4.23
Cura screenshot showing sliced part.

During slicing the following points should be remembered:

- Increasing slice height will increase the volumetric error and cusp height.
- Increasing the slice height will reduce the build time.
- Increasing slice height will increase the surface roughness.
- Slice height should be kept according to the layer thickness-related limitations of the 3D printer.

4.4.2 Tool Path

The deposition path provides the necessary information regarding material deposition for the machine. This path can be expressed using G- and M-codes or any other machine-specific syntax and format. When combined with additional parameters such as print speed and temperature, the deposition path forms the complete set of instructions sent to the machine.

Several area-filling strategies can be employed to fill the space inside the contours, and these strategies inform the material deposition plan. Some area-filling methods include grid cutting, pixel, mask, raster, direction, contour, spiral and hybrid, among others. Below are descriptions of a few of these methods:

- **Grid cutting:** in laminated object manufacturing (LOM), raw material is utilized in the form of sheets. Two sheets are bonded together using an adhesive activated by heat and pressure, followed by a cutting process that separates the area associated with the part from the remainder (as shown in Figure 4.24). Grid cutting is used as an area-filling method to remove unwanted portions of the material.

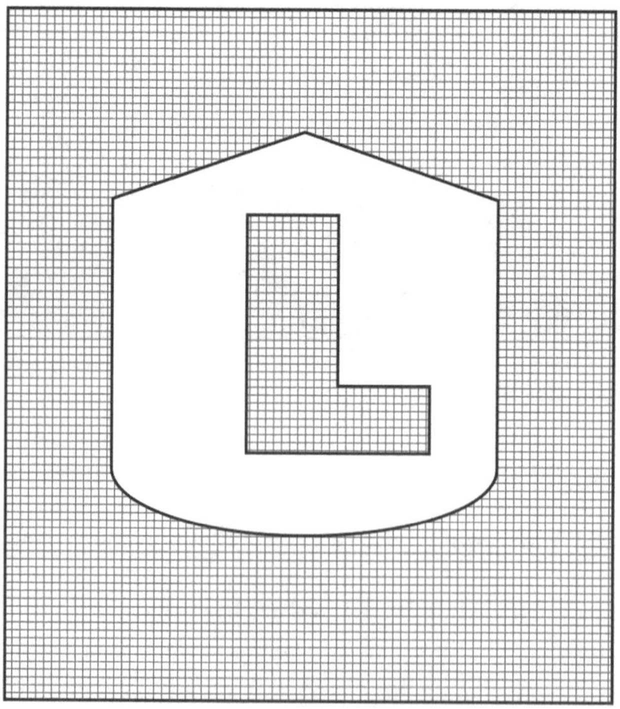

FIGURE 4.24
Grid cutting method.

- **Pixel:** in digital light processing (DLP)-based vat photopolymerization, binder jetting and material jetting processes, actions such as light projection, binder injection and material injection occur at the pixel level within the area defined by the object's boundary. The accuracy of material addition or removal is primarily determined by the pixel resolution.

  In these additive manufacturing techniques, the finer the pixel resolution, the more precise the material deposition or removal will be, resulting in higher accuracy and better surface finish of the final product. The pixel resolution plays a crucial role in governing the overall quality of the manufactured object, as it directly impacts the level of detail that can be achieved during the fabrication process.

  DLP-based vat photopolymerization employs light projection to selectively cure photopolymer resin in a layer-by-layer fashion, whereas binder jetting and material jetting processes involve the injection of binder or material onto the build platform to create a solid object. In all these methods, the pixel resolution dictates the level of precision and accuracy that can be attained, making it a critical factor in the success of the manufacturing process.
- **Mask:** as its name suggests it generates a mask (as shown in Figure 4.25) to obstruct light that helps to cure the selective region of the vat that is not masked and allows the light to go through the glass/screen.
- **Contour path:** the contour path is generated by offsetting the geometry inward direction. All the closed contours in a slice are collected together in a pattern to be offset.
- **Spiral path:** the spiral path is popular in NC machining and it also used for some specific 3D printing process.
- **Path-based area filling:** path-based filling defines the path, which should be followed by the material deposition nozzle like in FDM. Several patterns are used in the path-based area filling, i.e., raster, filling, contour filling, Hilbert curve filling, etc. It is up to the machine and user, which method is used to fill the area.

The path-based filling pattern with process parameters forms a toolpath that contains instructions related to axis movement and deposition control. Some other parameters are as follows:

- infill density varies road gap;
- nozzle diameter that helps to decide layer height;
- bed temperature;
- nozzle temperature to fuse the filament;
- filament diameter;

FIGURE 4.25
Mask to cure the Vat photopolymer.

- feed rate that helps to maintain a constant volume of material deposition;
- print speed;
- raster angle, etc., are the other parameters that we need to specify.

To generate the tool path code for fused deposition modelling (FDM), several critical variables must be defined. One of these variables is layer height, which is used when slicing the STL model of the object intended for printing, as well as for depositing material at the specified height. The total number of slices is determined based on the layer height.

Another important variable is the raster angle, which significantly affects the mechanical properties of the finished part by defining the angle of the raster fill. Raster angle values can range from 0° to 90°, and users have the flexibility to input their preferred angle in the provided field. Additional required parameters include nozzle diameter and bed temperature, both essential for maintaining the appropriate temperatures for the extruder and the build platform.

Filament diameter is used to calculate the feed speed in relation to the material deposition rate. Each printer command line includes a variable *E*, which refers to the extrusion parameter; this parameter is also influenced by the distance between two consecutive points in the nozzle path. In the case of pallet-based extrusion, the filament diameter is not a necessary parameter and is redundant when determining the tool path for that specific extrusion system.

There are two distinct variables to define: print speed and travel speed. Print speed relates to the rate of material deposition, while travel speed pertains to the nozzle's speed when idle. The nozzle must move faster during idle times, requiring a higher pulse feed rate compared to when it is actively printing. The nozzle diameter is another critical parameter that impacts the calculation of material extrusion (*E*) and is directly linked to the road gap of the tool path. The offset value indicates the distance between the innermost wall of the outer shell and the raster of the object, ensuring proper overlap and bonding between the object's outer boundary and the raster fill. Conversely, the offset number refers to the number of outer boundaries for material deposition. Additionally, RPM is solely relevant for the pallet-based extrusion system.

The process parameters can vary depending on the specific machine model. In Cura software, key process parameters that must be defined prior to tool path generation include infill density, infill pattern and shell thickness.

Screen capture of a customer-made process planning software is shown in Figure 4.26.

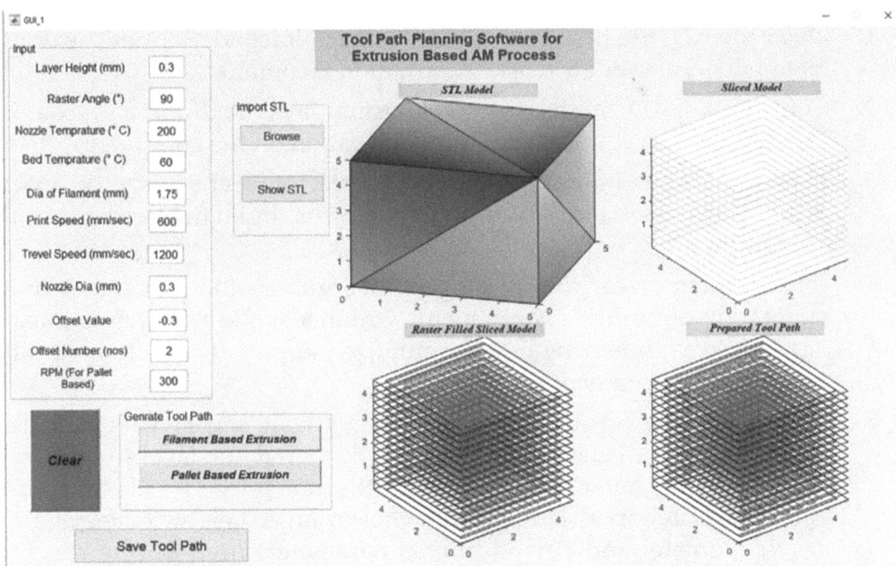

FIGURE 4.26
GUI of a toolpath generation program showing important process parameters of FDM-based 3D printing.

In various 3D printing processes, specific parameters can differ significantly. However, slice height is a common parameter present across multiple 3D printing techniques. While this course does not cover the intricate details of toolpath generation for each printing process, understanding the basic concepts of toolpaths is crucial for interpreting the variations found in different methods.

In conclusion, slicing and deposition path generation are vital components of the 3D printing workflow, responsible for converting a CAD model into a physical object with precision and accuracy. By recognizing the significance of these processes and implementing various optimization strategies, one can ensure that the final printed product is structurally sound and faithful to the original design.

As 3D printing technology advances, improvements in slicing algorithms and deposition techniques will enhance the capabilities of additive manufacturing, opening new avenues for design, innovation and production across diverse industries. This chapter lays the groundwork for understanding these essential concepts and encourages further exploration into the dynamic field of 3D printing.

Self-Check Exercises

1. Open the STL file in a text editor (such as Notepad, Sublime Text, or Notepad++) and analyze the information it contains.
2. Convert a CAD model into STL using fine, medium and coarse resolutions, and check the memory allocation on the computer.
3. How can different part orientations affect the efficiency of a 3D printing process in terms of support structures, build time and material consumption?
4. What strategies can be employed to optimize the placement and orientation of multiple components within a single print job to maximize build volume usage and minimize support material and post-processing requirements?
5. Design a simple 3D object using a CAD software and save it as an STL file. Then, use a slicing software to slice the object into layers according to the specified layer height. Analyze the generated G-code and provide a brief report discussing the chosen layer height, raster angle, nozzle diameter and any additional parameters used during slicing. Include screenshots of the slicing process and the final sliced object, along with your observations on the effect of these parameters on the print quality and structural integrity of the final 3D printed object.

Discussion Topics

1. Discuss the effect of different part orientations on the efficiency of a 3D printing process in terms of support structures, build time and material consumption.
2. Discuss the strategies that can be employed to optimize the placement and orientation of multiple components within a single print job to maximize build volume usage and minimize support material and post-processing requirements.

Multiple Choice Questions

1. What are the two ways in which STL files can store information?
   (a) Binary encoding and ASCII encoding
   (b) Geometric encoding and algebraic encoding
   (c) Compressed encoding and uncompressed encoding
   (d) Facet encoding and vertex encoding
2. What is the primary advantage of using binary format over ASCII format for 3D printing?
   (a) Binary files are smaller and easier to share
   (b) Binary files are visually easy to read and check
   (c) Binary files are more suitable for manual inspection
   (d) Binary files have better compatibility with CAD software
3. Which of the following is an advantage of using STL files in 3D printing?
   (a) Ability to store metadata and colour information
   (b) Universally recognized and supported by most 3D printers
   (c) Easy to modify and edit
   (d) Accurate representation of curved surfaces without approximation
4. What is a recommended setting for chordal deviation when exporting STL files for good quality prints and file size?
   (a) 0.1 to 0.01 mm
   (b) 0.01 to 0.001 mm
   (c) 0.001 to 0.0001 mm
   (d) 0.1 to 1 mm

5. What is the main disadvantage of using STL files compared to OBJ files in 3D printing?
   (a) STL files are larger in size
   (b) STL files are more complex to share and publish
   (c) STL files lack color and texture information
   (d) STL files are not suitable for 3D scanned objects
6. Which of the following is NOT a step in a typical STL file repair workflow?
   (a) Auto-repair
   (b) Separating shells
   (c) Adjusting print resolution
   (d) Closing holes, bridging gaps
7. What is the main advantage of using native file formats in CAD software?
   (a) They are universally compatible
   (b) They can take full advantage of specific program features
   (c) They have smaller file sizes
   (d) They are suitable for 3D scanned objects
8. What is the main purpose of STL viewers?
   (a) To modify the STL data
   (b) To passively view STL data
   (c) To generate support structures
   (d) To create new STL files
9. What is a common feature in advanced STL viewers?
   (a) Data modification
   (b) Remeshing
   (c) Serial number addition
   (d) All of the above
10. Which file format is specifically used for 3D scanned objects?
   (a) OBJ
   (b) STL
   (c) PLY
   (d) IGES
11. Why is part orientation important in the 3D printing process?
   (a) It affects build quality, processing and post-processing
   (b) It has no impact on support generation
   (c) It reduces the efficiency of the slicing process
   (d) It decreases the use of the build volume
12. What are the two categories of methods used to align the proper part orientation in 3D printing process planning software?
   (a) Manual and automatic methods
   (b) Geometric and algebraic methods
   (c) Proportional and non-proportional methods
   (d) Linear and non-linear methods

13. What transformation operations are used in manual placement?
    (a) Translation, rotation, mirroring and scaling
    (b) Rotation, scaling and reflection
    (c) Translation, scaling and projection
    (d) Mirroring, scaling and extrusion
14. Which type of scaling maintains the object's proportions?
    (a) Uniform scaling
    (b) Non-uniform scaling
    (c) Anisotropic scaling
    (d) Isotropic scaling
15. What is the purpose of the mirror transformation in 3D printing process planning software?
    (a) To create a symmetric copy of the object
    (b) To enlarge or shrink the object
    (c) To change the orientation of the object
    (d) To translate the object to a new position
16. In 3D printing, what is the primary benefit of using automatic placement methods?
    (a) They require no user input
    (b) They position the part based on specific parameters for optimal results
    (c) They are faster than manual methods
    (d) They always produce perfect prints
17. Which of the following is NOT a popular model for achieving optimal part orientation in 3D printing software?
    (a) Cost minimization model
    (b) Support minimization model
    (c) Multi-objective model
    (d) Surface area maximization model
18. Why are support structures necessary in additive manufacturing?
    (a) To create overhanging features
    (b) To improve the appearance of the final product
    (c) To speed up the printing process
    (d) To reduce material usage
19. In Cura software, what are the two types of support structures available?
    (a) Grid-type support and layer support
    (b) Column-type support and tree support
    (c) Honeycomb support and lattice support
    (d) Triangle support and square support

20. How does the support placement setting in 3D printing software affect support structures?
    (a) It adjusts the positioning of support structures on the model or build plate
    (b) It changes the density of the support structures
    (c) It modifies the material used for support structures
    (d) It determines the height of support structures
21. In 3D printing, what are the two types of deposition paths?
    (a) Primary material deposition and secondary material deposition
    (b) Support material deposition and contour boundary deposition
    (c) Primary material deposition and support material deposition
    (d) Contour boundary deposition and slicing material deposition
22. Which method of slicing in 3D printing involves slicing a part obtained in any CAD format without the need for an intermediate format?
    (a) Direct Slicing
    (b) Indirect Slicing
    (c) Uniform Slicing
    (d) Adaptive Slicing
23. Which file format is commonly used as an intermediate format in indirect slicing?
    (a) OBJ
    (b) STL
    (c) PLY
    (d) 3MF
24. Which type of slicing adjusts layer thickness across the build direction to achieve certain objectives?
    (a) Direct Slicing
    (b) Indirect Slicing
    (c) Uniform Slicing
    (d) Adaptive Slicing
25. Which effect can be mitigated by using adaptive slicing techniques?
    (a) Staircase effect
    (b) Warping effect
    (c) Overhang effect
    (d) Elephant foot effect
26. What is the user-defined parameter that dictates how the CAD model is sliced?
    (a) Slice width
    (b) Slice height
    (c) Slice depth
    (d) Slice angle

27. What is the main goal of contour organization in the slicing process?
(a) To reduce build time
(b) To identify exterior and interior contours
(c) To minimize the staircase effect
(d) To increase the surface finish

28. Which of the following area-filling strategies is used in laminated object manufacturing (LOM) to eliminate unwanted portions of the material?
(a) Grid cutting
(b) Pixel
(c) Mask
(d) Contour path

29. In digital light processing (DLP)-based vatphotopolymerization, binder jetting and material jetting processes, what directly impacts the level of detail that can be achieved during the fabrication process?
(a) Slice height
(b) Infill density
(c) Pixel resolution
(d) Raster angle

30. Which area-filling method involves offsetting the geometry in the inward direction and is used in some specific 3D printing processes?
(a) Path-based area filling
(b) Spiral path
(c) Contour path
(d) Grid cutting

Answers

1. (a) Binary encoding and ASCII encoding
2. (a) Binary files are smaller and easier to share
3. (b) Universally recognized and supported by most 3D printers
4. (b) 0.01 to 0.001 mm
5. (c) STL files lack color and texture information
6. (c) Adjusting print resolution
7. (b) They can take full advantage of specific program features
8. (b) To passively view STL data
9. (d) All of the above
10. (c) PLY
11. (a) It affects build quality, processing and post-processing
12. (a) Manual and automatic methods
13. (a) Translation, rotation, mirroring and scaling

14. (a) Uniform scaling
15. (a) To create a symmetric copy of the object
16. (b) They position the part based on specific parameters for optimal results
17. (d) Surface area maximization model
18. (a) To create overhanging features
19. (b) Column-type support and tree support
20. (a) It adjusts the positioning of support structures on the model or build plate
21. (c) Primary material deposition and support material deposition
22. (a) Direct slicing
23. (b) STL
24. (d) Adaptive slicing
25. (a) Staircase effect
26. (b) Slice height
27. (b) To identify exterior and interior contours
28. (a) Grid cutting
29. (c) Pixel resolution
30. (c) Contour path

5

Preprocessing – Selecting the Material for 3D Printing

In manufacturing, materials play a crucial role as they significantly impact the performance of the final part. Similarly, in 3D printing, the choice of material is pivotal as it determines the outcome of the printing process. Materials serve as the foundation of 3D printing, influencing the functionality and applicability of the printed parts across various industries.

3D printing processes are inherently material-dependent, meaning that each process is compatible with a specific range of materials. The selection of material is contingent upon the intended use and application of the 3Dprinted parts. This chapter explains the importance of material selection for 3D printing thereby enabling the learner to select suitable material for the given 3D printing process. It also covers the material properties for various 3D printing techniques, applications and limitations of 3D printing materials. The chapter comprises the following main sections:

5.1 Materials: Types and Selection Criteria
5.2 Materials for Solid-Based 3D Printing Processes
5.3 Materials for Powder-Based 3D Printing Processes
5.4 Materials for Liquid-Based 3D Printing Processes

After reading through the material, self-check exercises, objective questions and discussion topics presented in this chapter, the learner will be able to do the following:

(a) Explain the importance of material selection for 3D printing.
(b) Select the suitable material for the 3D printing process.
(c) Evaluate the material properties for various 3D printing techniques.
(d) Examine the applications and limitations of 3D printing materials.

DOI: 10.1201/9781003562344-5

5.1 Materials: Types and Selection Criteria

The choice of materials in manufacturing and 3D printing has a significant impact on the quality and performance of the final products. Different 3D printing processes are compatible with specific materials, and the selection of materials depends on the intended use of the printed parts. These parts have a wide range of applications across various industries, including rapid prototyping, biomedical, electronics, aerospace, jewelry manufacturing and automotive sectors (refer to Figure 5.1).

Unlike traditional manufacturing methods, the properties of 3Dprinted parts are influenced not only by the raw material but also by the printing parameters and build direction. It is important to consider that properties achieved in the *XY* plane may differ from those in the *Z* direction, resulting in anisotropic characteristics. This inherent anisotropy emphasizes the vital role of material selection in 3D printing to ensure optimal performance and functionality of the printed parts.

5.1.1 Types of Materials for 3D Printing

The different types of materials used in 3D printing can be categorized into polymers, metals, ceramics, and composites due to advancements in 3D printing technology as illustrated in Figure 5.2.

Polymers, which are extensively used in 3D printing, have large molecular structures composed of repeated subunits. They can be categorized as thermoplastic, thermoset and elastomers, all of which are used in 3D printing. These materials offer a good strength-to-weight ratio and low densities, and they can be printed at relatively low temperatures compared to metals. Thermoplastic polymers can be softened when heated and solidified when cooled, making them suitable for processes like fused deposition modelling (FDM). Common thermoplastic polymers used in the 3D printing process are Acrylonitrile butadiene styrene (ABS), Polylactic acid (PLA), Polyamides

FIGURE 5.1
Various applications of 3D printing.

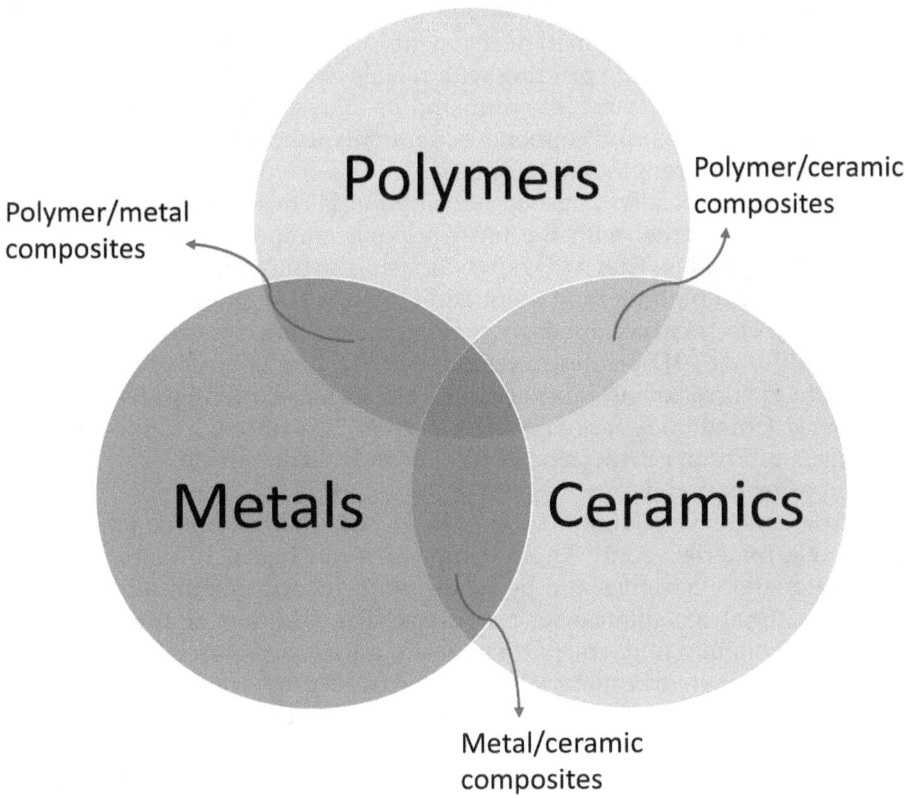

FIGURE 5.2
Material types for 3D printing.

(nylon), Polycarbonate (PC), Polyvinyl alcohol (PVC), High impact polystyrene (HIPS), High-density polyethylene (HDPE), acrylonitrile styrene acrylate (ASA), Polypropylene (PP), Thermoplastic polyurethane (TPU) and Polyethylene terephthalate glycol-modified (PETG).

On the other hand, thermosets are network polymers that do not soften on heating, making them harder and stronger than thermoplastics. They are suitable for high-temperature applications and are commonly used in electronics and household appliances. Epoxies and acrylic are frequently used thermosets for 3D printing.

Elastomers are known for their elastic behaviour, elasticity and resilience, and they have good thermal and electrical insulating properties. They are commonly used in engineering applications, such as gaskets, seals and hoses, with polyurethane and silicon being the most widely used elastomers for 3D printing.

Metals can be used in 3D printing processes to create parts with strong and dense properties. Unlike polymers and ceramics, metals have highly ordered atomic arrangements and can conduct heat and electricity well. They are composed of one or more metallic elements and also non-metallic elements in small amounts, and 3D printing with metals can include both ferrous and non-ferrous materials. Steel, titanium and its alloys, inconel, aluminum and its alloys, super alloys, and copper are commonly used metals for 3D printing due to their high strength and toughness.

Ceramics are typically made up of compounds consisting of metallic and non-metallic elements, with the most popular compositions being oxides, nitrides and carbides. They are known for being brittle and prone to breaking. Ceramics have a high melting point and are quite strong, but they have low thermal and electrical conductivity, making them useful for insulation. When produced through 3D printing, ceramics are ideal for components that need to withstand corrosion and mechanical wear, especially in high-temperature conditions. Common types of ceramics used in 3D printing include alumina, zirconia, aluminum nitride, silicon carbide and silicon nitride.

Composite materials are created by combining two or more different materials to achieve a combination of properties that cannot be attained by any single material alone. These composites can be made using combinations of metals, ceramics and polymers, with reinforcements added to the matrix material to enhance its properties. The addition of high-modulus reinforcements like fibres or particulates to a low-modulus matrix material results in a host matrix material with improved properties, including high strength-to-weight and stiffness-to-weight ratios. These enhanced properties expand the potential applications of 3D printed parts. The fibres can be added to the matrix in chopped form (short fibers) or used as continuous fibres. Additionally, a comparison of tensile strength for metals, ceramics, polymers and composites is illustrated in Figure 5.3.

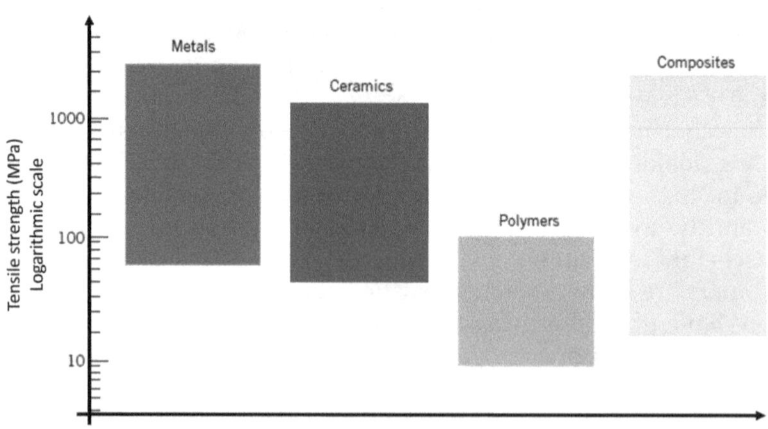

FIGURE 5.3
Variation in tensile strength of polymers, metals, ceramics, and composites.

5.1.2 Materials Selection Considerations

When using 3D printing, choosing the right material is a critical part of the product development process. Various factors, such as the intended application of the part, must be taken into account when selecting a material. It is important to ensure that the chosen material meets the necessary certifications for the project. For instance, if a 3Dprinted polymer part needs to be sterilized using an autoclaving process, the material should be able to withstand high temperature and pressure without deforming. If biocompatibility is important for a specific application, the chosen 3D printing material must possess this characteristic. Additionally, certain applications may require flame, smoke and toxicity retardant materials, so the selected material must meet these requirements.

The functionality of the 3D printed parts also depends on the chosen material, as well as the environment in which the printed part will be used. Factors such as the level of load applied or the temperature to which it will be subjected need to be considered when selecting a material.

The geometry of the printed part, including intricate features and minimum resolution, must also be considered when choosing a material. It is important to note that the characteristics of materials are closely tied to the corresponding 3D printing technology. For example, a liquid-based 3D printing process may offer higher resolution compared to a solid-based or powder-based 3D printing process. Figure 5.4 shows the influence of material on 3D printed parts' key factors.

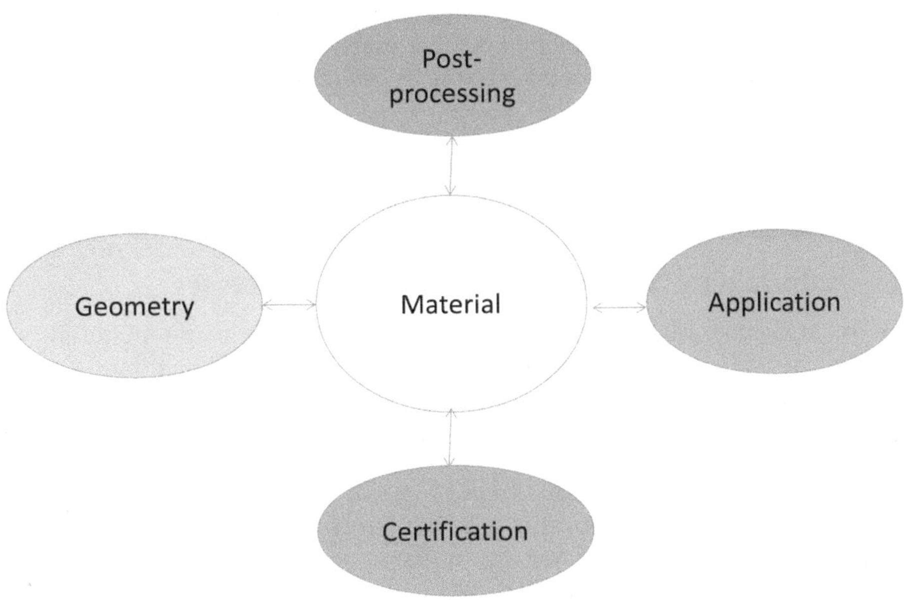

FIGURE 5.4
Impact of materials on parts key factors.

In many 3D printing processes, post-processing is often necessary. Different materials are suitable for specific post-processing techniques. For example, if there is a complex geometric component with many support structures that are hard to remove manually, a material that can be dissolved in a specific medium may be used for easy support removal without damaging the part. Additionally, some materials may require curing after printing, such as photopolymer, or heat treatment, like stainless steel. Therefore, material selection should consider these factors.

5.1.3 3D Printing Technology Specific Materials

3D printing methods are specific to the materials they can handle. The material chosen directly impacts how well the part will be printed and its functionality. Different materials may require specific temperature controls, nozzle sizes or layer adhesion techniques during printing. Understanding the properties and behaviour of different materials is crucial for selecting the right printing process and achieving the desired results.

Different 3D printing techniques can only handle a limited variety of materials. For instance, the vat polymerization process can use UV-curable resins, material jetting can handle waxes, ceramics and UV-curable resins, binder jetting can process polymers, ceramics and metals, while material extrusion systems like FDM can use thermoplastics. 3D printing processes that use material in powder form can process thermoplastics and metals. Sheet lamination techniques can use papers, metals and thermoplastics, while direct energy deposition techniques can process metallic materials. The list below shows the types of processable materials for various 3D printing techniques. Figure 5.5 shows the various types of materials compatible with 3D printing processes.

5.2 Materials for Solid-Based 3D Printing Processes

We have discussed how various 3D printing methods use materials in different ways. One common form of material used in 3D printing is solid. Two 3D printing technologies that use solid materials are fused deposition modelling (FDM) and laminated object manufacturing (LOM). FDM is the more widely used process because it offers several advantages, including ease of material change, material handling and better properties. To use the FDM process, the material needs to be in filament form. The filament, made of thermoplastic material, must be prepared with the right diameter and mechanical properties for processing. The filaments are melted and extruded to create the part layer by layer. On the other hand, for the LOM process, the

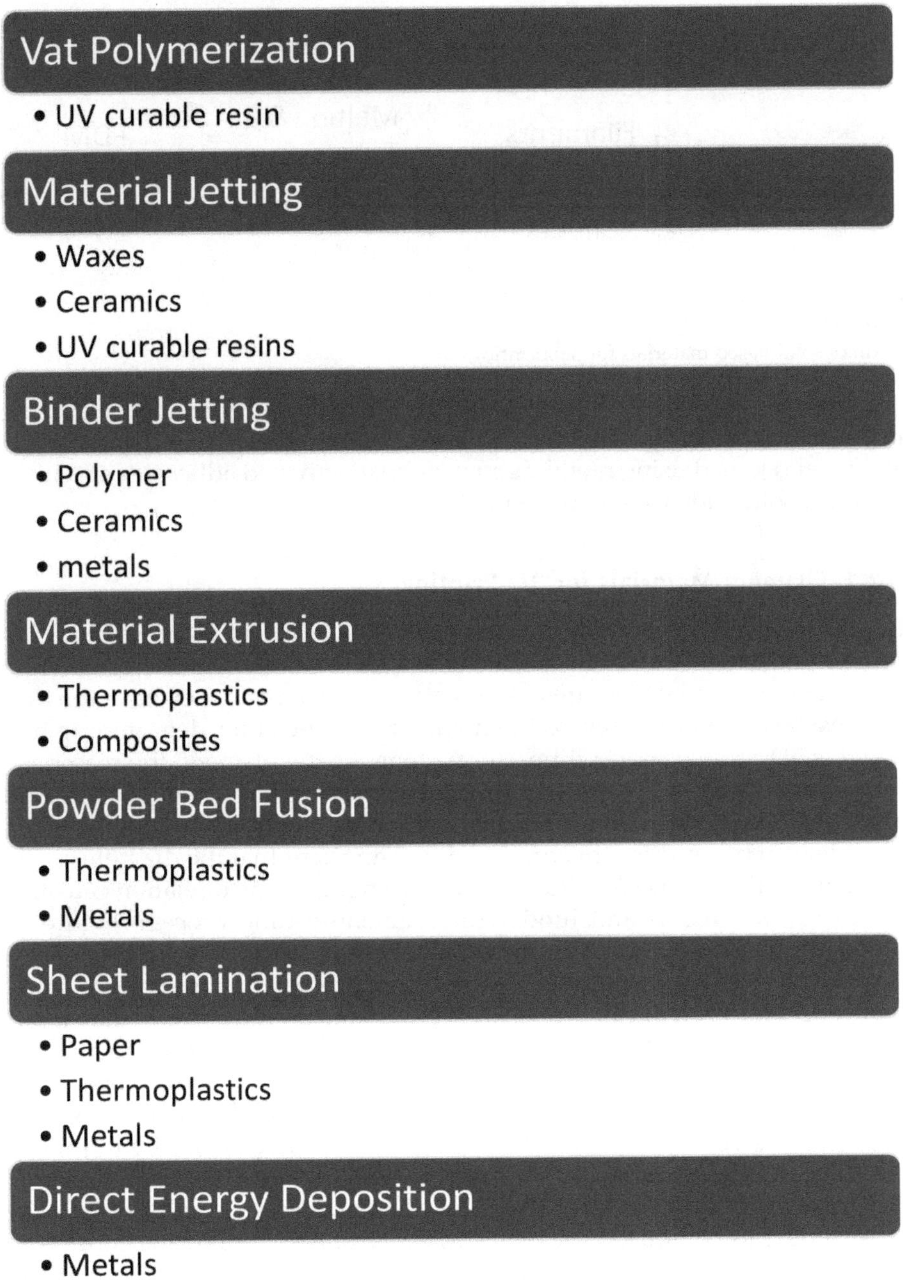

FIGURE 5.5
Type of materials compatible with 3D printing processes.

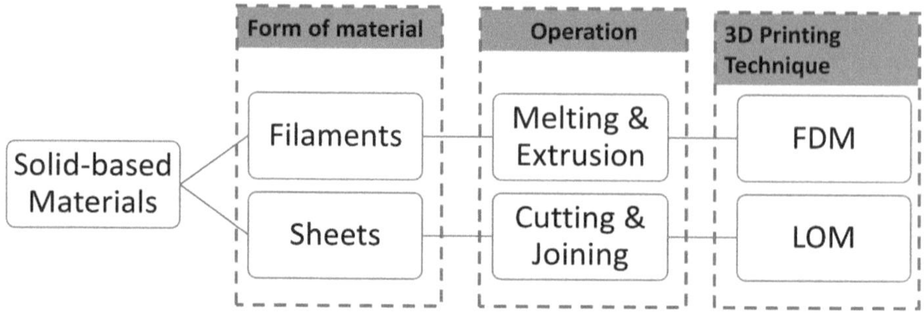

FIGURE 5.6
Form of solid-based materials for 3D printing.

materials need to be in sheet form. These sheets are then cut into the required shapes and joined using methods such as heat-activated adhesives and ultrasonic consolidation (see Figure 5.6).

5.2.1 Filament Materials for 3D Printing

In order to use FDM 3D printing, material needs to be transformed into filament form. This filament is similar to a wire, with a circular cross-section. Creating the filaments requires careful attention to ensure they have the right dimensions. The properties of the filament are crucial for determining how well the 3D printed parts will perform. To make the filament, the material is first extruded. Most 3D printers use filaments with a diameter of 1.75 mm, although 3 mm filaments are also available. A single- or double-screw extruder can be used for this process. It is important that the diameter of the filament remains consistent throughout its length, as any deviation can affect the extrusion process and hinder the filament-feeding process. Figure 5.7 illustrates the schematic diagram of filament fabrication.

When 3D printing using the FDM process, it is essential to use a filament with specific mechanical properties. The filament must have the appropriate stiffness to facilitate successful extrusion during printing. Insufficient stiffness could lead to buckling prior to entering the heating zone, negatively impacting the printing process. Additionally, the filament should not be excessively brittle, but should display desirable flexibility to prevent breakage during printing.

A variety of commercially available filaments, both soluble and semi-crystalline, can be utilized for the FDM process. Amorphous thermoplastics are preferred for material extrusion due to their melt characteristics compared to semi-crystalline polymers. ABS, widely used in the FDM process, is a tough material with good impact and heat resistance.

Another commonly used material, PLA, is biodegradable and offers good rigidity but is prone to brittleness. PC is known for its durability and strength,

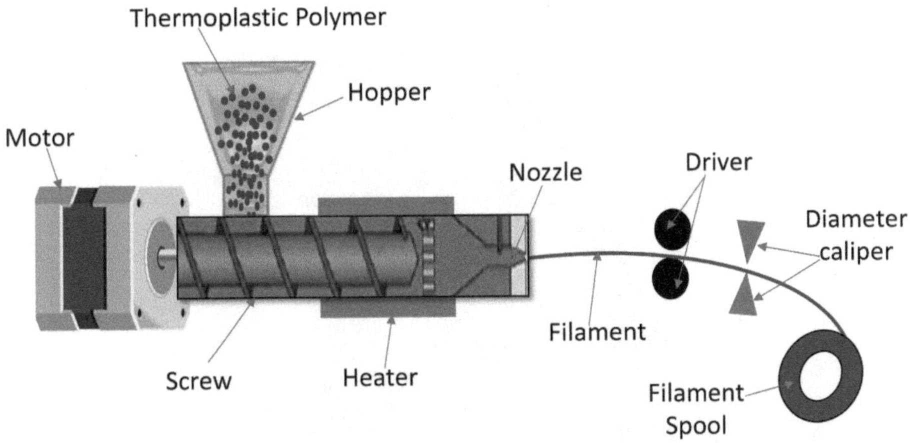

FIGURE 5.7
Schematic representation of a single screw extruder to prepare filament for 3D printer.

providing high impact and heat resistance, making it suitable for challenging environments. PETG, a chemical-resistant material with high transparency and water resistance, also offers excellent printability.

Nylon, a strong, durable and lightweight material, possesses good impact and heat resistance but is challenging to 3D print. TPU, a flexible material with properties of thermoplastic and polyurethane, exhibits flexibility and high impact strength, offering superior vibration damping. PVA, a water-soluble material, is widely used as a support material. HIPS, another soluble material, is used as support material with ABS.

ASA, known for its resistance to UV light and impact loading, is ideal for outdoor applications. HDPE is recognized for its high strength-to-density ratio, weather resistance, and being food-safe, although it is difficult to 3D print. PP offers good fatigue resistance, making it suitable for high-cycle, low-strength applications. PEEK, a semi-crystalline polymer, boasts good mechanical and chemical resistance. Additionally, composite filaments with reinforcing materials such as carbon fibre, ceramics and metals are available. Figure 5.8 shows the commercially available filaments for 3D printing.

5.2.2 Temperature Selection for Filaments

Understanding the optimal temperature settings for 3D printing filament is crucial for successful printing. Each filament material has unique temperature requirements for the extruder and print bed, based on factors such as the material's glass transition temperature (T_g) and melting temperature. The T_g signifies the point at which a polymer transitions from a hard, glassy state to a softer, rubbery state. When setting the printing temperature, it is important to slightly exceed the T_g to ensure proper flow and layer adhesion, while

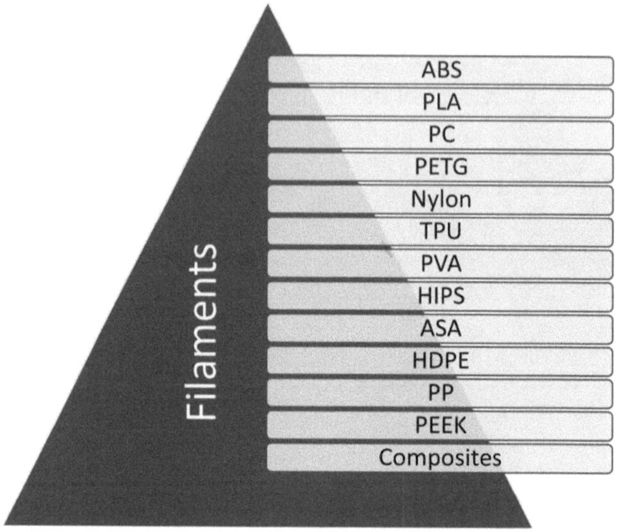

FIGURE 5.8
Commercially available filaments for 3D printing.

avoiding excessive temperatures that could lead to deformation. Melting temperature, on the other hand, represents the point at which a material shifts from a solid to a liquid state. For thermoplastic filaments like ABS, PLA and PETG, the printing temperature is generally set slightly below the melting temperature to facilitate controlled extrusion without causing material degradation.

Manufacturers of filaments typically offer recommended printing temperatures based on rigorous testing and analysis. These serve as a starting point for users and can help achieve optimal print quality. However, users can also experiment with the temperature settings to refine print quality based on factors such as layer adhesion, stringing and overhang performance. Even small adjustments in temperature can significantly impact print quality and success.

Bed temperature selection is equally important, as it directly influences adhesion between the filament and the print bed. Different filament materials necessitate specific bed temperatures tailored to their properties. For example, ABS filament typically adheres well to a heated bed, while PLA filament may require a lower bed temperature. Manufacturer recommendations and the type of print bed surface, such as glass or adhesive, further influence the required bed temperature. Higher bed temperatures can enhance layer adhesion and mitigate warping risks, especially for larger prints and materials prone to warping, like ABS.

Users can experiment with bed temperatures to identify the optimal setting for their specific printer and filament, ensuring consistent and successful

TABLE 5.1

Recommended Temperature Settings for Various Filament Materials for FDM

Filament material	Extruder Temperature	Bed temperature	Heated Bed
ABS	220–250°C	95–110°C	Required
TPU	225–245°C	45–60°C	Optional
PLA	190–220°C	45–60°C	Optional
HIPS	230–245°C	100–115°C	Required
PETG	230–250°C	75–90°C	Required
Nylon	220–270°C	70–90°C	Required
Composite (carbon fibre reinforced)	200–230°C	45–60°C	Optional
ASA	235–255°C	90–110°C	Required
PC	260–310°C	80–120°C	Required
PP	220–250°C	85–100°C	Required
Composite (metal filled)	190–220°C	45–60°C	Optional
Composite (wood filled)	190–220°C	45–60°C	Optional
PVA	185–200°C	45–60°C	Required
HDPE	230–260°C	60°C	Required
PEEK	390–400°C	130°C	Required

(*Note*: the temperature settings may vary depending upon the manufacturer's recommendation)

prints. For more detailed information, refer to Table 5.1 for a comparison of extruder and bed temperature recommendations for various commercially available filaments.

5.2.3 Material Properties of Filament Materials

The selection of filament materials can significantly impact the properties of 3D printed components. However, it is important to consider various process variables in addition to filament materials to achieve desired outcomes. Understanding the essential characteristics of filament materials is crucial for selecting the right materials for specific applications. For instance, ABS offers good impact resistance and durability with an ultimate tensile strength of 40MPa, while TPU provides flexibility and resilience with a range of ultimate strengths from 26 to 43MPa. PLA is a popular choice for biodegradable material in FDM, and HIPS is often used as a soluble support material due to its ultimate strength of 32MPa. PETG, nylon and carbon fibre-reinforced filaments each offer unique strengths and properties, catering to different application requirements. By thoroughly comparing filament attributes such as strength, flexibility, temperature resistance and printability, users can make informed decisions for their 3D printing projects, ensuring that the chosen materials align with the intended functionality and performance of the printed pieces for successful outcomes. Table 5.2 shows the tensile strength and density for various commercially available filament materials for FDM.

TABLE 5.2

Tensile Strength and Density for Various Filament Materials for FDM

Filament material	Ultimate Strength	Density
ABS	40MPa	1.04g/cm^3
TPU	26–43MPa	1.19–1.23g/cm^3
PLA	65MPa	1.24g/cm^3
HIPS	32MPa	1.03–1.04g/cm^3
PETG	53MPa	1.23g/cm^3
Nylon	40–85MPa	1.06–1.14g/cm^3
Composite (carbon fibre reinforced)	45–48MPa	1.3g/cm^3
ASA	55MPa	1.07g/cm^3
PC	72MPa	1.2g/cm^3
PP	32MPa	0.9g/cm^3
Composite (Metal filled)	20–30MPa	2–4g/cm^3
Composite (wood filled)	46MPa	1.15–1.25g/cm^3
PVA	78MPa	1.23g/cm^3
HDPE	29MPa	0.958 g/cm^3
PEEK	100 MPa	1.26 g/cm^3

5.2.4 Applications and Limitations of Filament Materials in 3D Printing

The filament materials used in FDM techniques offer great versatility for a wide range of applications. Recent advancements in filament development have expanded the options available, and commercially available thermoplastic filament materials are now suitable for various uses including wear-resistant parts, waterproofing, flexible prototypes, conceptual models, jigs, fixtures, and tooling in industries such as automotive, aerospace, and dental fields.

PLA is predominantly utilized in prototyping work because of fast design iteration, clear surface finish, and ease of printing. It is a material of choice for students and hobbyists, due to the fact that it is non-toxic and has little warpage when printing. In medicine, PLA is used for biodegradable implants, surgical guides, and eco-friendly packaging. It enjoys popularity in the making of artistic and architectural models, sculpture making, ornamentation, and conceptual architectural schemes. However, PLA presents itself with many drawbacks, chief among them being low heat resistance since PLA deforms and gets soft at low temperatures and thus is not applicable in heat environments. It is also a brittle material and has low impact resistance; this restricts its application in functioning parts that are under mechanical stress. It is also not very weather-resistant since it deteriorates after prolonged exposure to UV radiation and moisture. ABS, on the other hand, has been popularly and extensively utilized in the automotive industry in making components of dashboards, interior panels, and functional models. In consumer electronics, it is employed in making casings for remote controls, game controllers, and so on. Because of its mechanical strength and wear-resistance, ABS is also applicable in industrial cases such as gears, brackets, and boxes. Yet with these advantages, ABS introduces several issues, notably its high-temperature requirement while printing, needing a heated bed, which is above 100°C quite often, and an enclosed chamber to lessen warping. In truth, during the printing of ABS, some volatile organic compounds are emitted, yielding some styrene fumes that require proper ventilation. Another disadvantage to this statement is its sensitivity to UV rays.

PETG is used for several applications in the food-safe capabilities. PETG also performs well in outdoor conditions because it has UV and water resistance flexibilities, which make it suitable for outdoor signs and equipment. PETG is also used in prototyping for industrial applications with average impact and chemical resistance properties. However, PETG has its weakness such as oversized stringing and oozing when printing that requires meticulous retraction settings. Also, it has poor adhesion and considerable weakness due to delamination under high mechanical stress in some few other materials. Flexible filaments like TPU are very efficient in applications requiring elasticity and also resistance to impact. They find their way in wearable technology, like soft wristbands, watch straps, and different sports equipment. In medicine, TPU is used in prosthetic limbs, shaped to fit the individual, orthopedic insoles, and rehabilitation devices. It also finds industrial applications in seals and gaskets for automobiles and mechanical works, where it is important to have chemical resistance and, more importantly, flexibility. The core issue here is that TPU is very hard to print, as these work best when printed at very low speeds on direct-drive extruders in order to avoid bunching the filament. High levels of stringing and oozing threaten the TPU and require either instance optimized retraction or cooling control. Finally, since TPU is nonrigid, it limits the use of optional load-bearing applications.

Nylon has found application in all fields, from automobiles to aerospace, for making low-weight and high-strength parts: wheels, hinges, and mechanical assemblies, among others. Nylon has also found uses in sporting and textile applications, like semi-flexible shoe soles or sporting equipment. However, Nylon has notable limitations, such as high water absorption, that will need them stored in dry conditions to make sure no defects come up in prints. Additionally, this material has poor bed adhesion and will need to be printed with special adhesive or in a heated enclosure to control the warping. Widely used in the manufacture of safety goggles, helmets, and impact shields in the field of personal protection, PC is an important plastic in the industrial field, where highly organized and high-temperature strength components are needed. The aerospace and defense sector use polycarbonate in UAV parts and aircraft interiors. Unfortunately, due to the high-temperature requirement for printing, a hot end with an all-metal design, and a heated chamber would need to be in place. Polycarbonate, meanwhile, is also prone to hydrolysis, whereby moisture absorption can further compromise the quality of the resulting print. It also shows some degree of brittleness making it susceptible to fracture under extreme mechanical stress.

Despite the advances made, filaments still face certain challenges. Maintaining a constant diameter along the length of the filament is very hard, and a breaking of the filament during printing will interrupt the process. Moreover, moisture absorption into the filaments would adversely affect the extrusion process, thus causing print defects like bubbling, poor layer adhesion, and nonuniform extrusion. The porosity inherent in FDM parts also can limit the performance of printed parts such as mechanical strength, airtightness, and overall durability. Warping and shrinking are also common problems, especially in the case of ABS and Nylon materials that require controlled environments for the best printing outcomes. Research into materials formulations, drying methods, and print capabilities aims at mitigating such restrictions so that filaments-based 3D printing becomes real dependable for industries and commercial applications. Choosing a filament for use in FDM 3D printing highly influences the performance, longevity, and usability of the finished part.

5.3 Materials for Powder-Based 3D Printing Processes

Various 3D printing methods make use of material in powder form, such as selective laser sintering (SLS), selective laser melting (SLM), direct energy deposition (DED) when powder is used, and binder jetting (BJ). The powder can be stored in a container or dispensed through a nozzle during printing. A variety of materials including metal, polymer and ceramic powders are utilized in these processes, making the application domain of powder-based 3D printing wider than other methods. The quality of the printed parts produced through powder-based techniques relies on the chosen material and the bonding of powder particles, which can be sintered, melted or consolidated using binders for layer-by-layer manufacturing. Figure 5.9 illustrates the powder-based materials and associated 3D printing technique.

In the SLS method, the powder particles are bonded together through a process called sintering. Lasers play a key role in this method, as they are used to sinter and fuse the powder particles. When the laser beam interacts with the powder, it raises the temperature to the point near melting, causing the particles to fuse together with the previous layer, ultimately creating a solid object.

For SLM and DED, the approach involves melting powder particles to fuse them, which is typically used for metal powders. The powdered materials are heated to their melting point to fuse them together, resulting in the creation of fully dense objects. For a visual representation, please refer to Figure 5.10.

In the process of binder jetting, the powder particles stick to the binder at the printing spot. The excess powder remains loose and provides a foundation for the subsequent layers to be printed. The size of the powder particles is a critical factor in determining the level of detail achievable in the 3D printed components.

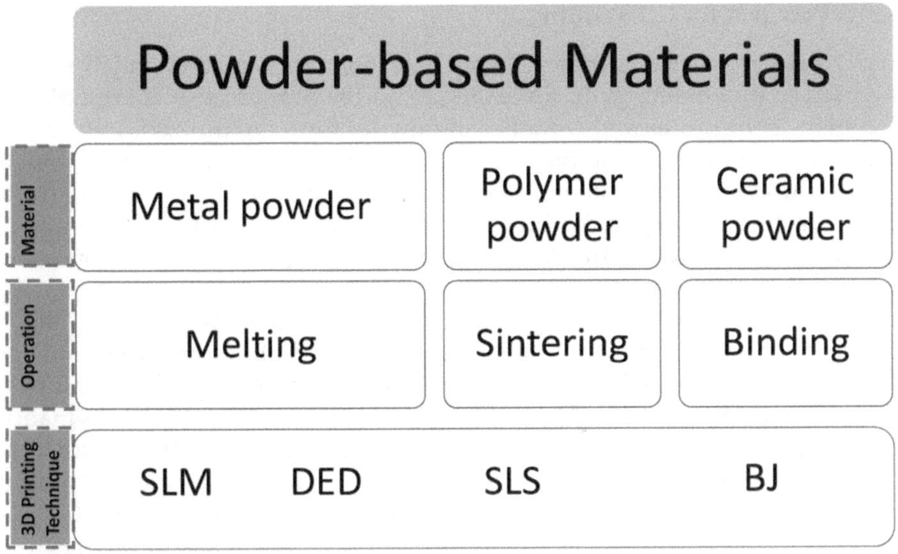

FIGURE 5.9
Powder-based materials and associated 3D printing technique.

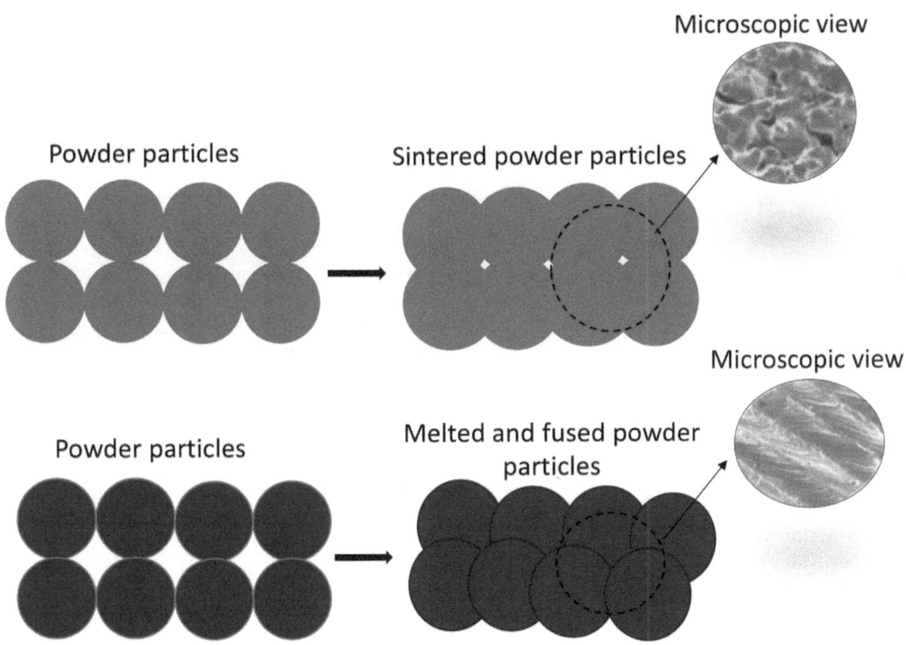

FIGURE 5.10
Sintering and melting of powder particles during the 3D printing process.

5.3.1 Powders for 3D Printing

In 3D printing, various materials such as polymers, metals and ceramics are used in powdered form. For example, in the SLS process, thermoplastic polymers, composites and ceramics in powder form are utilized. Polyamide and glass-filled polyamide powders are common choices for creating tough and stiff polymer components, while thermoplastic elastomers are used for flexible, rubber-like parts. Additionally, materials like polycarbonate powder, stainless steel infused with bronze, and copper-polyamide powder are used for specific applications.

In the SLM process, metal powders such as ferrous, non-ferrous, alloys and ceramics are employed. Steel- and iron-based alloys are preferred for their strength and endurance, particularly in various industrial applications. Nickel-based alloys are highly valued in aerospace and automobile manufacturing due to their corrosion resistance and high-temperature tolerance. Titanium-based alloys are in high demand for their strength and lightweight properties, especially in aerospace and medical industries. Similarly, aluminum alloys are widely used in automotive and aerospace industries due to their exceptional heat conductivity and low density. Commercial SLM systems typically use powders with particle sizes ranging from 20 to 50 μm. Table 5.3 shows some available metal powder for the SLM process.

TABLE 5.3

Some Metal Powder for the SLM Process

Materials	Some Alloy Examples
Steel and Iron-based alloys	Fe-Ni Fe3AI Fe-Ni-Cr stainless steel tool steel Ultra-high carbon steel
Titanium-based alloys	Ti-6A1-4V Ti6A14V ELI
Nickel-based alloys	Inconel Chromel Ni-Ti
Other Metals (Aluminium, Copper, Magnesium, Cobalt-Chrome, Tungsten, Gold, Silver)	A16061 AlSi10Mg CoCr 24 Carat gold Tungsten
Ceramics	Glass-ceramics Alumina Silica Yttria stabilized zirconia Alumina zirconia Mixture Porcelain

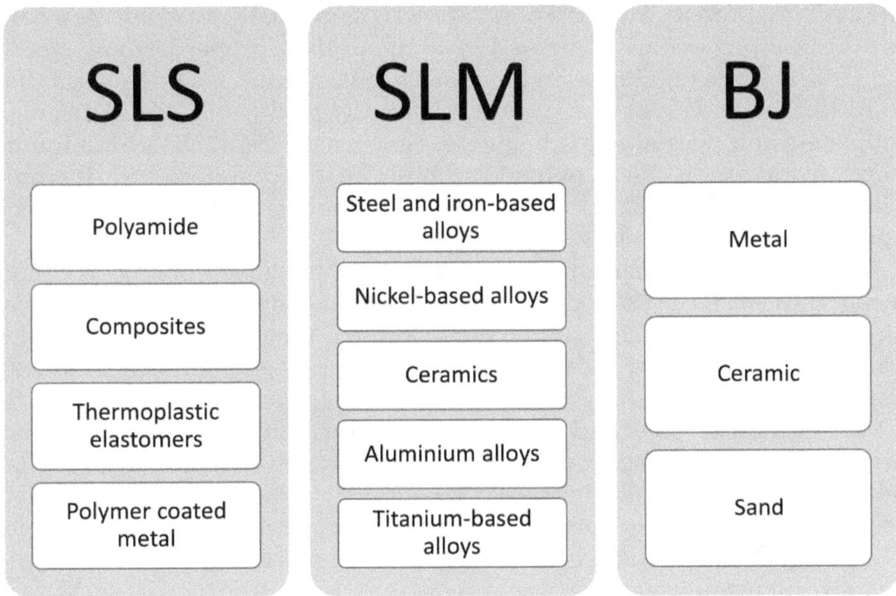

FIGURE 5.11
Various materials used for the common powder-based 3D printing process.

The binder jetting (BJ) process is a type of 3D printing that uses a powder bed and binder material to build parts layer by layer. This method is versatile and can be used with various materials including metals, sands and ceramics. Unlike other powder bed fusion (PBF) 3D printing techniques, binder jetting does not use heat during the build process, which helps prevent residual stresses from forming in the final parts. Different types of binders such as polymer binders, water-based binders, solvent-based binders, furan binders, silicate binders, phenolic binders and UV binders are utilized in binder jetting devices. These binders are chosen based on the specific application and are combined with powdered materials to create the desired parts. Each type of binder has its own method of removal, such as heat treatment for removing polymer binders or sintering for water-based binders. Additionally, UV binders offer high resolution and quick curing periods, but have limitations such as the need for UV light exposure during printing and a limited range of materials. Figure 5.11 provides a comparison of various powder-based methods and the materials they can process.

5.3.2 Material Properties of Powder Materials

In order to achieve optimal results in 3D printing, the characteristics of the powder materials used are extremely important. Properties such as mechanical properties, particle size, flowability and density all have an impact on the printing process. For instance, various commercially available powders for the SLS process are compared in terms of their properties. Polyamide 12, known for its toughness and resistance to chemicals, is widely used due to its durability. Polyamide 11, derived from vegetable oil, exhibits similar toughness and resistance to chemicals. Other materials such as aluminum-filled nylon, glass-filled nylon, carbon-fiber filled nylon and thermoplastic polyurethane (TPU) offer specific properties suitable for different applications.

Furthermore, commercially available metal powders for 3D printing, including steel, iron-based alloys, titanium-based alloys, nickel-based alloys, aluminum, copper, magnesium, cobalt-chrome, tungsten, gold and silver, have unique properties such as corrosion resistance, strength, hardness and temperature resistance.

Ceramics, on the other hand, are known for their high melting temperatures, brittle nature and surface roughness. Understanding these properties is vital when selecting materials for specific 3D printing applications. To achieve the best print result, the properties of the powder materials play a crucial role. The properties that influence the process are their mechanical properties, particle size, flowability and density.

The mechanical properties of common powder materials used for the SLS process are given below in Figure 5.12.

Figure 5.12 shows a comparison of various commercially available powders for the SLS process and their respective properties. Polyamide 12, a general-use plastic, is renowned for its toughness, high impact strength and resilience against chemicals and weathering. Polyamide 11, derived from vegetable oil, exhibits similar toughness and resistance to chemicals, making it popular in various industries. Aluminum-filled nylon, combining nylon with aluminum dust, boasts a high density and mechanical strength, suitable for robust components. Glass-filled nylon, comprising 70% nylon and 30% glass

Material	Polyamide 12	Polyamide 11	Aluminum-Filled Nylon	Glass-Filled Nylon	Carbon-Fiber Filled Nylon	TPU
Description	• General-use plastic • Popular due to its toughness, high impact strength, and ability to resist both chemicals and weather	• Polyamide bioplastic is produced from vegetable oil • Can absorb lower amount of water, PA 11 can accept the high loading of fillers and resist chemicals • High elasticity	• Produced by combining polyamide/nylon and aluminum dust • Alumide has a much higher density than other SLS 3D printing materials • Exceptional thermal and conductive properties	• PA-GF is 70% nylon and 30% glass fiber • Stiff powdered material • Very good mechanical strength • Its effectiveness in resisting high temperature and wear makes PA-GF one of the most preferred material	• PA-FR is 70% nylon and 30% carbon fiber • Widely in manufacturing and molding processes due to its high-strength-to-weight ratio • Utilized for components for automobiles and airplanes	• High elongation • High tear strength and elongation for strong, durable, elastomeric parts

FIGURE 5.12
Various materials properties of SLS materials.

fibre, offers stiffness and strength ideal for structural parts. Carbon-fibre filled nylon, with its exceptional thermal and conductive properties, finds applications in demanding environments like automotive and aerospace sectors. Lastly, thermoplastic polyurethane (TPU) stands out for its high elongation and tear strength, making it a preferred choice for elastomeric parts. These materials cater to a wide range of applications, from manufacturing to moulding processes, across diverse industries.

Steel- and iron-based alloys exhibit high corrosion resistance, strength and micro-hardness. Titanium-based alloys offer superior shear strength and low porosity, making them ideal for various applications. Nickel-based alloys boast high-temperature resistance, fatigue strength, excellent corrosion resistance, wear resistance and good weldability. Other metals such as aluminium, copper, magnesium, cobalt-chrome, tungsten, gold and silver are known for their high strength, with aluminium reaching up to 400 MPa and increased hardness when powder copper is added.

Ceramics, on the other hand, are characterized by high melting temperatures, a brittle nature, and high surface roughness. Understanding these properties is crucial for selecting the appropriate material for specific 3D printing applications.

5.3.3 Applications and Limitations of Powder Materials in 3D Printing

Materials commonly used in SLM, such as steel- and iron-based alloys, are widely applied in various industries. In the medical and biomedical fields, these materials are frequently used for implants and orthodontic products due to their durability and biocompatibility. Additionally, alloys like steel and iron are favoured for their thermal conductivity and resistance to corrosion in heat exchangers. They are also used in lightweight structures, where their strength-to-weight ratio is advantageous. Titanium-based alloys find extensive applications in medical and dental settings due to their biocompatibility and lightweight nature, while nickel-based alloys are essential in aircraft engines and combustion chambers because of their high-temperature resistance and corrosion properties. Other metals such as aluminum, copper and gold are used in biomedical, dental and industrial applications. Ceramics, on the other hand, are used in medical and dental fields for dental restorations and bone implants, as well as for various industrial purposes. The choice of materials plays a crucial role in diverse applications across industries.

The powder utilized in SLS serves various purposes, including creating conceptual models, functional prototypes and working models with properties similar to final products. It is also crucial for pattern creation in investment casting and facilitates rapid tooling due to its durability and compatibility with manufacturing processes. The selection of SLS materials depends on properties such as strength, durability and heat resistance, and significantly impacts product development and manufacturing workflows.

Materials used in the BJ process have wide-ranging applications across various industries, from automotive and aerospace to art and design. They are adaptable for creating prototypes, aerospace components, cores, moulds, tooling, casting patterns, intricate jewellery pieces and decorative objects, offering full-colour prints.

Most of the PBF 3D printing process requires sintering or melting of the powder, leading to high power consumption due to the laser's high wattage. The process is typically carried out in an inert environment, requiring the 3D printing system to accommodate inert gas tanks. Moreover, the large particle size of the powder results in a poor surface finish.

5.4 Materials for Liquid-Based 3D Printing Processes

The process of 3D printing using liquid materials is commonly used to create parts. These liquid materials are mainly thermoset polymers, with some usage of thermoplastics. Materials such as epoxies, acrylates and polyamides are frequently used in liquid-based 3D printing processes like stereolithography (SLA), digital light processing (DLP) and material jetting (MJ). These processes use photopolymers that cure or solidify when exposed to UV light, resulting in excellent surface finishes compared to other 3D printing methods. The resins used in liquid-based 3D printing are typically UV-curable photopolymers, which can be modified to meet specific mechanical and chemical requirements. Figure 5.13 illustrates the liquid-based materials and associated 3D printing techniques.

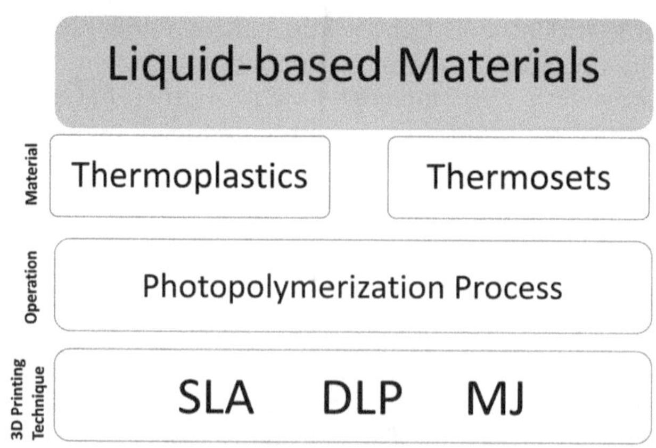

FIGURE 5.13
Liquid-based materials and associated 3D printing techniques.

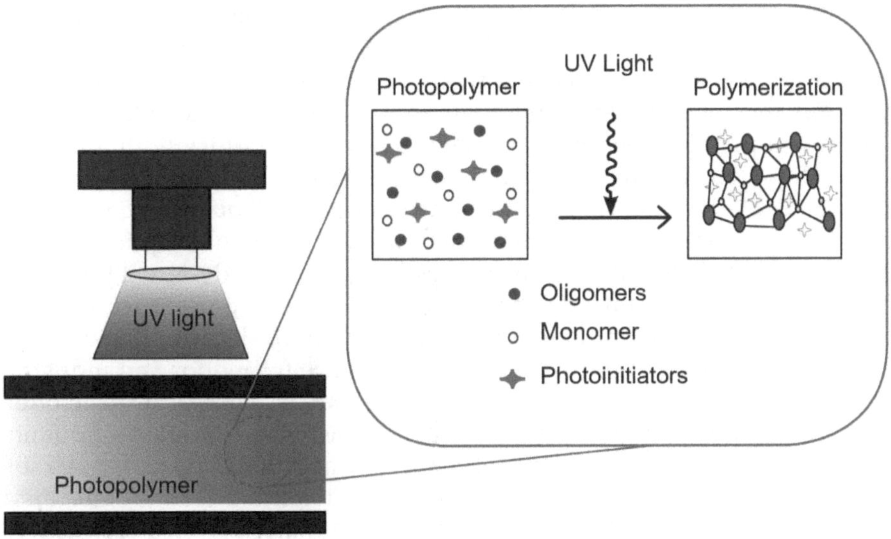

FIGURE 5.14
Schematic of the photopolymerization process during the 3D printing process.

The photopolymerization process, which is central to these techniques, can be activated by a photochemical process, often involving energy from a radiation source. This process can be free-radical or cationic. Liquid-based 3D printing methods are known for producing parts with the highest resolution, accuracy, clear details and smooth surface finish compared to other plastic 3D printing processes. Figure 5.14 shows the schematic of the photopolymerization process during the 3D printing process.

5.4.1 Liquid Materials for 3D Printing

There are various commercially available liquid-based materials for the 3D printing process. Understanding these proprietary materials is important for choosing the right one for particular applications. Resins offered by Formlabs©, esun© and Phrozen Technology are some of the resin manufacturers. The discussion in this section covers some of the commercially available resins for 3D printing. Standard resins offer a smooth, matte surface finish and great resolution, making them suitable for various 3D printing applications. Clear resin is the only material that is transparent for plastic 3D printing and can be polished to achieve optical transparency. Draft resin is one of the fastest materials for 3D printing, with rates up to four times quicker than ordinary resins and ten times faster than FDM printing. Resins with excellent compression, stretching, bending and impact resistance are available, making them practical and durable. Rigid resins offer highly filled, robust and stiff materials that are resistant to bending, chemicals, heat and dimensional changes under load. Polyurethane resins are known for their long-term durability, stability in UV, temperature and humidity, flame retardancy, serializability and resistance to chemicals and abrasion. High-temperature resin stands out for its precision and tolerance to high temperatures. Elastic and flexible resins can sustain compression, bending and flexing without rupturing, giving them a flexibility similar to that of rubber. Silicone 40A Resin is the first commercially available 100% silicone 3D printing material with qualities comparable to cast silicone. Medical and dental resins offer biocompatible options for creating medical and dental appliances. Jewellery resins are designed for investment casting and vulcanized rubber molding, offering complex features and excellent shape retention. Flame retardant (FR) resin delivers flame retardancy, heat resistance, stiffness and creep resistance for indoor and industrial situations with high temperatures or ignition sources, while ESD resin offers ESD-safe qualities to enhance electronics manufacturing workflows. Alumina 4N Resin, made of 99.99% pure alumina technical ceramic, offers remarkable thermal, mechanical and conductive qualities suitable for a range of technical applications. The commercially available liquid materials used for the DLP 3D printing process are classified as hard plastics, high-temperature plastics, elastomers and castable resin. Hard plastic is a versatile material used for functional prototyping as well as producing end-use parts, providing high-performance qualities with a high degree of elongation and exceptional flexural and tensile properties. High-temperature resin is designed for interior applications in automobiles, tooling and moulds with a heat deflection temperature (HDT) of 100°C. Elastomers enable the production of flexible end products and parts with isotropic qualities, maintaining mechanical properties and resolution while having a smooth surface finish. Castable resin is a significant advancement in high-speed casting model printing, producing incredibly crisp details with the greatest wax content available for 3D printing.

Material jetting uses UV-curable photopolymers and involves three main types of materials: base, composite and support materials. Base resins offer diverse characteristics such as castable, strong, flexible, transparent, biocompatible and all-purpose. Each has its own set of advantages and limitations for specific applications. Composite resins combine different base resins, requiring a thorough understanding of their individual properties to create the desired combination. The efficient handling of various material containers by the machine also plays a role in material selection. Support materials, typically made of dissolvable materials, are designed for easy removal from prints and vary in solubility depending on the material used.

5.4.2 Material Properties of Liquid-Based Materials

The available liquid materials for the SLA process offer a wide range of properties, allowing for diverse applications. Resins with high tensile strength and moderate tensile modulus are tough and durable, providing resilience to impact and bending forces. They also offer decent elongation at break, ensuring flexibility and preventing premature failure under stress. Resins with high heat deflection temperature (HDT) are ideal for applications requiring stability under elevated temperatures. Ceramic-reinforced resins offer superior tensile modulus and flexural modulus, making them suitable for demanding structural applications where strength and rigidity are essential. Standard and clear resins provide versatility and clarity, allowing for a wide range of aesthetic and functional uses. Each material's distinctive properties cater to specific requirements in 3D printing applications.

The range of photopolymers available for DLP 3D printing is quite diverse, and understanding their properties is essential for choosing the right materials. Hard plastics, for example, offer excellent strength and stiffness, with a tensile strength of 60 MPa and a tensile modulus of 1890 MPa, as well as an elongation at break of 47% and impact strength of 53.8 J/m. However, their heat deflection temperature of 56°C limits their use in high-temperature environments. On the other hand, high-temperature plastics provide a balance of strength and heat resistance and are capable of withstanding elevated temperatures. Elastomers offer flexibility and resilience, while castable resins provide versatility for applications requiring mouldability and precision. Each material type has unique properties suited for specific 3D printing needs.

5.4.3 Applications and Limitations of Liquid Materials in 3D Printing

The range of materials available for liquid-based 3D printing offers diverse applications. Standard resins are commonly used for creating concept models and prototypes, while clear resin is ideal for parts requiring optical transparency. Draft resin enables quick iterations in the early stages of prototyping, and tough resins provide durability for various components. Rigid resins are suitable for making robust parts, while polyurethane resins cater to demanding applications in automotive, aerospace and machinery components. High-temperature resin is suitable for applications requiring heat resistance, and flexible resins find use in consumer products, robotics and medical equipment. Medical and dental resins ensure precision and biocompatibility, while jewellery resins allow for intricate designs. ESD and flame retardant resins cater to specialized industries, and alumina 4N resin offers dependability in demanding applications.

However, there are limitations to liquid-based 3D printing materials. They often require support structures for overhangs and undercuts, leading to complex printing processes. Post-processing, including support removal and post-curing, can be time-consuming and carries the risk of damaging the printed model. These limitations necessitate careful consideration when choosing materials for 3D printing projects.

Self-Check Exercises

1. How do the properties of 3D printed parts differ between the *XY* plane and the *Z* direction?
2. Explain the role of material selection in achieving optimal performance in 3D printing.
3. Describe the differences between thermoplastic and thermoset polymers used in 3D printing.
4. What are the advantages of using metals in 3D printing, and which metals are commonly used?
5. Discuss the applications and limitations of ceramic materials in 3D printing.
6. How do composite materials enhance the properties of 3D printed parts?
7. What factors should be considered when selecting a material for 3D printing?
8. Describe the importance of temperature settings in filament-based 3D printing.
9. Explain the significance of filament properties in the performance of 3D printed parts.
10. What challenges are associated with using filament materials in 3D printing?
11. What factors should be considered when selecting a binder for binder jetting 3D printing?
12. How does the particle size of powder materials affect the quality of 3D printed components?
13. How do the mechanical properties of polyamide 11 differ from polyamide 12?
14. What role do material properties like flowability and density play in the powder-based 3D printing process?
15. Compare the use of nickel-based alloys and titanium-based alloys in 3D printing. What are their respective advantages?

Discussion Topics

1. Discuss the key factors influencing material selection in 3D printing, such as strength, flexibility and applications.

2. Discuss the comparison between the characteristics and applications of solid, powder and liquid materials used in 3D printing, considering factors like printing techniques and post-processing requirements.

Multiple Choice Questions

1. Which material type is commonly used in FDM?
   (a) Thermoplastics
   (b) Ceramics
   (c) Metals
   (d) Composites
2. What is a characteristic of thermoset polymers?
   (a) They soften when heated.
   (b) They are suitable for high-temperature applications.
   (c) They are commonly used for flexible parts.
   (d) They have low molecular weight.
3. Which material is known for its biodegradability in 3D printing?
   (a) ABS
   (b) PLA
   (c) Nylon
   (d) PC
4. Which material is commonly used as a support material and is water-soluble?
   (a) ABS
   (b) HIPS
   (c) TPU
   (d) PVA
5. What property of metals makes them suitable for 3D printing?
   (a) Low density
   (b) High strength and toughness
   (c) Brittle nature
   (d) Low thermal conductivity
6. What is the typical diameter of filaments used in FDM 3D printing?
   (a) 0.5 mm
   (b) 1.75 mm
   (c) 2.5 mm
   (d) 3.5 mm
7. Which material is known for its resistance to UV light and impact loading?
   (a) PLA
   (b) ASA
   (c) Nylon
   (d) PETG

8. What does the glass transition temperature (T_g) signify for a polymer material?
   (a) The point at which it melts
   (b) The point at which it solidifies
   (c) The point at which it transitions from a hard to a rubbery state
   (d) The point at which it decomposes
9. What is a common use for elastomer materials in 3D printing?
   (a) High-temperature applications
   (b) Flexible engineering applications
   (c) Electrical conductivity
   (d) Structural components
10. What is a critical factor in powder-based 3D printing methods like SLS and SLM?
    (a) Filament diameter
    (b) Powder particle size
    (c) Liquid viscosity
    (d) Sheet thickness
11. Which material is ideal for components needing to withstand corrosion and mechanical wear?
    (a) Ceramics
    (b) Thermoplastics
    (c) Thermosets
    (d) Metals
12. What is the primary advantage of using composites in 3D printing?
    (a) Low melting point
    (b) High strength-to-weight ratio
    (c) High thermal conductivity
    (d) Low cost
13. Which process uses filaments with specific mechanical properties for 3D printing?
    (a) Binder jetting
    (b) Direct energy deposition
    (c) Fused deposition modelling
    (d) Selective laser sintering
14. What is a limitation of FDM parts related to material porosity?
    (a) Reduced flexibility
    (b) Limited performance
    (c) Increased weight
    (d) Higher cost
15. What process involves the sintering of powder particles using a laser?
    (a) Binder jetting
    (b) Selective laser sintering
    (c) Vat polymerization
    (d) Material extrusion

16. What are the four primary 3D printing methods that utilize powder-based materials?
(a) Fused deposition modeling, stereolithography, digital light processing, binder jetting
(b) Selective laser sintering, selective laser melting, direct energy deposition, binder jetting
(c) Fused deposition modeling, selective laser sintering, selective laser melting, digital light processing
(d) Stereolithography, digital light processing, binder jetting, direct energy deposition

17. In the SLS method, what process bonds the powder particles together?
(a) Melting
(b) Sintering
(c) Deposition
(d) Jetting

18. Which materials are commonly used in the SLM process for 3D printing?
(a) Polymers
(b) Metals
(c) Ceramics
(d) Composites

19. Name three types of binders used in the binder jetting process.
(a) Water-based, solvent-based, UV-based
(b) Metal-based, polymer-based, ceramic-based
(c) Powder-based, liquid-based, gas-based
(d) Organic-based, inorganic-based, hybrid-based

20. What is a critical factor in determining the level of detail achievable in 3D printed components using powder-based materials?
(a) Colour of the powder
(b) Size of the powder particles
(c) Temperature of the environment
(d) Speed of the printer

21. What are the main materials used in liquid-based 3D printing?
(a) Metals
(b) Polymers
(c) Ceramics
(d) Composites

22. In which 3D printing method are UV-curable photopolymers commonly used?
(a) Fused deposition modelling
(b) Selective laser sintering
(c) Stereolithography
(d) Direct energy deposition

23. Name a common application for titanium-based alloys in 3D printing.
    (a) Jewellery making
    (b) Aerospace components
    (c) Clothing design
    (d) Food packaging
24. What properties make nickel-based alloys suitable for aerospace applications?
    (a) Low density and flexibility
    (b) High-temperature resistance and corrosion resistance
    (c) Transparency and biocompatibility
    (d) Colour variety and affordability
25. Which material used in SLA is known for its optical transparency?
    (a) Standard resin
    (b) Clear resin
    (c) Draft resin
    (d) Rigid resin

Answers

1. (a) Thermoplastics
2. (b) They are suitable for high-temperature applications
3. (b) PLA
4. (d) PVA
5. (b) High strength and toughness
6. (b) 1.75 mm
7. (b) ASA
8. (c) The point at which it transitions from a hard to a rubbery state
9. (b) Flexible engineering applications
10. (b) Powder particle size
11. (d) Metals
12. (b) High strength-to-weight ratio
13. (c) Fused deposition modelling
14. (b) Limited performance
15. (b) Selective laser sintering
16. (b) Selective laser sintering, selective laser melting, direct energy deposition, binder jetting
17. (b) Sintering
18. (b) Metals

19. (a) Water-based, solvent-based, UV-based
20. (b) Size of the powder particles
21. (b) Polymers
22. (c) Stereolithography
23. (b) Aerospace components
24. (b) High-temperature resistance and corrosion resistance
25. (b) Clear resin

6

Printing Using Solid-Based Raw Material

Based on the raw material used the additive manufacturing processes can be categorized as solid-based, liquid-based and powder-based 3D printing processes. This chapter outlines the different solid-based 3D printing processes and their working principles. It covers fused deposition modelling (FDM) technology, fused filament fabrication (FFF) and direct energy deposition (DED) solid-based 3D printing processes along with their field of application, advancements and future trends. The chapter comprises the following main sections:

6.1 Overview of Solid-Based 3D Printing Processes
6.2 Fused Deposition Modelling 3D Printing Process
6.3 Direct Energy Deposition (With Wire) 3D Printing Process
6.4 Sheet Lamination 3D Printing Process
6.5 Applications of FDM, DED and SL
6.6 Advanced Concepts and Future Trends

After reading through the material, self-check exercises, objective questions and discussion topics presented in this chapter, the learner will be able to do the following.

(a) Outline different solid-based 3D printing processes.
(b) Explain the working principle of FDM technology, FFF and DED (using wire as the raw material).
(c) Identify the key components of an FDM printer.
(d) Use FDM-based 3D printing processes to develop products.
(e) List the applications of the FDM 3D printing.
(f) Outline the advancements and future trends in FDM based 3D printing.

DOI: 10.1201/9781003562344-6

6.1 Overview of Solid-Based 3D Printing Processes

Based on the raw material used the additive manufacturing processes can be categorized as follows:

1. **Solid-based 3D printing processes:** these processes utilizes solids as the primary medium (raw material) to create the part or prototype. The raw material may be in the form of wire or sheets.
    - FDM or FFF (with wire as raw material),
    - DED (with wire as raw material) and
    - sheet lamination process (with sheet as raw material)
2. **Liquid-based 3D printing processes:** these processes utilizes liquid as the primary medium (raw material) to create the part or prototype.
    - Photopolymerisation techniques: steriolithography (SLA), digital light processing (DLP), liquid crystal display (LCD), continuous liquid interface printing (CLIP)
    - Material jetting techniques: continuous ink jetting (CIJ), drop on demand (DoD)
    - Direct ink writing (DIW)
3. **Powder-based 3D printing processes:** these processes utilize powders of different materials as the primary medium (raw material) to create the part or prototype.
    - Selective laser sintering (SLS)
    - Selective laser melting (SLM)
    - Direct metal laser sintering (DMLS)
    - Electron beam melting (EBM)
    - Binder jetting, multi-jet fusion
    - DED (with powder as raw material)

6.1.1 Overview of Fused Deposition Modelling

As we have already discussed in Chapter 1, Section 1.1 that FDM, also known as the material extrusion additive manufacturing technique or fused filament fabrication, utilizes polymers as the raw material (filament in the form of wire) for creating 3D objects. The filament is usually heated to a molten state and then extruded through the nozzle of the machine (3D printer). The nozzle head can move in three Cartesian (translatory) degrees of freedom (DoF) to deposit the extruded polymer on the build plate layer by layer as per the G-code of digital model. Low operation time and cost, easily understandable printing technique, small equipment size, simple and easy changing of print material, and comparably low-temperature process are the main pros of this process. FDM finds applications across various industries, ranging from automotive and aerospace to healthcare and consumer goods. In automotive manufacturing, FDM is employed for rapid prototyping of parts and components, facilitating iterative design improvements and reducing time-to-market.

6.1.2 Overview of Laminated Object or Sheet Lamination 3D Printing Process

Sheet lamination is a 3D printing technology that involves stacking and laminating ultra-thin sheets of material. Initially, a thin layer of material is rolled out or put on the building platform. The procedure for connecting the layers varies depending on the type of sheet lamination utilized. SDL and ultrasonic additive manufacturing (UAM) (refer to Figure 6.1) entail gluing the layers together first, followed by cutting the 3D form. In contrast, computer-aided manufacturing of laminated engineering materials (CAM-LEM) cuts each layer to form before gluing them together. This procedure is repeated until all layers are completed, resulting in the entire height of the item. When the layers are finished, the print block is removed, and any undesirable exterior edges are cut away, revealing the final 3D item. A multitude of technologies, including heat and sound, can be used to fuse the material layers together.

Sheet lamination generates items with lesser additive precision than other additive manufacturing (AM) processes, but its benefits include

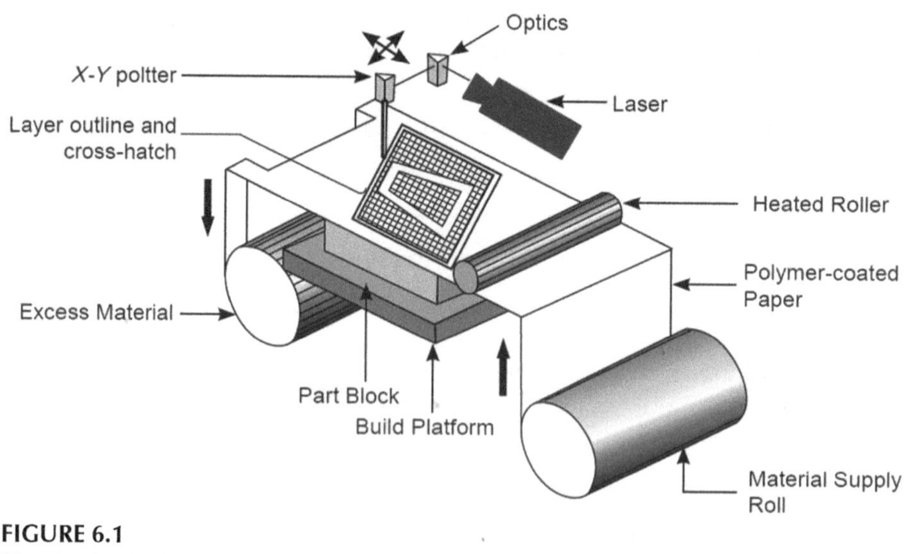

FIGURE 6.1
Sheet lamination 3D printing process.

cost-effectiveness and shorter manufacturing time. This enables product designers to produce low-fidelity prototypes with easily available, low-cost materials.

6.1.3 Overview of Directed Energy Deposition 3D Printing Process (Wire as Raw Material)

DED is an advanced 3D printing technique primarily used for repairing or adding material to existing components, although it can also be employed to construct items from scratch. It is often applied in industrial settings, such as fixing damaged turbine blades or propellers. DED employs a focused energy source, such as a laser, electron beam or plasma, to melt the material. The key difference lies in the delivery of the material itself, which is either in wire or powder form. As the material exits the nozzle, it melts immediately, enabling the creation of complex structures. The process continues, with the layers solidifying until the object is either formed a new or repaired. Overall, DED serves as a versatile and efficient method for repairing and building parts in various industrial applications.

While DED itself has been in use for several years, the specific use of wire feedstock (refer to Figure 6.2) with a laser energy source is a relatively recent development and has gained increasing attention in various industries due to following reasons:

- Material versatility: the use of wire feedstock offers a wider range of material options compared to other DED feedstock options, such as powder. This grants the deposition of a broader variety of alloys, making it well-suited for a diverse range of applications across different industries.

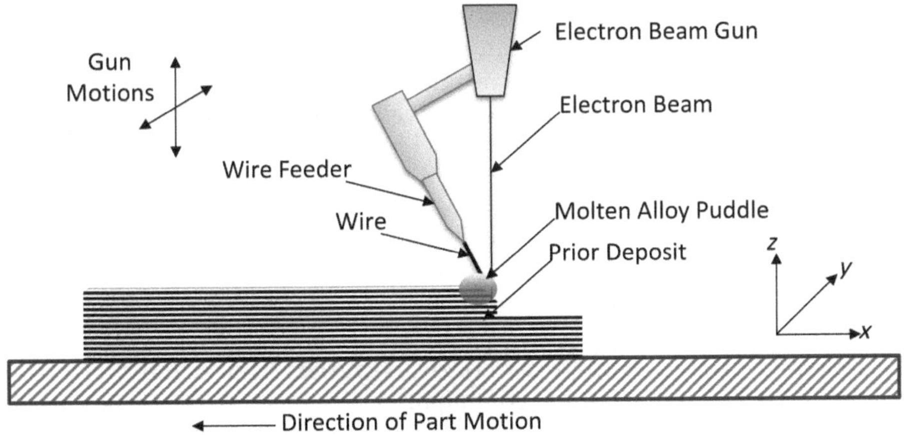

FIGURE 6.2
Wire fed direct energy deposition 3D printing process.

- Continuous deposition: the continuous feed of wire feedstock during the DED process allows for a seamless and uninterrupted additive build, resulting in increased efficiency and reduced production time. This continuous deposition also eliminates the need for frequent interruptions and manual handling, further streamlining the process.
- Large-scale applications: this combination is particularly well-suited for producing large-scale components due to the continuous feed and reduced need for support structures. It allows for the creation of sizable parts with reduced production time and cost-effectiveness.
- Material change: switching between different materials is often more straightforward with wire-based additive manufacturing. The ability to swap wire types with relative ease can support the production of diverse components without significant downtime and also allow the printing of dual material parts.

6.2 Fused Deposition Modelling 3D Printing Process

6.2.1 Working Principle of FDM

The process begins with a computer-aided design (CAD) file, which is sliced into thin horizontal layers. FDM printers employ two types of materials: modelling material for the end product and support material, which serves as a temporary scaffold throughout the printing process. A spool(s) of thermoplastic filament(s), typically made of materials like ABS or PLA, is then fed into a heated nozzle, where it melts to a semi-liquid state. When the printer achieves the desired nozzle temperature, it extrudes the filament through an extrusion head and nozzle. The nozzle heated the filament to glass transition temperature before extruding it onto the build platform or table. The controller directs and coordinates the nozzle and base during printing layer by layer according to the deposition path that also known as the tool path as per G-code. As each layer is deposited, it quickly cools and solidifies, fusing with the previous layers to form a solid part. When a layer is completed, the build platform moves down or the nozzle is elevated to maintain the gap between the deposited layer and the nozzle, and the process begins again with the next layer. This continuous process of material extrusion and layer-by-layer deposition continues until the entire object is fabricated (refer to Figure 6.3).

The success of FDM is dependent on robust layer adhesion. As molten thermoplastic passes through the nozzle, it pushes against the previously printed layer due to gravity. The combination of high temperature and gravitational force causes the layer to cross-link of polymer, allowing it to merge with the layer underneath it. Alternatively, the extrusion head might travel

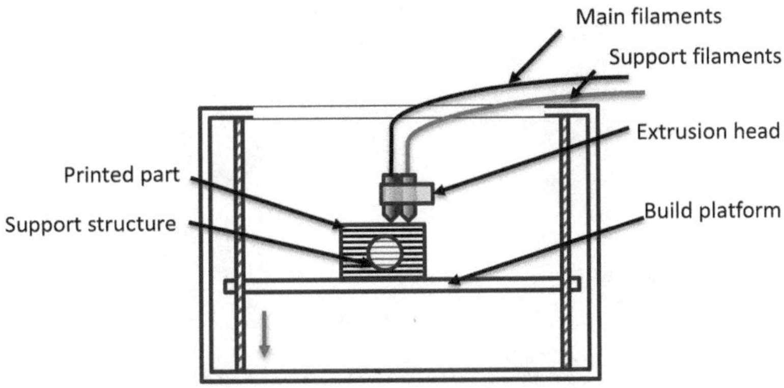

FIGURE 6.3
Fused deposition modelling 3D printing process.

upward rather than downward. This procedure continues until the last component is finished.

After removing the object from the FDM printer, the support materials are removed either by soaking it in a water and detergent solution or by physically breaking off the thermoplastic supports. Sanding, milling, painting and plating can all be used to improve the usefulness and look of an object.

FDM offers several advantages, including cost-effectiveness, simplicity and the ability to create complex geometries with ease. Despite its simplicity, achieving high-quality prints requires careful consideration of parameters such as layer thickness, nozzle temperature and printing speed to ensure optimal results.

6.2.2 Components of FDM Printer

We have already discussed the different components of a 3D printer in Chapter 2, Section 2.1. Here are the main components of an FDM printer (refer to Figure 6.4):

- **Frame and structure** – the frame provides the structural support for the entire printer. Itis usually made of metal or plastic to ensure stability during the printing process.
- **Print bed** – the print bed is the surface where the object is built layer by layer. It can be heated or unheated depending on the printer model and the type of filament being used. Heated beds help in preventing warping and improve adhesion of the first layer.
- **Extruder assembly** – the extruder assembly consists of a motor-driven gear system and a heated nozzle. It feeds the filament into the hot end where it is melted and extruded onto the print bed. Some printers have

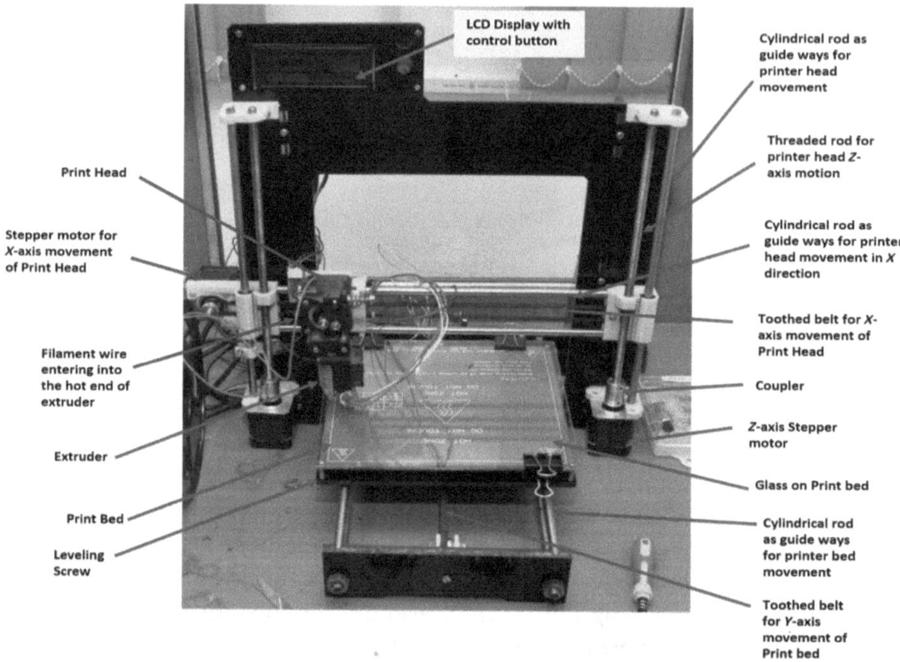

FIGURE 6.4
Assembled open-source FDM type 3D printer.

a single extruder, while others may have dual or more, allowing for multi-material or multi-colour printing.

- **Hot end** – the hot end is where the filament is melted and deposited onto the print bed. It typically consists of a heating element (like a resistor or a cartridge heater) and a temperature sensor (usually a thermistor or thermocouple).The nozzle size can vary depending on the desired print quality and speed.
- **Filament** – FDM printers use thermoplastic filament as the raw material for printing. Common types of filament include PLA (polylactic acid), ABS (acrylonitrile butadiene styrene), PETG (polyethylene terephthalate glycol), and more. Filament comes in spools and is available in various colours and diameters.
- **Control board and electronics** – the control board is the brain of the 3D printer, responsible for interpreting the G-code instructions and controlling the movement of the printer components. It typically includes a microcontroller, stepper motor drivers and other necessary electronics.
- **Stepper motors** – stepper motors are used to precisely control the movement of the printer's components, including the extruder, print bed and possibly other axes if it is a multi-axis printer. They provide

accurate positioning and are capable of both linear and rotational motion.

- **Cooling fans** – cooling fans are used to cool down the printed layers quickly to prevent warping and improve print quality. They are often positioned near the hot end and the printed object.
- **End stops and sensors** – end stops are sensors that detect the physical limits of the printer's axes. They are used to home the printer (establish a reference point) and prevent the motors from moving beyond their mechanical limits, which could damage the printer.
- **Display and user interface** – many FDM printers come with a display screen and user interface for controlling the printer settings, initiating prints and monitoring the printing progress. These components work together to enable the FDM printer to create three-dimensional objects layer by layer from digital designs.

6.2.3 Selection Criteria for FDM Materials

(This is also covered Chapter 5, Section 5.2.)

When it comes to selecting materials for FDM, there are several important criteria to consider (refer to Table 6.1). FDM is a type of 3D printing that melts thermoplastic filament and deposits it layer by layer to create objects (refer to Figure 6.5).

TABLE 6.1

List of Materials and Properties

Material Name	Properties
ABS	+ Good strength + Good temperature resistance - More susceptible to warping
PLA	+ Excellent visual quality + Easy to print with - Low impact strength
NYLON	+ High strength + Excellent wear and chemical resistance - Low humidity resistance
PETG	Food Safe* + Good strength + Easy to print with
TPU	+ Very flexible - Difficult to print accurately
PEI	+ Excellent strength to weight + Excellent fire and chemical resistance - High cost

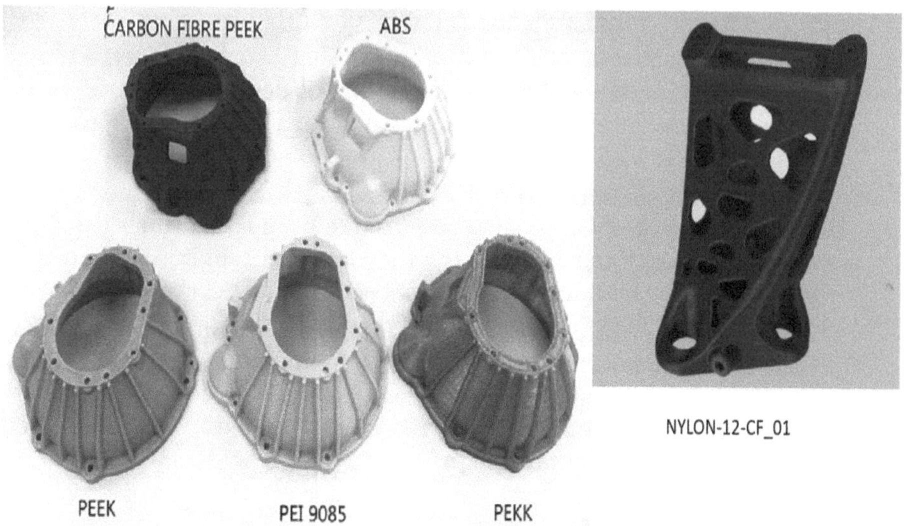

FIGURE 6.5
FDM 3D printed components with different FDM materials.

- **Melting point** – since FDM involves heating the filament to melt it, the material must have a melting point within a suitable range. If the melting point is too low, the object may deform during printing, while if itis too high, the filament may degrade or not extrude properly.
- **Material's strength and durability** – the chosen material should be able to withstand the stresses and environmental conditions. For example, if you are making a prototype for a mechanical part, so we need material with high tensile strength and resistance to impact.
- **Material's flexibility or stiffness** – some applications may require a flexible material that can bend without breaking, while others may need a rigid material for structural integrity.
- **Compatibility** – compatibility with post-processing techniques is also essential. Certain materials may be easier to sand, paint or glue than others.
- **Cost and availability** – some materials may be more affordable or easier to source than others, which can significantly impact your project's budget and timeline. Ultimately, the selection of FDM materials depends on a balance of these factors, and specific requirements of the project or application. By carefully considering the melting point, strength, flexibility, post-processing compatibility, cost and availability, we can choose the most suitable material to bring your 3D printing projects to life.

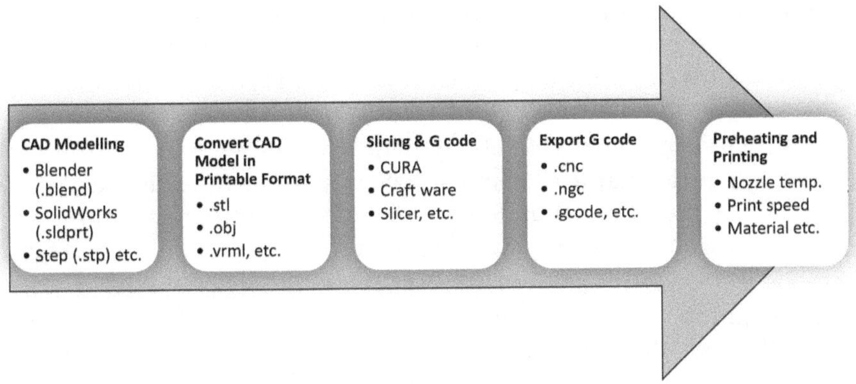

FIGURE 6.6
Flow diagram of FDM process.

6.2.4 Steps in FDM 3D Printing

Preparing 3D models for printing involves several steps to ensure a successful 3D Printing. Figure 6.6 depicts the same using a flowchart and name of the software available.

- **Select or design a 3D model** – the first step is to either select a pre-existing 3D model from online website OR create new design using 3D modeling software like TinkerCAD, Blender or AutoCAD OR perform the 3D scanning of the existing product OR CT scan data of any human body part. Ensure the model is suitable for FDM printing, meaning it has no overhangs beyond the printer's capabilities and is manifold to avoid printing errors. (This is covered in detail in Chapter 3, Section 3.1.)
- **Export or save as STL file** – once the 3D model is ready, it needs to be saved or exported in the STL (stereolithography) file format. STL files represent the surface geometry of the 3D model using triangular facets. (This is covered in detail in Chapter 4, Section 4.2.)
- **Import into slicing software** – import the STL file into slicing software such as Cura, Simplify3D or Prusa Slicer. Slicing software divides the 3D model into thin horizontal layers (slices) and generates the tool paths necessary for the printer to recreate each layer. Set parameters like layer height, infill density and print speed according to your desired print quality and material. (This is covered in detail in Chapter 4, Section 4.3.)
- **Slice and generate G-code** – once the settings are adjusted, slice the model within the software. This process generates G-code, a set of instructions that tells the 3D printer how to move and extrude filament to create the object layer by layer. (This is covered in detail in Chapter 4, Section 4.4.)

- **Transfer G-code to the printer** – save the sliced G-code file onto an SD card or connect the computer directly to the printer if supported. Transfer the file to the printer's interface. Some printers allow for direct communication with slicing software, enabling a seamless transfer process. (This is covered in detail in Chapter 4, Section 4.4.)
- **Prepare printer and filament** – before starting the print, ensure the printer is properly calibrated and levelled. Load the appropriate filament into the printer's extruder, taking care to follow manufacturer guidelines regarding temperature settings and filament type.
- **3D printing** – with the printer and filament prepared, initiate the print job through the printer's interface. Monitor the print progress periodically to catch any issues early on, such as warping, layer adhesion problems or filament jams.
- **Post-processing** – once the print is complete, remove it from the print bed carefully, depending on the model and material used, post-processing steps like sanding, painting, or assembly may be necessary to achieve the desired finish. These are the steps to be followed to effectively prepare 3D models for printing using FDM printers, ensuring successful and high-quality prints. (This will be covered in detail in Chapter 9.)

6.2.5 Slicing Software and Parameters

Slicing software is a crucial component in the 3D printing process. It takes a 3D model file (typically in .STL format) and generates a set of instructions (G-code) that tells the 3D printer how to build the object layer by layer. Figure 6.7 presents the parameters related to geometry, process and structure.

- Importing the 3D model: the first step is to import the 3D model into the slicing software. The slicing software then analyzes the geometry of the model to determine how it should be printed.
- Setting parameters: before slicing the model, various parameters need to be configured. These parameters include layer height, infill density, print speed, temperature, support structures and more. Each parameter affects the quality, strength and appearance of the final print.
- Layer height: layer height refers to the thickness of each individual layer of filament that the printer will deposit. A smaller layer height results in finer details but increases print time, while a larger layer height speeds up printing but may sacrifice surface quality.
- Infill density: infill density determines the internal structure of the printed object. Higher infill percentages result in stronger prints but require more material and time. Lower infill percentages reduce material usage but may compromise structural integrity.

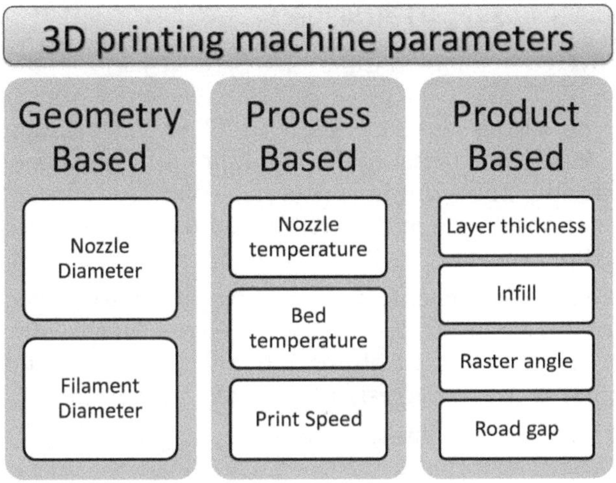

FIGURE 6.7
Parameters influencing the fused deposition modelling process.

- Print speed and temperature: print speed and temperature settings dictate how quickly the printer moves and how hot the extruder and heated bed are during printing. Optimal settings depend on the material being used and the desired balance between speed and quality.
- Support structures: models require support structures to prevent overhangs and bridges from collapsing during printing. Support structures are generated by the slicing software based on the model's geometry and can be customized in terms of density and placement.
- Slicing the model: once all parameters are set, the slicing software generates the G-code instructions based on the configured settings. This G-code contains instructions for the printer's movements, extrusion rates, temperatures and other parameters required to reproduce the 3D model layer by layer.
- Previewing and exporting: after slicing, the software typically provides a preview of the sliced model, showing each layer and any support structures. Once satisfied, the G-code is exported to an SD card or sent directly to the 3D printer for printing.
- Printing and fine-tuning: finally, the sliced model is printed on the 3D printer. Throughout the printing process, users may monitor the print and make adjustments to parameters if necessary to optimize print quality. By understanding and adjusting these slicing parameters, users can achieve the desired balance between print quality, strength and speed for their FDM 3D prints.

6.2.6 Importance of Support Structures

(This is also covered in detail in Chapter 4, Section 4.3.)

- Support structures are important in FDM 3D printing for several reasons. Firstly, they ensure the stability and accuracy of the printed object during the printing process. Without support structures, overhangs and bridges in the design would collapse or deform, leading to a failed print.
- Support structures enable the printing of complex geometries that would otherwise be impossible to achieve. By providing temporary support for overhanging features, intricate designs with intricate details or protrusions can be successfully printed layer by layer.
- Support structures facilitate the printing of parts with large overhangs or steep angles, improving the overall quality of the final print. Support structure help in maintaining the integrity of the model by preventing sagging or warping, resulting in smoother surfaces and sharper edges.
- Support structures play a vital role in minimizing material wastage and optimizing print time. By strategically placing supports only where needed, excessive use of filament can be avoided, reducing costs and enhancing efficiency.
- Furthermore, support structures aid in post-processing tasks such as removing support material and finishing the printed object. While some support structures are designed to be easily removable, they still provide necessary stability during printing and can be cleaned up afterward.
- Support structures are indispensable in FDM 3D printing, ensuring the successful fabrication of complex designs, maintaining print quality, reducing material wastage, and facilitating post-processing tasks.

6.3 Direct Energy Deposition (With Wire) 3D Printing Process

DED is an advanced 3D printing technology commonly used in industrial applications, particularly for repairing or adding material to existing components. While DED can be employed to fabricate parts from scratch, it is more frequently used for tasks such as repairing damaged turbine blades or propellers.

In principle, DED is similar to certain powder bed fusion (PBF) methods, like laser powder bed fusion (LPBF) or electron beam melting (EBM), as it also uses a concentrated energy source (laser or electron beam) to melt material. However, in DED, the key difference is that the material is melted as it is deposited through a nozzle, rather than being selectively melted from a pre-existing powder bed. This positions DED at the intersection of material extrusion and powder bed fusion techniques. The energy source and the material feed nozzle are controlled using either a gantry system or a robotic arm, which directs the deposition process. Additionally, in hybrid manufacturing setups, the substrate can be moved during deposition to enable the creation of more complex geometries. This ability to manipulate both the material feed and the substrate makes DED a versatile process in manufacturing and repair operations.

6.3.1 Types

DED technology, though capable of creating parts from metals, ceramics and polymers, is primarily used for manufacturing metal components. Based on the energy source used to melt the material, DED can be classified into three major types:

Laser-based DED: utilizes a high-power laser as the heat source, commonly applied to process metals and alloys like titanium.

Electron beam-based DED: uses an electron beam to melt materials such as titanium alloys and high-temperature alloys.

Plasma/electric arc DED: employs plasma or an electric arc to melt steels, titanium, and nickel alloys.

DED can also be categorized based on the material feedstock used to fabricate parts:

Powder-based DED: uses either a laser or electron beam as the energy source and is suited for processing metals and alloys like cobalt-chrome.

Wire-based DED: utilizes a laser, electron beam or arc for melting materials such as titanium, aluminum and steel.

This breakdown shows how DED technology can be tailored for different applications depending on the choice of energy source and material feedstock, making it highly versatile for various industrial uses.

6.3.2 Laser-Based DED

Laser-based DED utilizes a high-power laser as the heat source to melt materials during the deposition process. This method is particularly valued for its precision and ability to work with a range of metals and alloys, such as titanium, steel and inconel. One of its key strengths lies in the ability to produce complex geometries with high accuracy and excellent surface finishes, making it a popular choice for applications that demand fine detail and exacting tolerances.

However, despite its many advantages, laser-based DED does have certain limitations. High costs are often associated with this technology due to the complexity of the equipment and the energy-intensive nature of the process. Additionally, it is best suited for materials that can efficiently absorb laser energy, which narrows the range of applicable materials.

An example of a widely used laser-based DED system is Optomec's laser engineering net shape (LENS). This system exemplifies the merits of laser-based DED, offering high precision, the capability to fabricate complex parts, and superior surface finishes. Yet, like other laser-based systems, it comes with the challenges of higher operational costs and material-specific limitations.

6.3.3 Electron Beam-Based DED (EBAM)

Electron beam-based directed energy deposition (EBAM) uses a concentrated electron beam to liquefy materials while depositing them, making it well-suited for applications involving high temperatures and reactive materials. Commonly utilized materials are titanium alloys and high-temperature alloys. This technique offers high deposition rates, is suitable for manufacturing large-scale parts and results in low residual stress. However, it necessitates a vacuum environment, involves high equipment expenses, and is constrained in terms of material variety.

Electron beam-based DED systems, like Sciaky's electron beam additive manufacturing (EBAM), use a focused electron beam to melt powdered material during deposition. This process provides benefits such as fast deposition rates, the ability to produce large-scale components, and reduced residual stress. However, it also has drawbacks, including the requirement for a vacuum environment, high equipment costs and a limited range of compatible materials.

6.3.4 Plasma or Electric Arc-Based DED

Plasma or electric arc-based DED utilizes a plasma arc or electric arc as the heat source to melt materials during the deposition process. This technology is particularly well-suited for repair applications and the production of large parts, where the method's high deposition rates and cost-effectiveness make it a preferred choice in industries such as aerospace, automotive and shipbuilding.

Commonly processed materials include steel, titanium and nickel alloys, all of which are commonly used in industrial applications that require durable, high-strength components. Plasma or electric arc-based DED provides high deposition rates, making it ideal for large-scale manufacturing. Its relative cost-efficiency, particularly when producing bulky components, and its ability to process a range of metals give this method robust versatility.

However, this method comes with certain limitations. Due to the nature of the plasma or electric arc, the process tends to have lower precision compared to laser or electron beam-based systems. As a result, parts often require post-processing to achieve a smooth finish, as the initial deposition can leave behind rougher surface textures. Additionally, the process can lead to high residual stresses within the part, which may require further treatment to relieve.

This balance of high deposition efficiency and cost-effectiveness versus lower precision and surface quality makes plasma or electric arc-based DED an attractive option for certain large-scale manufacturing and repair applications, where speed and cost are prioritized over fine detail.

6.3.5 Powder-Based DED

Powder-based DED involves feeding powder material into the deposition area, where it is melted by a heat source, typically a laser or electron beam. Examples of this technology include LENS and laser metal deposition (LMD). These processes, which will be explored in more detail in Chapter 8, are highly valued for their precision and ability to repair or add features to existing parts with intricate details.

Materials commonly used in powder-based DED include metals and alloys such as titanium, cobalt-chrome and stainless steel, making it versatile for a wide range of applications. The method offers several benefits, including the following:

- **High precision** – allowing for detailed repairs and part modifications.
- **Flexibility in material choice** – supporting various metals and alloys.
- **Effectiveness in complex repairs** – making it ideal for high-value or critical components.

However, there are challenges associated with powder-based DED, such as the following:

- **High cost** of powder feedstock, which can limit its economic feasibility for certain applications.
- **Powder handling issues**, including storage, contamination and consistency, which may affect process efficiency and part quality.

6.3.6 Wire-Based DED

Wire-based DED uses a metal wire as the feedstock, which is melted by a heat source, such as a laser or an electric arc, and deposited layer by layer (refer to Figure 6.8). This method is known for its efficiency and cost-effectiveness, especially when constructing large structures where precision is less critical.

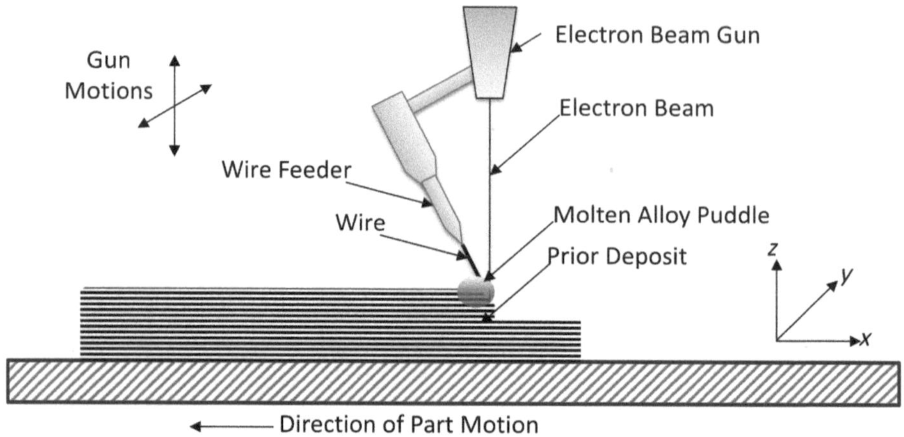

FIGURE 6.8
Schematic of wire feed DED.

Materials commonly employed in wire-based DED include titanium, aluminum and stainless steel, making it ideal for producing large parts in industries like aerospace and automotive. The key advantages of wire-based DED include the following:

- **High material utilization** – reducing waste compared to powder-based methods.
- **Lower costs** – as wire feedstock is generally more economical than powder.
- Particularly suited for **large-scale part fabrication.**

Despite its benefits, wire-based DED has some limitations:

- **Lower precision** compared to powder-based alternatives, making it less suitable for intricate parts.
- **Limited to wire feedstock**, which restricts material choices compared to powder-based methods.

These characteristics make wire-based DED a practical option for large-scale, cost-sensitive manufacturing, but less ideal for applications requiring fine detail or complex geometries.
Steps followed in 3D printing of components using DED (with wire) process:

1. **Model design**:
   ✓ Create a 3D model of the component using CAD software.
   ✓ Slice the model into layers with slicing software.

- ✓ Generate a .gcode file using 3D printer process planning software.

2. **Machine preparation**:
   - ✓ Load the wire feedstock into the DED machine.
   - ✓ Ensure the machine is equipped with an energy source (laser, electron beam, or plasma arc).
3. **Material deposition**:
   - ✓ Feed the wire through a nozzle.
   - ✓ Melt the wire using the focused energy source.
   - ✓ Deposit the molten material onto the substrate or baseplate according to the .gcode coordinates.
4. **Layer-by-layer construction**:
   - ✓ Lower the baseplate after each layer is deposited.
   - ✓ Repeat the process until the entire component is constructed.
5. **Cooling and solidification**:
   - ✓ Allow each deposited layer to cool and solidify before adding the next layer.
   - ✓ Ensure a strong bond between layers during this process.
6. **Post-processing**:
   - ✓ Conduct post-processing steps as needed, such as heat treatment, machining, or surface finishing.
   - ✓ Achieve the desired properties and dimensions for the final component.

This structured approach facilitates the precise construction of components through the DED wire process, covering all stages from initial design to final adjustments.

6.3.7 Materials for the DED (Wire) Process

DED primarily uses metals in wire or powder form, but it can also process polymers and ceramics. For example, Arevo manufactures composite frames using polymer and carbon fiber filaments. DED can work with a wide variety of metals, as any weldable metal is suitable for creating parts with this technology. Commonly used materials include the following:

- **Stainless steels**: known for its corrosion resistivity, strength and durability, it is widely used in aerospace components, automotive parts, medical implants, tooling and machinery.

- **Titanium alloys**: known for high strength-to-weight ratio, corrosion resistance, and biocompatibility, these alloys are ideal for aerospace and defence applications as well as medical implants.
- **Aluminium alloys**: lightweight with high thermal conductivity and corrosion resistance, alloys are perfect for aerospace structures, automotive components, consumer goods and heat exchangers.
- **Nickel alloys** (e.g., inconel): these are excellent choices for aerospace components, gas turbine engine parts, chemical processing equipment and high-temperature applications due to their strength and resistance to corrosion and heat.
- **Tungsten and molybdenum**: these materials, characterized by high melting points and heat resistance, are used in aerospace components, electrical contacts, high-temperature furnace parts and radiation shielding.
- **Cobalt-chromium alloys**: ideal for dental and orthopaedic implants, aerospace components and turbine engine parts, these alloys offer high strength, wear resistance and biocompatibility.
- **Tool steel**: known for their hardness, toughness and heat resistance, these steels are perfect for tool and die components, moulds, cutting tools and wear-resistant parts.
- **Ceramic materials** (e.g., zirconia, alumina, silicon nitride): suitable for aerospace components, medical implants, cutting tools and wear-resistant applications, these materials are prized for their high hardness, corrosion resistance and thermal stability.
- **Intermetallics** (e.g., titanium aluminides, nickel aluminides): required in aerospace components, turbine engine parts and automotive exhaust systems, these materials offer exceptional high-temperature strength and corrosion resistance.

6.3.8 Advantages and Disadvantages of Directed Energy Deposition

DED technologies are primarily utilized in metal additive manufacturing, owing to their unique process characteristics. They are particularly well-suited for repairing and augmenting existing components.

6.3.8.1 Advantages

- **High build rates**: DED offers higher deposition rates at relatively low resolution, resulting in faster build times compared to other metal additive manufacturing methods.

- **Dense and strong parts**: the process produces parts with high density and mechanical properties comparable to those of cast or wrought materials.
- **Near net shape production**: DED enables the fabrication of parts that are near net shape, requiring minimal post-processing.
- **Repair capabilities**: this technology is ideal for applications that involve adding material to existing parts, making it particularly useful for repair applications.
- **Multi-material capability**: advanced DED systems can accommodate multiple powders or wire types, allowing for the creation of components with customized alloy compositions.
- **Large parts production**: DED can fabricate significantly larger parts; for example, Sciaky's EBAM printers can produce components several meters in height.
- **Easy material change**: the process allows for on-demand material feeding from separate containers, facilitating easy refilling or switching of materials during operation.
- **Reduced material waste**: DED minimizes material waste by depositing only the necessary amount during fabrication, unlike processes such as powder bed fusion (selective laser sintering and direct metal laser sintering), which require a complete build platform filled with metal powder.

6.3.8.2 Disadvantages

- **High capital costs**: the initial investment for DED systems is generally higher than that for other metal additive manufacturing technologies.
- **Lower build resolution**: parts produced using DED tend to have lower resolution and poorer surface finishes, often resembling sand or investment castings. This necessitates additional secondary processing, such as machining or aqua blasting, which can increase both time and costs.
- **No support structures**: DED does not utilize support structures during the build process, limiting the ability to create features like overhangs, which can be challenging to implement.

6.3.9 Factors for Choosing a DED Process

- Material compatibility
- Build size and volume
- Precision
- Print resolution

- Layer thickness
- Surface finish
- Deposition rate
- Multi-material requirement
- Post-processing requirements

6.4 Sheet Lamination 3D Printing Process

Sheet lamination is one of the seven types of additive manufacturing processes defined by ISO/ASTM. It involves building a 3D object by stacking and laminating thin sheets of material. The lamination method can involve bonding, ultrasonic welding or brazing, while the final shape is achieved by laser cutting or CNC machining. Out of all the seven types of additive manufacturing technologies, sheet lamination produces parts with the lowest additive resolution. However, its low cost and faster manufacturing time allow product designers to create low-fidelity prototypes from standard, easily available, low-cost materials.

6.4.1 Types

Sheet lamination can be classified based on the type of build material employed, including options like paper, plastic, metal or woven fiber composites. It can also be categorized according to the methods used for shaping, such as CNC milling, laser cutting or aqua blasting. Additionally, sheet lamination can be distinguished by the techniques utilized to bond the sheets, including adhesive bonding, thermal bonding and ultrasonic welding. There are variations in the sequence of forming and bonding: in some cases, sheets are shaped first and then bonded to create a 3D structure, as seen in the CAM-LEM process. In contrast, processes like UAM bond the sheets together before cutting them into the desired shapes.

Given these variations, sheet lamination can be further divided into several categories:

- Laminated object manufacturing (LOM)
- Selective lamination composite object manufacturing (SLCOM)
- Plastic sheet lamination (PSL)
- Computer-aided manufacturing of laminated engineering materials (CAM-LEM)
- Selective deposition lamination (SDL)
- Composite-based additive manufacturing (CBAM)
- Ultrasonic additive manufacturing (UAM)

Each type of sheet lamination operates with slight differences, though they all adhere to the same fundamental principles. Figure 6.1 provides a schematic overview of the sheet lamination 3D printing process, which includes several essential steps:

1. **3D model creation**: the process begins by creating a digital 3D model using specialized software or by scanning an existing object.
2. **Slicing**: the digital model is then sliced into thin layers through specialized software, which will form the basis for the final object.
3. **Material selection**: next, the appropriate material for printing is chosen. Sheet lamination can accommodate a variety of materials, including paper, plastic and metal.
4. **Printing**: after material selection, a printer or cutting device applies the layers onto sheets of the chosen material, using a laser or another cutting tool to shape each layer accordingly.
5. **Bonding**: once the layers are cut, they are bonded together using an adhesive, ensuring that the layers remain securely attached. Depending on the type of sheet lamination, the next layer may or may not be bonded to the previous one. For instance, SDL and UAM bond the layers before cutting the final 3D shape, while CAM-LEM cuts the layers first and then bonds them together. This process continues until all layers are assembled to achieve the desired height. Afterward, the print block is removed, and any excess material is trimmed away to reveal the finished 3D object.
6. **Stacking**: the layers are stacked on top of one another, with adhesive used to secure each layer to the one below.
7. **Cutting**: once the stacking is complete, the final object is cut from the material stack using a laser or other cutting device, a crucial step that determines the object's appearance and size.
8. **Post-processing**: the final step involves cleaning the object and removing any excess material to achieve a smooth and precise finish.

In sheet lamination, the thickness of each layer corresponds to that of the thin sheets used, significantly influencing the final quality. The layer thickness is also dependent on the specific machine and process employed.

6.4.2 Material for Sheet Lamination Process

Sheet lamination is versatile and can utilize various materials, including paper, polymers, ceramics and metals, each employing distinct binding methods (refer to Table 6.2). The most commonly used material is paper,

TABLE 6.2

Sheet Lamination Material Options

Process	Material Options
LOM	Papers
PSL	Most polymers
SLCOM	Carbon fibre, fibreglass, aramid fibre (i.e., Kevlar), PBO (i.e., Zylon), along with metal fibres like steel, aluminium or titanium
CAM-LEM	Alumina other ceramics available in our processes include zirconia, zirconia-toughened alumina (ZTA), silica, and a variety of oxide and non-oxide ceramics.
SDL	Standard paper
CBAM	Long-fibre fabrics bonded with thermoplastic matrix materials. Carbon fibre combined with either Nylon12, PEEK and glass fibre with Nylon 12 and PEEK
UAM	A wide range of material combinations can be printed as any material that can be welded in a single part opening the design space to enable property gradients and engineered material properties for increased performance.

which typically comes with a pre-applied adhesive. In this case, heat and pressure activate the adhesive layer to bond the sheets together.

For polymers, the process relies on heat and pressure to melt the sheets without needing an adhesive. Metal sheets, on the other hand, are bonded using ultrasonic welding. Meanwhile, materials like fibre-based composites and ceramics are combined through thermal energy, often achieved by oven baking the layers together.

6.4.3 Advantages and Disadvantages of Sheet Lamination

6.4.3.1 Advantages

LOM offers several advantages that make it an appealing choice for various applications. One of the primary benefits is its faster print time, although post-processing is often necessary to achieve the desired finish. The process allows for the integration of hybrid manufacturing systems and facilitates easy handling of materials. LOM can produce components from ceramics and composite fibres, expanding its utility. Additionally, original equipment manufacturer (OEM) components, such as sensors and wires, can be embedded during the layering stage, depending on the specific sheet lamination technique and materials used. LOM is relatively cost-effective since it employs standard materials, and it does not require support structures, which can simplify the production process. Furthermore, LOM boasts a larger working area compared to many other additive manufacturing technologies, enabling the creation of larger parts. It also supports full-colour printing, allowing for a wide spectrum of colours. In some techniques, the material state remains unchanged during or after the process, and multi-material layers can be incorporated. Additionally, in certain cases, cut materials can be easily recycled, although this may pose challenges if the part is smaller than the sheet or build area.

6.4.3.2 Disadvantages

LOM also has its drawbacks. One significant limitation is that the layer height cannot be adjusted without changing the thickness of the sheets, which directly affects the part's resolution along the vertical build axis. The finish quality can vary based on the materials used, such as paper or plastic, often requiring additional post-processing to achieve a satisfactory surface. Moreover, the range of materials available for LOM is limited. The process can be time-consuming and challenging when it comes to removing excess material after lamination, leading to higher waste generation compared to other additive manufacturing methods. Creating hollow parts or internal cavities can be particularly difficult in certain LOM processes, such as those that bond first and then form. Additionally, the bonding strength is contingent on the specific laminating technique employed; in some instances, adhesive bonds may not provide sufficient durability for long-term product integrity. Finally, material waste can be substantial if the part dimensions are smaller than the sheet size or build area, further complicating the process.

6.5 Applications of FDM, DED and SL

6.5.1 Industrial Applications of FDM

- Prototyping: FDM is widely used for rapid prototyping across various industries, including automotive, aerospace, and consumer goods. Car manufacturers use FDM to create prototype parts such as engine components or interior features to assess their performance and design.
- Tooling and jigs: FDM is employed to produce custom tooling, fixtures and jigs used in manufacturing processes. These aids help streamline production, improve accuracy and reduce lead times.
- End-use parts: with advancements in materials and printer technology, FDM is increasingly used to manufacture end-use parts, particularly in low-volume production scenarios. Industries such as healthcare, electronics and industrial machinery benefit from the production of functional parts using FDM.
- Functional prototypes: FDM produced prototypes can be functional, allowing for testing of mechanical properties, fit and performance (refer to Figure 6.9).

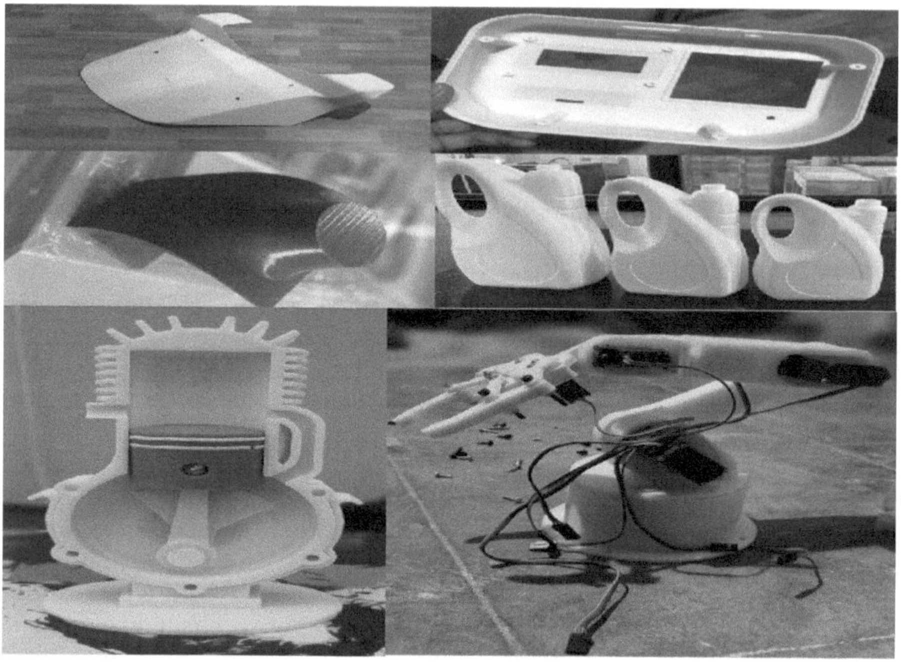

FIGURE 6.9
FDM 3D printing components.

FIGURE 6.10
FDM 3D printing architectural model.

- Production tooling: FDM is utilized in the fabrication of moulds, dies and patterns used in traditional manufacturing processes like injection moulding and casting. This approach reduces lead times and costs associated with traditional tooling methods.

- Architecture: there was a time when architectural models of residential complexes (refer to Figure 6.10), power plants, stadiums and the like were made by using multiple materials like wood, polystyrene, acrylic, etc. These were not only more costly to make but were also difficult to transport in addition to the long manufacturing lead time. With the availability of FDM 3D printing technology, the architectural models can now be rapidly printed, transported easily in pieces and easily assembled on location. Additionally, they resemble the original structure to the point.
- Motorsport: Formula 1 teams and others are seeing the impact that 3D printing can have on their performance and have heavily invested in fused deposition modelling to get ahead of their competitors. The use of FDM and 3D printing in motorsports allows teams to shave those vital seconds from their lap times by improving aerodynamics by creating lightweight pieces through 3D printing. Doing so will help reduce drag and increase speed, which can make a big difference when it comes to race time. Motorsport teams are now using these industrial machines to design functional prototypes and test these parts to analyze their effectiveness before introducing them on a larger scale.

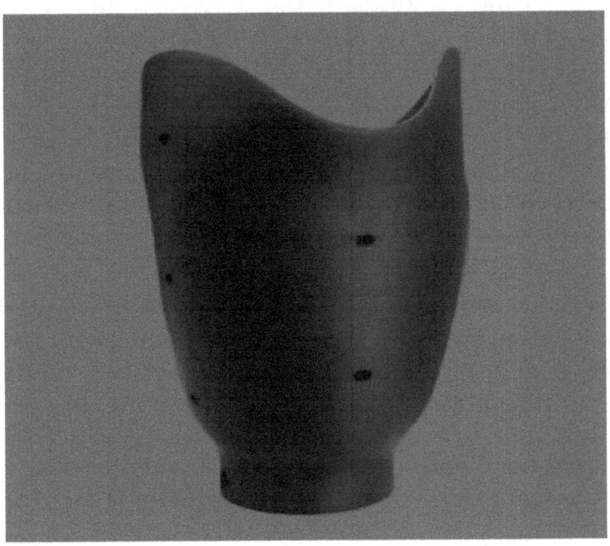

FIGURE 6.11
Arm prosthetics FDM 3D printed components.

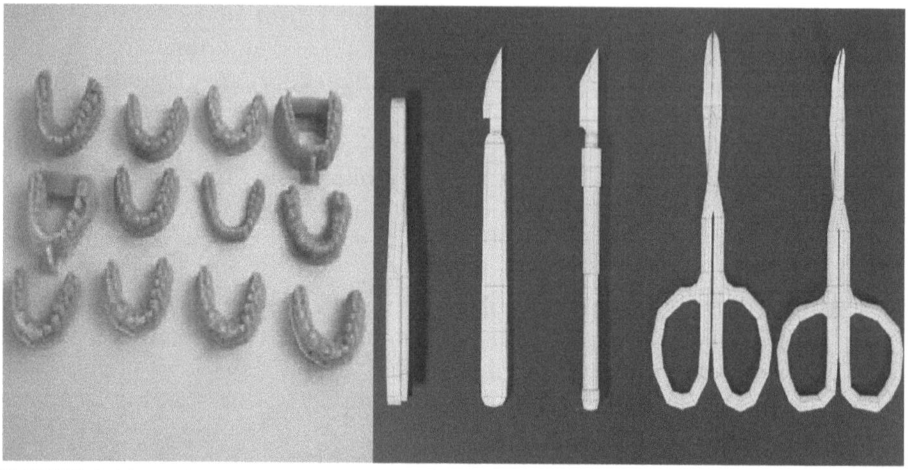

FIGURE 6.12
3D printed tooth and implant surgeries.

6.5.2 Medical Applications of FDM

In the medical field, FDM technology has gained significant attention due to its versatility, cost-effectiveness and ability to produce complex geometries with ease.

- FDM has been employed in the production of custom prosthetics (refer to Figure 6.11) and orthotics. By precisely controlling the deposition of materials, prosthetic limbs and orthotic devices can be tailored to fit individual patients, enhancing comfort and functionality.
- Customized surgical instruments – FDM can be utilized to create patient-specific surgical guides for orthopedic procedures such as joint replacements. These guides help surgeons accurately position implants, resulting in better alignment and longevity of the implant, ultimately reducing the risk of complications and improving patient recovery (refer to Figure 6.12).
- Biomedical research and education: FDM technology also plays an important role in biomedical research and education. Researchers can utilize 3D printing to create anatomically accurate models of organs, tissues and pathological conditions for studying disease mechanisms and testing new treatments. Additionally, medical students can benefit from hands-on learning experiences with 3Dprinted anatomical models, enhancing their understanding of complex physiological concepts.

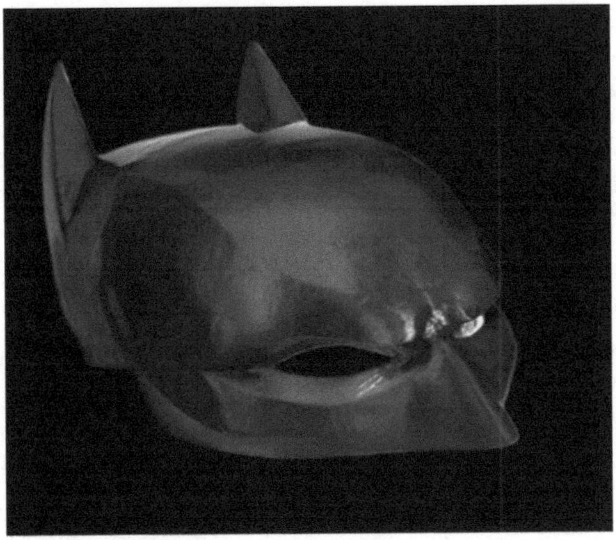

FIGURE 6.13
Props and Cosplay FDM 3D printed part.

6.5.3 Consumer Products Produced by FDM

FDM is a widely used additive manufacturing process that is particularly adept at producing consumer products.

- Customized phone cases: FDM enables the production of personalized phone cases tailored to individual preferences, including unique designs, textures and even integrated features such as card holders or kickstands.
- Jewellery and accessories: from intricate pendants to custom-fit bracelets, FDM allows jewellery designers to explore new creative avenues and offer consumers one-of-a-kind pieces that reflect their personal style.
- Home decor: FDM is increasingly being used to produce decorative items for the home, including lampshades, vases and ornamental sculptures. With FDM, designers can experiment with complex geometries and artistic forms to create visually striking pieces.
- Household items: by rapidly manufacturing replacement products like buttons, hooks, clips, shower heads, customized luxury items like mobile covers, lamps, and simple goods like pen holders, planters, stackable drawers, keychains, etc., the 3D printing can help the household a lot.
- Consumer electronics accessories: FDM can be utilized to manufacture a variety of accessories for consumer electronics, such as tablet

stands, headphone holders and cable organizers. These products can be customized to fit specific devices and user preferences.

- Props and cosplay: earlier it was difficult to manufacture cosplay products but FDM 3D printing has helped fans to recreate popular items from their favourite games, TV shows, movies and more. Artists, enthusiasts and hobbyists have used 3D printing to create such models (refer to Figure 6.13).

6.5.4 Advantages of FDM in Consumer Product Manufacturing

- Customization: the advantages of FDM for consumer products are its ability to facilitate customization. FDM allows manufacturers to cater to individual preferences.
- Rapid prototyping: FDM enables rapid prototyping, allowing designers and entrepreneurs to quickly iterate on product designs without the need for expensive tooling or lengthy manufacturing processes.
- Cost-effectiveness: compared to traditional manufacturing methods, FDM can be highly cost-effective, particularly for small-batch production runs. With minimal material waste and no tooling expenses.
- Complex geometry: FDM producing intricate geometries and complex shapes that may be challenging or impossible to achieve using conventional manufacturing techniques.
- Material versatility: FDM supports a wide range of thermoplastic materials, including PLA, ABS, PETG.

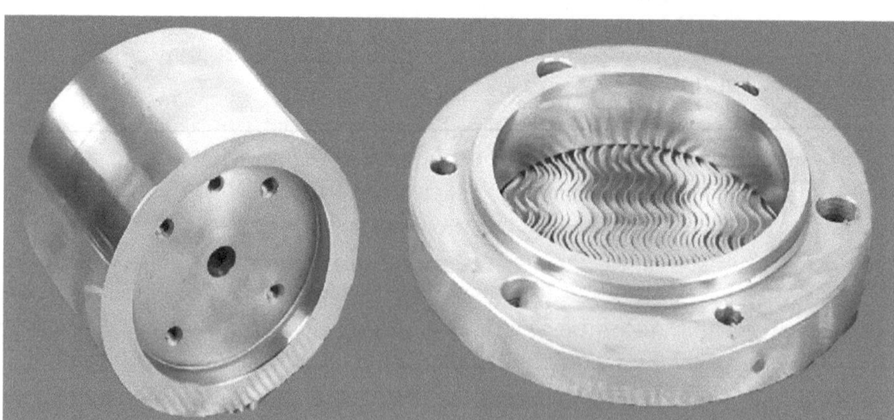

FIGURE 6.14
DED 3D printed part.

TABLE 6.3

Materials and Applications Related to DED

Material Type	Applications
Stainless Steel	Aerospace components, automotive parts, medical implants, tooling, machinery components
Titanium Alloys	Aerospace and defence components, medical implants
Aluminium Alloys	Aerospace structures, automotive parts, consumer goods, heat exchangers
Nickel Alloys (Inconel)	Aerospace components, gas turbine engine parts, chemical processing equipment, high-temperature applications
Refractory Metals	Aerospace components, electrical contacts, high-temperature furnace parts, radiation shielding
Cobalt-Chromium Alloys	Dental implants, orthopaedic implants, aerospace components, turbine engine parts
Tool Steel	Tool and die components, moulds, cutting tools, wear-resistant parts
Tungsten	Aerospace components, electrical contacts, radiation shielding, high-temperature applications
Ceramic Materials	Aerospace components, medical implants, cutting tools, wear-resistant parts
Metal Matrix Composites	Aerospace structures, automotive parts, high-performance machinery components
Inter-metallics	Aerospace components, turbine engine parts, automotive exhaust systems
Precious Metals	Jewellery, luxury goods, electronics, dental components

6.5.5 Applications of DED

Direct energy deposition is used in various industries, including aerospace, marine, oil and gas, defense and architecture. Aerospace manufacturers use DED to produce structural parts for satellites and military aircraft like titanium fuel tank domes for satellites (refer to Figure 6.14). DED is also used for structural parts for commercial aircraft as well. In addition to creating metal parts, DED technology is well-suited for repairing damaged parts. Due to the strong metallurgical bond and fine, uniform microstructures DED can create components like turbine blades and injection moulding tool inserts can be reconditioned. DED allows for the significant reduction of downtime and costs associated with part replacement while extending the part's life by repairing worn parts, moulds or dies. Furthermore, DED can be used to alter parts. For example, in using the method to deposit a wear-resistant hard-facing layer, a part's wear and corrosion resistance can be improved.

Like any other subtractive manufacturing process, parts made using DED can be heat-treated, hot-isostatic-pressed, machined or finished in any customary manner, which opens various new applications (refer to Table 6.3).

Most hybrid manufacturing systems use DED and have become increasingly popular.

6.5.6 Applications of Sheet Lamination

Different types of sheet lamination techniques are used for different purposes and are directly tied to the individual process. Sheet lamination can use a variety of materials such as paper, polymer and metal – but each requires a different method to bind the sheets of material together. Paper-based techniques such as LOM and SDL are used for full-colour prints, while metal-based sheet lamination is used in hybrid manufacturing. SLCOM can make production-quality composite fibre parts, while the CAM-LEM process can be used to make ceramic parts. Paper sheets are usually bound using heat and pressure to activate a layer of activated adhesive that is preapplied to the sheets. The range of materials developed by sheet lamination process includes paper manufacturing, foundry industries and embedded electronic devices, sensors, pipes and other features (refer to Figure 6.15).

FIGURE 6.15
Sheet lamination 3D printed part.

TABLE 6.4

Traditional Materials

Material Name	Properties	Industrial Application
ABS	Tough and strong	Automotive, aerospace, medical-device
ASA	Mechanical strength and UV stability	Functional prototyping from brackets and electrical housings to automotive prototypes and practical production parts for outdoor use under the sun
Nylon12	Good chemical resistance,high fatigue resistance and high impact strength	Ideal material for applications that demand impact-protective components and high fatigue endurance, including antenna covers, custom production tooling, friction-fit inserts and snap fits in automotive and aerospace industries
PC	High tensile and flexural strength	Functional prototypes, tooling and fixtures,blow-molding master in automotive and aerospace industries
PPSF/PPSU	Excellent chemical and heat resistance and mechanical strength	PPSF/PPSU can with stand various sterilization methods including ethylene oxide, auto claving, and radiation. Sterilizable medical devices, automotive prototypes, and tooling for demanding applications in a variety of industries
PEI or ULTEM	Biocompatible, excellent mechanical, chemical and thermal stability	Due to its high strength-to-weight ratio and existing certification, ULTEM is ideal for rapid prototyping and advanced tooling applications in aerospace,automotive,medical and food-production industries
PLA	Good tensile strength and surface equality	Ideal for model and prototypes that require aesthetic detail and environmentally-friendly for both home and office
TPU	Excellent tear and wear resistance, high impact strength and hardness	Exceptional flexibility (i.e., elongation at break) and corrosion resistance to many common industrial chemicals and oils. Highly versatile material with the both rubber and plastics properties for variety of industrial application

6.6 Advanced Concepts and Future Trends

6.6.1 FDM Material Innovations

In recent years, significant advancements have been made in developing novel materials for FDM, enhancing the properties and expanding the potential applications of 3D printed parts. Fused deposition modelling is a popular additive manufacturing technology known for its versatility and ease of use. Understanding these materials' characteristics is crucial for selecting the most

TABLE 6.5

Advanced Materials for FDM

Advanced Material	Properties	Industrial Applications
Carbon Fiber-Reinforced PLA	High strength-to-weight ratio, stiffness, and durability	Aerospace, automotive, sports equipment
Conductive PLA	Electrical conductivity, suitable for circuits	Electronics, prototyping
TPU (Thermoplastic Polyurethane)	Flexibility, elasticity, impact resistance	Footwear, medical devices, seals
Metal-infused PLA	Metallic appearance, higher density, conductivity	Prototyping, decorative items
Flame-Retardant ABS	Reduced flammability, compliance with safety standards	Electronics, automotive, aerospace

TABLE 6.6

Common Issues and Potential Solutions in FDM-Based 3D Printing

Common Issues	Solution
Poor Adhesion to Build Plate:	Ensure the build plate is clean and level. Apply adhesion aids like a thin layer of glue stick, painter's tape, or a specialized build surface.
	Adjust the nozzle-to-build plate distance (usually done through manual levelling or automatic bed levelling).
	Use a heated build plate and ensure it reaches the appropriate temperature for the filament being used.
Warped or Lifted Prints:	Ensure the build plate is heated to the proper temperature for the filament material. Apply a raft or brim in the slicer software to provide better adhesion.
	Use a heated enclosure to maintain a stable printing environment, especially for materials prone to warping like ABS.
Stringing or Oozing:	Adjust retraction settings in the slicer software to reduce the amount of filament being extruded when the nozzle moves between print locations. Increase travel speed to minimize the time the nozzle spends in open spaces.
	Check and clean the hotend to ensure proper filament flow and prevent clogs.
Layer Misalignment or Shifting:	Check that all belts and pulleys are properly tensioned and not slipping. Ensure the stepper motors are functioning correctly and not skipping steps.
	Reduce printing speed if the printer is struggling to keep up with the movements.
Under-Extrusion or Over-Extrusion:	Calibrate the extruder steps/mm to ensure the correct amount of filament is being extruded. Check for clogs or partial blockages in the hot end and clean as necessary.
	Adjust the printing temperature to better suit the filament material being used.

TABLE 6.6 (Continued)

Common Issues and Potential Solutions in FDM-Based 3D Printing

Common Issues	Solution
Nozzle Clogging:	Clean the nozzle with a cold pull or atomic pull method.Check the filament for any signs of moisture absorption and dry it if necessary. Use a filament filter to catch any debris or contaminants before they reach the hot end.
Inconsistent Extrusion:	Ensure the filament diameter is consistent and matches the settings in the slicer software. Check for partial blockages in the hot end and clear them if necessary. Calibrate the extruder steps/mm and flow rate to ensure consistent filament extrusion. Regular maintenance, such as cleaning the printer, checking for loose parts, and updating firmware, can also help prevent common issues from occurring. If problems persist despite troubleshooting, consulting the printer's user manual or seeking assistance from online communities or technical support.

suitable material for specific printing requirements. Tables 6.4 and 6.5 depict information about traditional and advanced materials for FDM, respectively.

6.6.2 Troubleshooting Common Issues

When a physical part is printed, in some cases the material flow can be interrupted due to some special issues therefore minor and major defects can occur. 3D printing machines involves diagnosing and resolving problems that may arise during the printing process. Table 6.6 presents some common issues and their potential solutions.

6.6.3 Future Trends in FDM 3D Printing Techniques

Future advancements in FDM will focus on expanding the range of printable materials beyond traditional thermoplastics.

- Introduction of composite materials, metal filaments and even bio-based polymers will enhance the functional properties of printed parts, making them suitable for a wider array of applications. Integration of additives such as carbon fibres or nanoparticles will improve mechanical strength, thermal conductivity and other performance metrics of printed objects.
- In one of the studies a hybrid 3D printing approach has been developed that combines FDM and DLP techniques for fabricating bioscaffolds,

enabling rapid mass production. The FDM technique fabricates outer moulds, while DLP prints struts for creating penetrating channels.

- Hybrid machining combines additive and subtractive manufacturing methods in one single solution. The concept was around since the 1990s but it was not commercially popular until recently. Today, more and more manufacturers are developing hybrid 3D printers as the demand for such machines keeps growing. Some manufacturers are offering even more functions than 3D printing and CNC – laser engraving and cutting and paste extrusion are often added to offer a versatile solution for the customer.
- Multi-material printing – emerging technologies aim to enable multi-material printing within a single FDM system. This capability will allow the fabrication of complex, multi-component parts with varying material properties in a single print run, enhancing design flexibility and reducing assembly requirements.
- High-speed printing – efforts are underway to enhance FDM printing speeds significantly.
- Innovations such as improved extrusion mechanisms, advanced slicing algorithms and optimized machine designs will contribute to faster print times without compromising part quality.
- Increased resolution and surface finish – future FDM systems will feature improved precision and finer layer resolution, resulting in smoother surface finishes and higher detail reproduction.
- Enhanced control over deposition parameters and the development of novel nozzle designs will contribute to superior print quality, making FDM more competitive with traditional manufacturing methods for high-precision applications.
- Integration of Industry 4.0 Technologies – adoption of Industry 4.0 principles will further automate and streamline the FDM manufacturing process.
- Integration of IoT sensors, machine learning algorithms and cloud-based monitoring systems will enable real-time quality control, predictive maintenance and remote operation of FDM printers, enhancing productivity and reliability.
- Sustainable manufacturing practices – with growing environmental concerns, the future of FDM will prioritize sustainability. Development of recyclable filaments, waste-reducing printing strategies, and energy-efficient manufacturing processes will minimize the environmental impact of FDM while maintaining its cost-effectiveness and accessibility.

6.6.4 Future Trends in Direct Energy Deposition 3D Printing Techniques

- Increased speed and efficiency: future advancements in DED technology are likely to focus on improving printing speed and overall efficiency. This involves enhancements in laser power, deposition rates and process optimization to reduce print times and increase throughput.
- Enhanced material options: future trends may include the development of novel metal alloys with improved properties, as well as the adaptation of DED for other materials beyond metals, such as ceramics and composites, carbon fibre, titanium and nickel-based super-alloys.
- Hybrid additive manufacturing (HAM) – process, which integrates DED with other processes like subtractive manufacturing (SM), mitigates these post-processing issues by enabling multitasking within an integrated system including uninterrupted manufacturing, reduced material movement, higher part complexity, decreased build time factor (BTF), space efficiency, operator simplicity, lower investment and reduced final part cost.
- Thermo-mechanical analyses to investigate the effects of deposition parameters, including the laser power, the scan speed and the feeding rate, on residual stress distributions of fabricated parts by wire laser additive manufacturing (WLAM) and laser additive manufacturing (LAM) processes.
- Improved precision and surface finish: efforts to enhance the precision and surface finish of DED-printed parts are expected to continue. Advancements in control systems, feedback mechanisms and real-time monitoring could contribute to finer resolution and smoother surface quality.
- Integration with CAD/CAM software: seamless integration with CAD and CAM software is crucial for streamlining the DED workflow. It involves the development of advanced software tools specifically tailored for DED processes, including features for automated path planning, optimization and simulation.
- *In situ* monitoring techniques are applied to DED processes for the sake of enhancing product qualities and decreasing material discontinuities through the process optimization. Visible cameras, pyrometers, infrared (IR) cameras, thermocouples (TCs), and X-rays are commonly employed for monitoring DED processes.

6.6.5 Future Trends in Sheet Lamination 3D Printing Techniques

Sheet lamination 3D printing techniques, also known as LOM, involve layering and bonding sheets of material together to create a 3D object. There are potential future trends that could shape its development:

- Topology optimization in 3D printing – topology optimization technique are needed for efficient material usage, appropriate structural design and for new design, analysis and optimization.
- Material diversity – future advancements in sheet lamination could involve the expansion of material options beyond paper and plastics. Researchers may explore new materials suitable for lamination, including composites, metals and ceramics, to enhance the strength, durability and functionality of printed parts.
- Improved resolution and surface finish – efforts to improve the resolution and surface finish of sheet lamination prints may lead to the development of finer and smoother layers. Innovations in printing technologies, such as precision cutting and bonding methods, could help achieve higher levels of detail and accuracy.
- Integration of functional elements – future trends may involve incorporating functional elements directly into the laminated sheets during the printing process. This could include embedding sensors, electronics or other components within the layers, enabling the creation of more complex and functional parts.
- Hybrid approaches – integration of sheet lamination with other manufacturing processes, such as CNC machining or post-processing techniques, could offer hybrid manufacturing solutions with enhanced capabilities and flexibility. Combining the strengths of different fabrication methods could result in more efficient production workflows and higher-quality parts.
- Smart lamination systems – advancements in automation, robotics and artificial intelligence may lead to the development of smart lamination systems capable of autonomous operation and real-time quality control. These systems could optimize printing parameters, detect defects and adjust processes on the fly to ensure consistent and reliable output.
- Recyclable and sustainable materials – with increasing emphasis on sustainability and environmental impact, future trends in sheet lamination may involve the use of recyclable and eco-friendly materials. Research into bio-based polymers, biodegradable composites, and other sustainable alternatives could drive the adoption of more environmentally responsible manufacturing practices.
- On-demand and customized production – sheet lamination technology has the potential to enable on-demand and customized production of parts and products. Future trends may involve the development of software tools and digital workflows that allow for easy customization and personalization of designs, catering to individual needs and preferences.
- Medical and biomedical applications – sheet lamination could find new applications in the medical and biomedical fields, such as printing

patient-specific anatomical models, surgical guides and implants. Advances in biocompatible materials and tissue engineering techniques may further expand the use of sheet lamination in healthcare-related applications.

Self-Check Exercises

1. Classify different 3D printing processes.
2. List solid-based 3D printing processes.
3. Write the steps to be followed to develop industrial components from FDM-based 3D printing process.
4. List the key components of an open-source FDM-based 3D printer.
5. Search and present case studies of FDM technology being used in automobile industries for manufacturing tools and parts.
6. Investigate the use of FDM in the medical field and propose potential applications for producing custom implants or prosthetics.
7. Analyze and design a prototype for a new consumer product utilizing FDM printing techniques.
8. Select a 3D model and optimize for printing, considering factors such as geometry, process parameter and based parameter.
9. Utilize slicing software to prepare any CAD model for printing, adjusting parameters to achieve desired print quality.
10. Develop the support structures for printing any complex industrial components and propose strategies for optimizing their placement and design.
11. List different methods used to perform post-processing on a FDM printed part.
12. Explores a novel post-processing technique for FDM parts. Explain the purpose of the technique and its potential benefits.
13. Discuss the different support structure generation options available in FDM software.
14. Describe the key slicer settings you would consider for a new FDM print and how they can impact the final product.
15. How can FDM functionalities help identify potential printing issues before starting the actual print?
16. Write the steps to be followed to develop components from direct energy deposition based 3D printing process.

17. Write the steps to be followed to develop industrial components from sheet lamination based 3D printing process.
18. List the areas of application of FDM, direct energy deposition and sheet lamination based 3D printing with examples.
19. List the areas of application of FDM-based 3D printing.
20. List the advantages of using FDM in consumer product manufacturing.
21. Comprehend the use of FDM in medical field with examples.
22. List the traditional and advanced materials for 3D printing techniques.
23. Comprehend the common issues and trouble shooting in FDM-based 3D printing technique.
24. Outline the future trends and advancements in FDM-based 3D printing technique.
25. Outline the future trends and advancements in direct energy deposition based 3D printing technique.
26. Outline the future trends and advancements in sheet lamination based 3D printing technique.

Discussion Topics

1. Discuss the effect of different part orientations on the efficiency of a 3D printing process in terms of support structures, build time and material consumption.
2. Discuss the latest advancements in eco-friendly materials for FDM printing and their impact on sustainability.
3. Explore how Industry 4.0 technologies, such as IoT and AI, are being integrated with FDM printing processes for enhanced efficiency and automation.
4. Investigate the latest developments in using FDM technology for bio printing applications, including tissue engineering and organ fabrication.
5. Discuss how FDM printers are being incorporated into smart manufacturing systems to enable real-time monitoring, predictive maintenance, and remote operation.
6. Explore the emerging trend of using FDM technology for printing metal parts and components, including advancements in metal filaments and hybrid FDM/AM systems.
7. Discuss the latest innovations in using FDM technology for 3D food printing, including personalized nutrition and culinary creativity.

8. Explore the regulatory landscape and challenges associated with using FDM technology for medical applications.
9. Investigate the growing trend of using FDM printing for producing custom-designed fashion and accessories, including sustainable and customizable clothing.

Multiple Choice Questions

1. What does FDM stand for in 3D printing technology?
   (a) Full dimension modelling
   (b) Fused deposition modelling
   (c) Fast design manufacture
   (d) Formative design mechanics
2. In FDM, the raw material used is in the form of:
   (a) Liquid resin
   (b) Powder
   (c) Filament
   (d) Gel
3. How many degrees of freedom does the nozzle head have in an FDM 3D printer?
   (a) 1
   (b) 2
   (c) 3
   (d) 4
4. Which part of an FDM printer is responsible for melting the filament?
   (a) Extruder assembly
   (b) Hot end
   (c) Control board
   (d) Stepper motors
5. The component that moves the build platform or the nozzle to add new layers in an FDM printer is called:
   (a) Cooling fans
   (b) Stepper motors
   (c) Filament
   (d) Hot end
6. What is the purpose of the heated print bed in an FDM printer?
   (a) To melt the filament before it reaches the nozzle
   (b) To prevent warping and improve the first layer's adhesion
   (c) To cool down the printed object faster
   (d) To heat the control board

7. What role do the end stops and sensors play in an FDM printer?
   (a) Melting the filament
   (b) Cooling the printed layers
   (c) Detecting the physical limits of the printer's axes
   (d) Extruding the filament
8. For a prototype requiring high tensile strength and impact resistance, which material property should be prioritized?
   (a) Flexibility
   (b) Strength and durability
   (c) Low cost
   (d) Easy to source
9. Which material would be best suited for an application requiring food safety?
   (a) NYLON
   (b) ABS
   (c) PLA
   (d) PETG
10. What does FDM stand for in the context of 3D printing?
    (a) Fused data modelling
    (b) Fused deposition modelling
    (c) Fixed dimension manufacturing
    (d) Formulated design manufacturing
11. Which of the following materials is commonly used in FDM printing?
    (a) Glass
    (b) PLA
    (c) Ceramic
    (d) Metal
12. FDM can create complex geometries that are difficult to produce using traditional manufacturing processes due to its ability to:
    (a) Mould materials
    (b) Conduct layer-by-layer deposition
    (c) Use a wide variety of metals
    (d) Perform high-speed machining
13. What is the primary function of the extruder assembly in an FDM printer?
    (a) To cool down the printed layers
    (b) To heat and extrude the thermoplastic filament
    (c) To move the build platform
    (d) To stabilize the printer structure
14. The nozzle head in an FDM printer moves in how many degrees of freedom to deposit the material?
    (a) 1
    (b) 2
    (c) 3
    (d) 4

15. Which component is responsible for melting the filament in an FDM printer?
    (a) Control board
    (b) Stepper motor
    (c) Hot end
    (d) Cooling fan
16. What characteristic is critical when selecting a material for a prototype that requires high tensile strength?
    (a) Flexibility
    (b) Material's strength and durability
    (c) Melting point
    (d) Colour availability
17. Which of the following materials is best suited for applications requiring flexibility in FDM printing?
    (a) ABS
    (b) PETG
    (c) TPU
    (d) NYLON
18. Which material is known for its excellent visual quality and ease of printing in FDM?
    (a) ABS
    (b) PLA
    (c) NYLON
    (d) PEI
19. Which of the following materials is best suited for applications requiring flexibility in FDM printing?
    (a) ABS
    (b) PETG
    (c) TPU
    (d) NYLON
20. What is the first step in preparing a 3D model for FDM printing?
    (a) Slicing the model
    (b) Exporting as an STL file
    (c) Selecting or designing a 3D model
    (d) Preparing the printer and filament
21. What file format is typically used to save a 3D model before importing it into slicing software?
    (a) PNG
    (b) STL
    (c) PDF
    (d) DOCX

22. What does the slicing software generate that instructs the 3D printer how to create the object?
    (a) STL file
    (b) JPEG image
    (c) G-code
    (d) PDF document
23. Which parameter does not directly influence the quality and speed of a 3D print according to the details given?
    (a) Layer height
    (b) Infill density
    (c) Colour of the filament
    (d) Print speed
24. What is the primary purpose of support structures in FDM 3D printing?
    (a) To increase printing speed
    (b) To support overhangs and prevent collapse
    (c) To enhance the colour quality of the print
    (d) To reduce the amount of filament used
25. In FDM 3D printing, what is meant by "infill density"?
    (a) The colour intensity of the printed object
    (b) The thickness of the printer's extruder nozzle
    (c) The internal structure's solidity of the print
    (d) The speed at which the filament is extruded
26. Which post-processing technique is mentioned for improving the appearance and texture of a FDM printed part?
    (a) Laminating
    (b) Sanding
    (c) Freezing
    (d) Electroplating
27. What is NOT a reason for the importance of support structures in FDM printing?
    (a) Enhancing the thermal stability of the printer
    (b) Allowing printing of complex geometries
    (c) Preventing material wastage
    (d) Maintaining the integrity of the model
28. Which of the following is not a step in the 3D printing process?
    (a) Exporting or saving as an STL file
    (b) Manual drawing of each layer
    (c) Importing into slicing software
    (d) Post-processing
29. What is the significance of ensuring a model is "manifold"?
    (a) To secure the model against hacking
    (b) To ensure the model can be successfully printed without errors
    (c) To enhance the model's colour fidelity
    (d) To reduce the model's file size for quicker slicing

30. What is the purpose of exporting a 3D model in STL format?
    (a) To enhance the model's resolution
    (b) To represent the model's surface geometry with triangular facets
    (c) To increase the model's color depth
    (d) To secure the model with encryption
31. Which parameter is NOT directly set in slicing software before generating G-code?
    (a) Colour of the model
    (b) Layer height
    (c) Infill density
    (d) Print speed
32. Higher infill density in a 3D printed object generally results in:
    (a) Lower strength but faster print time
    (b) Higher strength but more material usage
    (c) Lower material usage but higher flexibility
    (d) Faster print speed but lower quality
33. What is the primary function of G-code in 3D printing?
    (a) To design 3D models
    (b) To encrypt the 3D model file
    (c) To instruct the 3D printer during the printing process
    (d) To increase the 3D model's resolution
34. Which of the following is a key reason for using support structures in FDM 3D printing?
    (a) To enhance colour quality
    (b) To increase print speed
    (c) To support overhangs and prevent collapse
    (d) To reduce the amount of filament used
35. Why is post-processing necessary in FDM 3D printing?
    (a) To decrypt the G-code
    (b) To improve the printed object's appearance and functionality
    (c) To convert the model back to STL format
    (d) To prepare the model for slicing
36. Which of the following is NOT a typical post-processing technique?
    (a) Encrypting the G-code
    (b) Sanding the printed object
    (c) Applying a protective coating
    (d) Removing support structures
37. What role does layer height play in 3D printing?
    (a) Determines the colour intensity of the print
    (b) Affects the print's detail level and completion time
    (c) Controls the temperature of the heated bed
    (d) Specifies the printer's movement speed

38. How does infill density affect a 3D printed object?
    (a) Changes the colour properties of the filament
    (b) Directly affects the object's structural strength and weight
    (c) Modifies the printer's firmware settings
    (d) Alters the electrical conductivity of the print
39. Which factor is crucial for ensuring the 3D printer's readiness before starting a print job?
    (a) The printer's location
    (b) The level of printer calibration and bed leveling
    (c) The colour of the filament
    (d) The ambient temperature of the room
40. What is the layer resolution typically achieved in FDM printing?
    (a) 0.1–0.2 mm
    (b) 1–2 mm
    (c) 5–10 mm
    (d) 20–30 mm
41. In FDM, which material is commonly used for printing?
    (a) Aluminium
    (b) Nylon
    (c) Copper
    (d) Glass
42. Which of the following is a real-world industrial application of FDM?
    (a) Production of chocolate sculptures
    (b) Manufacturing custom prosthetics
    (c) Creating intricate jewelry
    (d) Designing paperweights
43. What is a significant advantage of FDM in industrial applications?
    (a) High cost-effectiveness
    (b) Low production speed
    (c) Limited material compatibility
    (d) Complexity limitations
44. Which material is commonly used in medical applications of FDM for producing implants?
    (a) PLA (polylactic acid)
    (b) ABS (acrylonitrile butadiene styrene)
    (c) PEEK (polyether ether ketone)
    (d) PETG (polyethylene terephthalate glycol)
45. What is a common challenge faced in FDM printing?
    (a) High-resolution output
    (b) Limited colour options
    (c) Poor adhesion between layers
    (d) Rapid printing speed

46. Which medical product can be manufactured using FDM technology?
    (a) Stethoscope
    (b) Blood pressure monitor
    (c) Dental implants
    (d) MRI machine
47. What is the main advantage of using FDM in the medical field?
    (a) Ability to produce non-customizable parts
    (b) Low precision
    (c) Customization according to patient anatomy
    (d) Limited material options
48. Which factor influences the choice of materials in FDM for medical applications?
    (a) Availability in local hardware stores
    (b) Aesthetics
    (c) Biocompatibility
    (d) Material cost
49. What is the primary concern when using FDM in medical applications?
    (a) Regulatory compliance
    (d) Material durability
    (c) Production speed
    (d) Surface finish
50. What is a critical consideration when selecting materials for FDM printing?
    (a) Material cost per unit
    (b) Availability of colour options
    (c) Ease of recycling
    (d) Heat resistance and mechanical properties
51. What material is commonly used in FDM for industrial applications requiring high strength and durability?
    (a) PLA (polylactic acid)
    (b) ABS (acrylonitrile butadiene styrene)
    (c) PETG (polyethylene terephthalate glycol)
    (d) Nylon
52. In FDM, what challenge is often faced in achieving intricate geometries and fine details?
    (a) Material shrinkage
    (b) Layer adhesion issues
    (c) Limited material availability
    (d) High operating costs
53. Which medical product commonly utilizes FDM for customized patient care?
    (a) Surgical implants
    (b) Dental aligners
    (c) Prosthetic limbs
    (d) Cardiac stents

54. What is a key advantage of FDM in consumer product manufacturing?
    (a) Low initial investment cost
    (b) High-speed production
    (c) Wide range of material options
    (d) Minimal post-processing required
55. Which material is often preferred in FDM for consumer products requiring flexibility and impact resistance?
    (a) Polycarbonate (PC)
    (b) TPU (thermoplastic polyurethane)
    (c) ASA (acrylonitrile styrene acrylate)
    (d) PEEK (polyether ether ketone)
56. What challenge is commonly faced in FDM when producing large-scale industrial components?
    (a) Limited material strength
    (b) Warping and distortion
    (c) High material cost
    (d) Inadequate resolution
57. In FDM, which parameter directly influences the speed of printing?
    (a) Layer height
    (b) Nozzle temperature
    (c) Bed temperature
    (d) Filament diameter
58. Which industry extensively uses FDM for rapid prototyping and functional testing of designs?
    (a) Automotive
    (b) Aerospace
    (c) Electronics
    (d) Textiles
59. Which additive manufacturing process is often compared with FDM due to its similar layer-by-layer approach?
    (a) Selective laser sintering (SLS)
    (b) Stereolithography (SLA)
    (c) Digital light processing (DLP)
    (d) Binder jetting
60. Which of the following is a common thermoplastic material used in FDM printing?
    (a) ABS (acrylonitrile butadiene styrene)
    (b) PVC (polyvinyl chloride)
    (c) PETG (polyethylene terephthalate glycol)
    (d) PTFE (polytetrafluoroethylene)
61. Which property of ABS makes it suitable for FDM printing?
    (a) Low melting point
    (b) High impact resistance
    (c) Transparency
    (d) Low cost

62. Which advanced FDM material is known for its flame-retardant properties?
    (a) Nylon
    (b) Carbon fibre
    (c) ULTEM™ (polyetherimide)
    (d) PLA (polylactic acid)
63. What common issue might occur during FDM printing due to improper bed levelling?
    (a) Warping
    (b) Layer delamination
    (c) Nozzle clogging
    (d) Poor surface finish
64. Which parameter should be adjusted to address over-extrusion in FDM printing?
    (a) Print speed
    (b) Layer height
    (c) Extrusion temperature
    (d) Cooling fan speed
65. What is a potential solution for improving adhesion between layers in FDM printing?
    (a) Increasing print speed
    (b) Decreasing nozzle diameter
    (c) Adding a heated build plate
    (d) Decreasing infill density
66. Which issue might occur if the filament diameter is inconsistent in FDM printing?
    (a) Over-extrusion
    (b) Under-extrusion
    (c) Warping
    (d) Layer shifting
67. What is a common cause of nozzle clogging in FDM printing?
    (a) High printing temperature
    (b) Low printing speed
    (c) Dust accumulation in the filament
    (d) Lack of maintenance
68. Which parameter should be adjusted to reduce stringing in FDM printing?
    (a) Extrusion multiplier
    (b) Retraction distance
    (c) Layer height
    (d) Cooling fan speed
69. How can layer adhesion issues be addressed in FDM printing?
    (a) Increasing print speed
    (b) Lowering extrusion temperature
    (c) Adjusting cooling fan speed
    (d) Increasing infill density

70. Which material is known for its biodegradability and is commonly used in FDM printing for eco-friendly applications?
(a) PLA (polylactic acid)
(b) ABS (acrylonitrile butadiene styrene)
(c) PETG (polyethylene terephthalate glycol)
(d) Nylon

71. Which advanced FDM material exhibits high strength, chemical resistance, and is often utilized in aerospace and automotive industries?
(a) Carbon fibrereinforced nylon
(b) Ultem (polyetherimide)
(c) PEEK (polyether ether ketone)
(d) TPU (thermoplastic polyurethane)

72. What common issue in FDM printing can occur due to insufficient bed adhesion?
(a) Warping
(b) Layer separation
(c) Over-extrusion
(d) Stringing

73. Which troubleshooting technique can help mitigate layer shifting during FDM printing?
(a) Adjusting nozzle temperature
(b) Increasing infill density
(c) Tightening belts and pulleys
(d) Decreasing print speed

74. What is a potential solution for improving bridging performance in FDM printing?
(a) Increasing layer height
(b) Lowering print speed
(c) Enabling support structures
(d) Adjusting fan speed

75. Which FDM material is known for its flexibility and elasticity, commonly used in applications requiring rubber-like properties?
(a) PLA (polylactic acid)
(b) ABS (acrylonitrile butadiene styrene)
(c) TPU (thermoplastic polyurethane)
(d) PETG (polyethylene terephthalate glycol)

76. How can under-extrusion issues in FDM printing be resolved?
(a) Increasing nozzle diameter
(b) Lowering print temperature
(c) Calibrating extruder steps/mm
(d) Increasing print speed

77. What is a potential solution for improving overhang performance in FDM printing?
    (a) Increasing layer height
    (b) Lowering print speed
    (c) Using soluble support materials
    (d) Adjusting retraction settings
78. What is a potential solution for addressing nozzle clogging issues during FDM printing?
    (a) Increasing extrusion multiplier
    (b) Decreasing print temperature
    (c) Using a filament dryer
    (d) Increasing print speed
79. Which FDM material is known for its excellent chemical resistance and is commonly used in applications requiring resistance to harsh chemicals?
    (a) PETG (polyethylene terephthalate glycol)
    (b) ABS (acrylonitrile butadiene styrene)
    (c) PEEK (polyether ether ketone)
    (d) Nylon

Answers

1. (b) Fused deposition modelling
2. (c) Filament
3. (c) 3
4. (b) Hot end
5. (b) Stepper motors
6. (b) To prevent warping and improve the first layer's adhesion
7. (c) Detecting the physical limits of the printer's axes
8. (b) Strength and durability
9. (d) PETG
10. (b) Fused deposition modelling
11. (b) PLA
12. (b) Layer-by-layer deposition
13. (b) To heat and extrude the thermoplastic filament
14. (c) 3
15. (c) Hot end
16. (b) Material's strength and durability
17. (c) TPU
18. (b) PLA

19. (c) TPU
20. (c) Selecting or designing a 3D model
21. (b) STL
22. (c) G-code
23. (c) Colour of the filament
24. (b) To support overhangs and prevent collapse
25. (c) The internal structure's solidity of the print
26. (b) Sanding
27. (a) Enhancing the thermal stability of the printer
28. (b) Manual drawing of each layer
29. (b) To ensure the model can be successfully printed without errors
30. (b) To represent the model's surface geometry with triangular facets
31. (a) Colour of the model
32. (b) Higher strength but more material usage
33. (c) To instruct the 3D printer during the printing process
34. (c) To support overhangs and prevent collapse
35. (b) To improve the printed object's appearance and functionality
36. (a) Encrypting the G-code
37. (b) Affects the print's detail level and completion time
38. (b) Directly affects the object's structural strength and weight
39. (b) The level of printer calibration and bed levelling
40. (a) 0.1–0.2 mm
41. (b) Nylon
42. (b) Manufacturing custom prosthetics
43. (a) High cost-effectiveness
44. (c) PEEK (polyether ether ketone)
45. (c) Poor adhesion between layers
46. (c) Dental implants
47. (c) Customization according to patient anatomy
48. (c) Biocompatibility
49. (a) Regulatory compliance
50. (d) Heat resistance and mechanical properties
51. (b) ABS (acrylonitrile butadiene styrene)
52. (b) Layer adhesion issues
53. (b) Dental aligners
54. (d) Minimal post-processing required

55. (b) TPU (thermoplastic polyurethane)
56. (b) Warping and distortion
57. (a) Layer height
58. (a) Automotive
59. (a) Selective laser sintering (SLS)
60. (a) ABS (acrylonitrile butadiene styrene)
61. (b) High impact resistance
62. (c) ULTEM™ (polyetherimide)
63. (a) Warping
64. (c) Extrusion temperature
65. (c) Adding a heated build plate
66. (b) Under-extrusion
67. (c) Dust accumulation in the filament
68. (b) Retraction distance
69. (d) Increasing infill density
70. (a) PLA (polylactic acid)
71. (c) PEEK (polyether ether ketone)
72. (a) Warping
73. (c) Tightening belts and pulleys
74. (d) Adjusting fan speed
75. (c) TPU (thermoplastic polyurethane)
76. (c) Calibrating extruder steps/mm
77. (c) Using soluble support materials
78. (c) Using a filament dryer
79. (c) PEEK (polyether ether ketone)

7

Printing Using Liquid-Based Raw Material

This chapter covers all the popular liquid-based 3D printing processes. It comprehends their principle of working and the key components of a DLP/SLA-based 3D printer. It also illustrates the use of DLP/SLA-based 3D printing processes to develop products along with their field of application, advancements and future trends. The chapter comprises the following main sections:

7.1 Overview of Liquid-Based 3D Printing Processes
7.2 Photopolymerization Techniques: SLA, DLP, LCD, CLIP
7.3 Material Jetting Techniques: DOD, CIJ
7.4 Direct Ink Writing: DIW
7.5 Applications of Liquid-Based 3D Printing
7.6 Advantages and Limitations of Liquid-Based 3D Printing
7.7 Troubleshooting Common Issues
7.8 Future Trends in Liquid-Based 3D Printing
7.9 Case Study for Prototyping Medical Devices

After reading through the material, self-check exercises, objective questions and discussion topics presented in this chapter, the learner will be able to do the following:

(a) Explain the working principle of various liquid-based 3D printing processes.
(b) Identify the key components of a DLP/SLA-based 3D printer.
(c) Use DLP/SLA-based 3D printing processes to develop products.
(d) List the applications of the liquid-based 3D printing.
(e) Outline the advancements and future trends in liquid-based 3D printing.

DOI: 10.1201/9781003562344-7

7.1 Overview of Liquid-Based 3D Printing Processes

Regardless of the type of 3D printing techniques used, the final outcome depends on how three key elements work together: the machine components, computer software and the resin materials used for 3D printing (refer to Figure 7.1). Software defines the shape of the object while the machine hardware builds the 3D printed object and the resin material defines the object. Just like other manufacturing processes, successful 3D printing needs well trained, skilled and experienced users.

Basic steps for creating objects (refer to Figure 7.2):

- As we have already studied that the 3D printing starts with making a 3D digital model using CAD software on a computer. This 3D model specifies the shape, size and structure of the object for printing. **(Design phase)**
- The digital model is then properly placed on the virtual build platform using 3D printer software followed by support system generation and setting of printing parameters using the software. The digital 3D model, along with its support system, is divided into thin horizontal layers using specialized slicing software. The complete information is saved as G-code file, which will be sent to the printer. **(Slicing phase)**

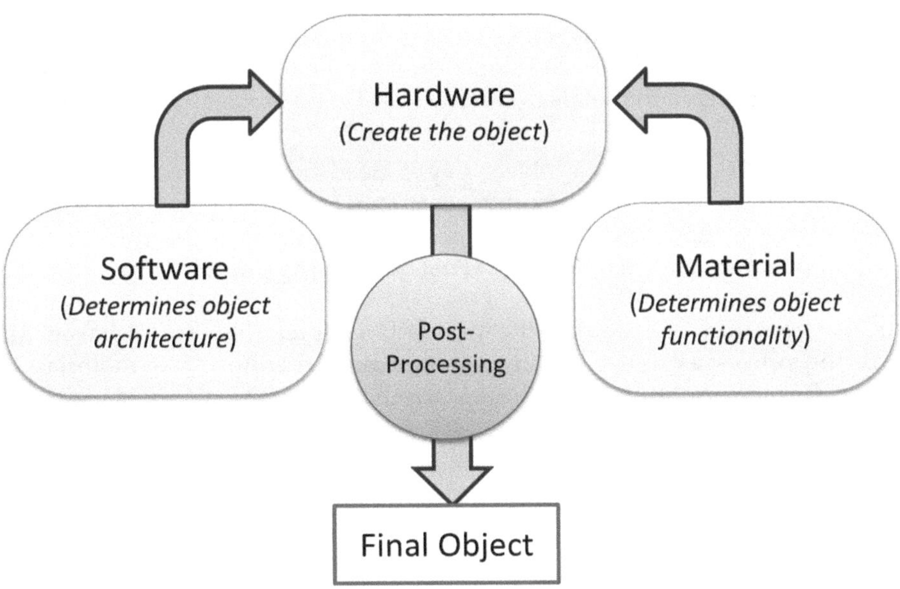

FIGURE 7.1
Three major components: hardware, software, and material.

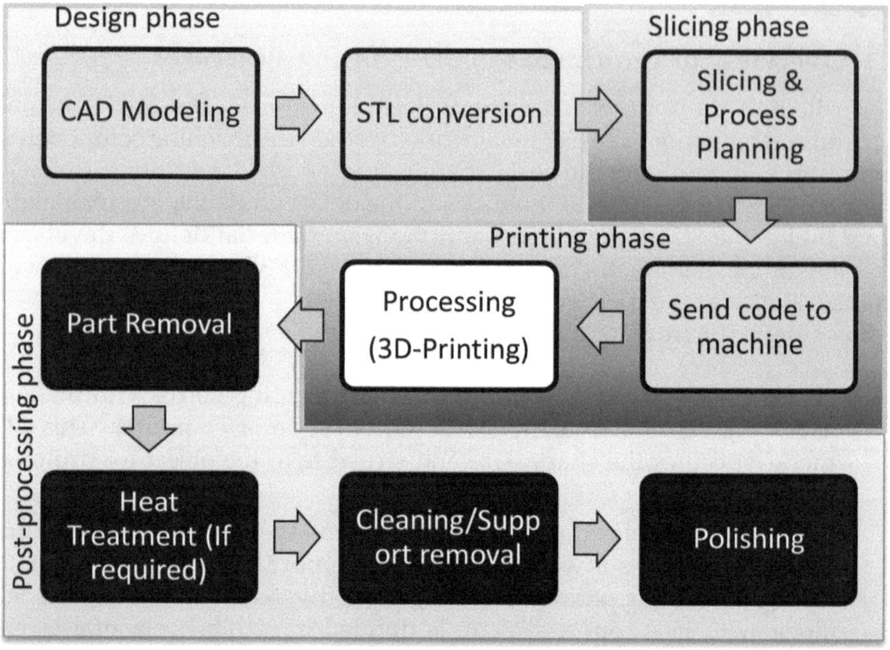

FIGURE 7.2
3D printing process steps.

- The 3D printer receives instructions in the form of sliced layers from the G-code file, then adds material layer by layer accordingly. The choice of printing material varies depending on the technology used, ranging from plastics and metals to ceramics and even biocompatible materials. **(Printing phase)**
- After 3D printing, the object might need additional post-processing steps such as removal of support structures, washing of resin material, curing, surface finishing and polishing to achieve the desired final appearance and functionality. **(Post-processing phase)**

As we have already seen in the previous chapter that **liquid-based 3D printing processes** utilize liquid as the primary medium (raw material) to create the part or prototype and categorized as follows:

- Photopolymerization techniques: steriolithography (SLA), digital light processing (DLP), liquid crystal display (LCD), continuous liquid interface printing (CLIP)
- Material jetting techniques: continuous ink jetting (CIJ), drop on demand (DOD)
- Direct ink writing (DIW)

7.2 Photopolymerization Techniques: SLA, DLP, LCD, CLIP

7.2.1 Principle of Photopolymerization and Processes

In photopolymerization, a light-sensitive resin called a photopolymer is used. This resin is placed in a vat container. Then, it is exposed to either visible or UV light. When this light strikes on the resin, it starts a chemical reaction called polymerization. This reaction causes the resin to turn into a solid by forming chains or cross-links. In the resin, there are monomers, oligomers and photoinitiators (refer to Figure 7.3). When the light hits the photoinitiators, they release reactive substances that help form chains between the monomers and oligomers.

This process is irreversible, meaning once the chains form; the resin cannot go back to being a liquid. By using this process, layers of resin are gradually built up to create the final product from a sliced STL file.

Photopolymerization is classified based on the method of curing (refer to Figure 7.4), which employ lasers (SLA), digital projection (digital light processing (DLP)), and light-emitting diodes (LEDs) and oxygen continuous digital light processing (CDLP)/continuous liquid interface production (CLIP).

There are two main approaches employed in the photopolymerization: the top-down and bottom-up methods. In the top-down method, the light

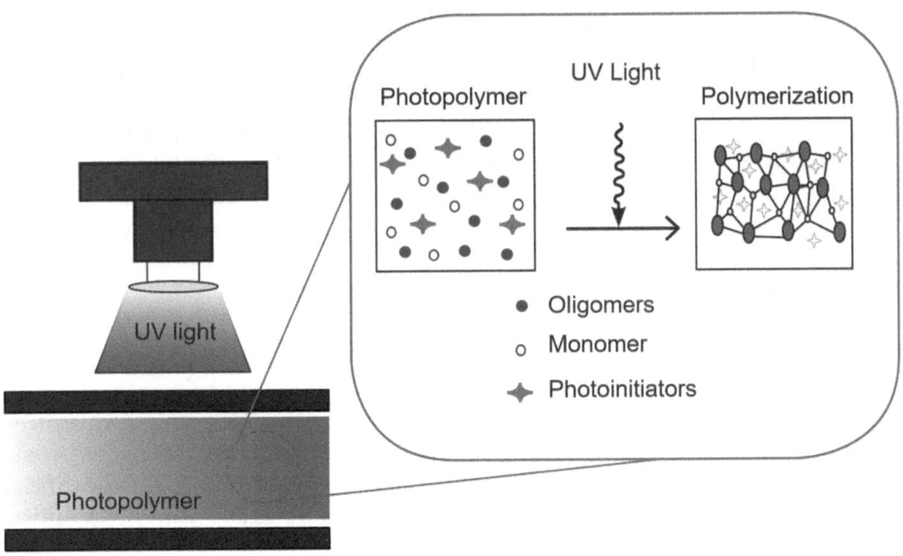

FIGURE 7.3
Liquid photopolymer (on the left), induced polymerization by light (small circle-monomer, large circle-oligomer, star-photo initiator).

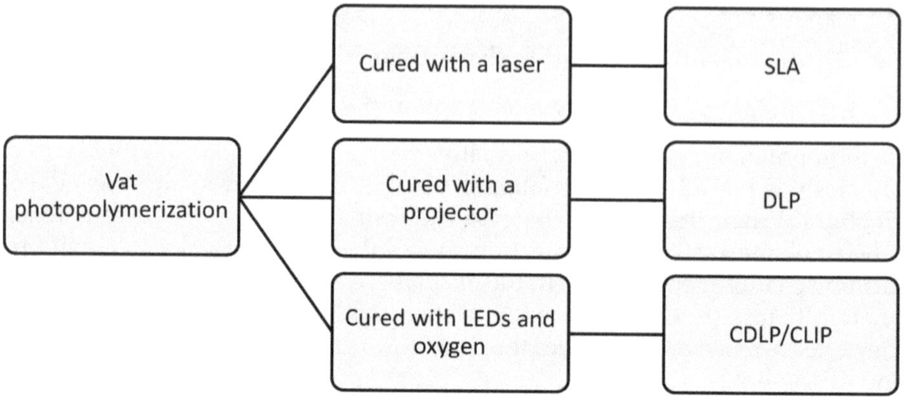

FIGURE 7.4
Different curing methods for vat photopolymerization.

source is placed under the resin vat, and the build platform rises gradually as the part is built upside down. In the bottom-up method, the light source is above the vat, and the build platform is dipped into the vat. Photopolymer resins come in various colours and have different properties for specific uses, such as tough, low residue (for investment casting), transparent and flexible resins.

7.2.2 Stereolithography (SLA)

SLA is a technique for building three-dimensional objects layer by layer, using a photopolymer resin that solidifies under specific ultraviolet (UV) light wavelengths, as shown in Figure 7.5. The SLA 3D printing process begins with a digital CAD model and this model is divided into a number of thin layers using specialized slicers software. Then, a UV laser beam solidifies the liquid resin layer by layer, following the pattern of the model on a platform submerged in a tank of resin. The platform moves either downwards or upwards (depending on the printer model) as each layer is cured, allowing subsequent layers to adhere to the previous ones, ultimately forming a complete object.

The precision of the laser ensures intricate details and fine features can be accurately reproduced, making SLA well-suited for applications requiring high accuracy and complex geometries, such as prototyping, jewellery production and dental prosthetics. After printing, excess resin is drained, and the object is typically rinsed in a solvent and cured under UV light to fully harden the material. Despite its exceptional resolution and surface finish, SLA may necessitate additional post-processing steps, like support removal and surface smoothing, to achieve the desired final outcome.

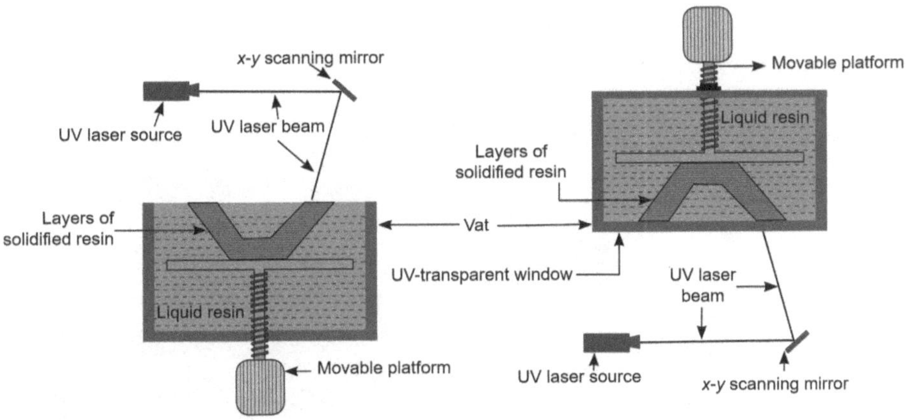

FIGURE 7.5
Schematic diagram of stereolithography based 3D printing technique.

7.2.3 Digital Light Processing (DLP)

DLP 3D printing (refer to Figure 7.6) consist of a high-resolution UV light projector, which focuses a UV light on a photosensitive liquid resin to polymerize the resin layer by layer, resulting in the creation of complex 3D printed objects. The process initiates as a digital model is sliced into thin horizontal layers, which are then sequentially projected onto a vat of UV-sensitive resin. Upon exposure to the intense UV light emitted by the projector, the resin solidifies, conforming to the shape of each layer. This precise curing mechanism enables the creation of intricate geometries and fine details with exceptional accuracy, down to the micron level.

The primary distinction between SLA and DLP 3D printing lies in their light sources and how they control the selective curing of resin. In SLA, a laser serves as the light source, guided by mirrors to specific points on the liquid's surface in the *x-y* plane (see Figure 7.6). These mirrors systematically guide the laser to cover the entire area of each layer of the object under construction. On the other hand, DLP utilizes a conventional light source, like an arc lamp, coupled with a liquid crystal display panel to project light (refer to Figure 7.6), akin to systems found in movie theatres. This projected light instantly cures the entire layer of the object being printed, providing a different approach for the solidification of resin material as compared to SLA technique.

Additionally, liquid-based 3D printing (DLP) exhibits high speed advantages over traditional additive manufacturing techniques due to its ability to cure entire layers simultaneously, reducing production times significantly. Furthermore, the utilization of UV light in the DLP process allows for the fabrication of objects with high structural integrity and smooth surface finishes, making it particularly well-suited for applications requiring

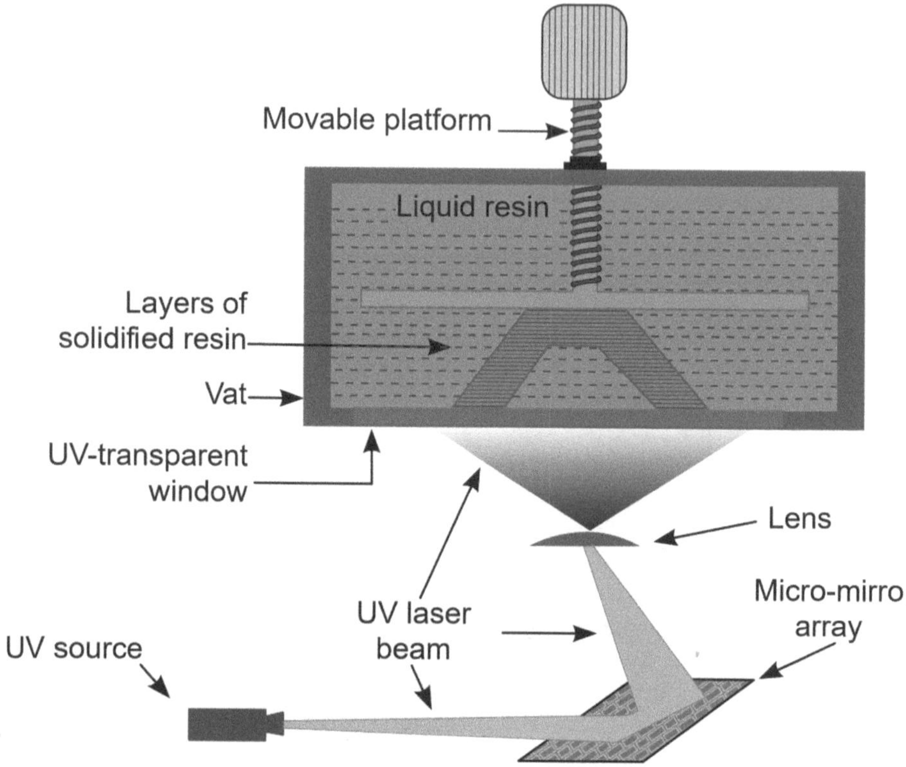

FIGURE 7.6
Schematic diagram of DLP based 3D printing technique.

intricate designs and fine surface quality. Despite its advantages, DLP 3D printing also presents challenges such as the need for post-processing to remove excess resin and the limited selection of materials compatible with the UV curing process.

7.2.4 Continuous Liquid Interface Printing (CLIP)

A CLIP 3D printing technique relies on several key components including photosensitive resin, an oxygen-permeable (catalyst), a dead zone, an ultraviolet (UV) projector and a build platform. It operates by precisely controlling the interaction between light and oxygen to continuously solidify liquid resin into desired objects (see Figure 7.7). Unlike traditional layer-based approaches, CLIP utilizes a continuous liquid interface, eliminating the need for layering and resulting in significantly faster printing speeds.

The CLIP vat photopolymerization process uses resin as the base material. Through a transparent projector on the bottom of the vat, ultraviolet light illuminates the specific cross-section of the object to be printed. During this

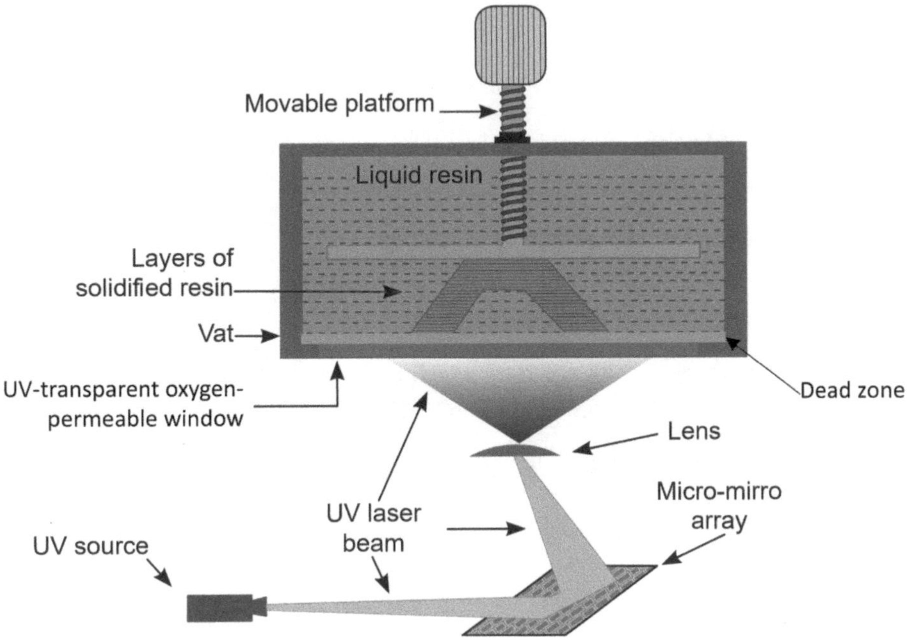

FIGURE 7.7
Schematic diagram of continuous liquid interface printing (CLIP) based 3D printing technique.

process, the resin solidifies, or photopolymerizes, as a result of the exposure to light. A thin layer of resin flows underneath the object as it gradually rises, keeping it in contact with the bottom surface. In this case, a dead zone is created by an oxygen-permeable membrane beneath the resin. This horizontal liquid layer stops resin from sticking to the window, which stops the process of the resin hardening between the window and the polymerizer.

Unlike standard SLA methods, CLIP's continuous 3D printing process boasts speeds up to 100 times faster. Moreover, it enables the production of intricate details and overhangs with minimal need for support structures, reducing post-processing requirements. The versatility of materials compatible with CLIP, ranging from tough engineering-grade resins to flexible elastomers, further broadens its application across various industries, such as aerospace, automotive and healthcare.

7.2.5 Liquid Crystal Display (LCD)

Instead of employing lasers or projectors for polymer curing, the DPP manufacturing technique utilizes LCD technology. Known as LCD 3D printing, this method utilizes unaltered LCD screens in conjunction with a specialized daylight polymer. In this innovative approach, the traditional micro-mirror and projector lens are replaced by an LCD screen driven by UV-emitting LEDs.

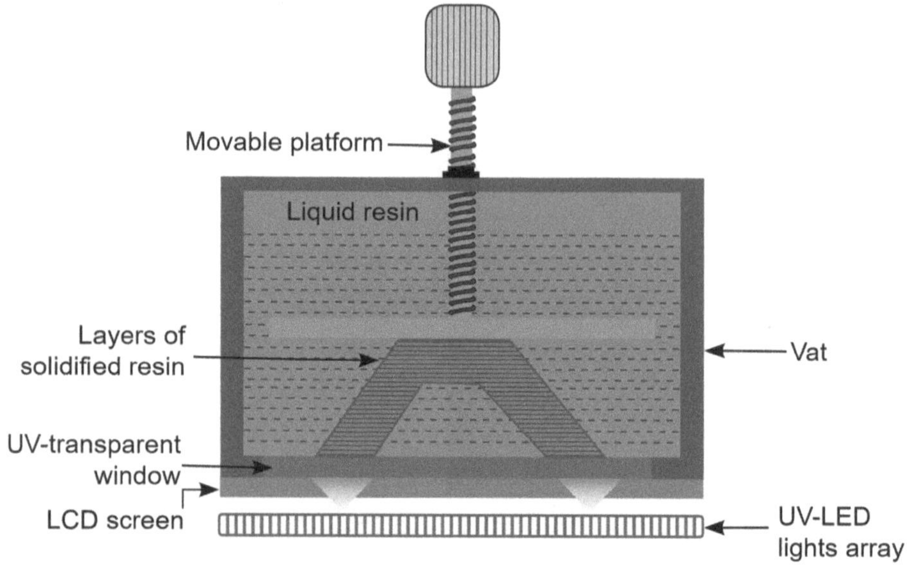

FIGURE 7.8
Schematic diagram of liquid crystal display (LCD) based 3D printing technique.

Similar to digital light processing technique, these LEDs can control the emission of UV lights for individual pixels. Consequently, as the LEDs create the complex shape of the desired object, the liquid crystal screen transmits this image through a UV-transparent projector into the resin material inside the vat of the machine. This process allows for simultaneous exposure and curing of the entire layer, thereby facilitating the gradual layer-by-layer construction of the part (see Figure 7.8).

7.2.6 SLA and DLP Technologies

SLA and DLP represent commonly employed techniques for 3D printing of resin materials. These approaches are popular for their ability to produce accurate, uniform and intricate prototypes using advanced materials.

SLA technology employs a laser beam, while DLP technology utilizes a UV light source to solidify selectively resin materials. In SLA, the laser emitted from a mirror focuses on various points within the liquid resin in the x-y plane, curing it. This process involves the laser continuously traversing the entire surface area of the resin layer being 3D printed (refer to Figure 7.5). Conversely, DLP operates by projecting UV light through a projector, illuminating the complete surface area of each layer of the printed components during the 3D printing process (refer to Figure 7.9).

DLP printing usually takes less 3D printing time to create an object because it does not have to raster scan each layer one by one, unlike some other

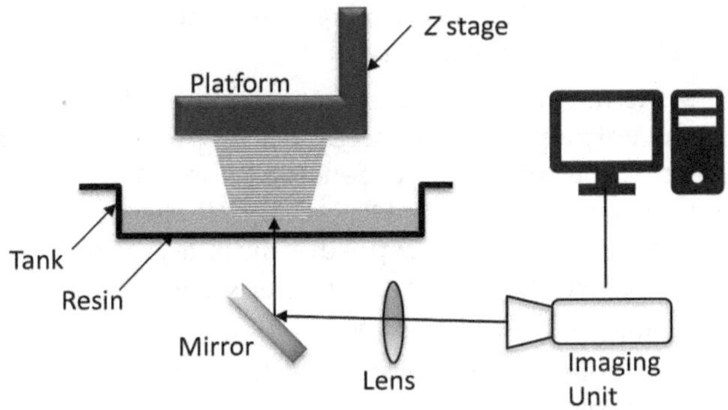

FIGURE 7.9
Schematic diagram of DLP based 3D printing process.

machines. A new and exciting DLP printing technology is the introduction of continuous liquid interface production (CLIP) by 3D printer manufacturer (Carbon 3D, Redwood City, CA). CLIP employs an innovative technique involving an oxygen-inhibiting layer positioned above a membrane within the resin vat. This layer, often referred to as a "dead zone," effectively prevents resin polymerization at the interface between the membrane and the object being printed. As a result, it streamlines the vat photopolymerization process by reducing mechanical steps.

7.2.7 Components of SLA/DLP 3D Printing Machine

The schematic diagram of SLA/DLP is shown in Figure 7.6.

- **Light source**: the heart of SLA printer is the laser source and controlled mirrors while in case of the DLP printing process it is the light source, typically a high-intensity projector or an array of UV LEDs. This light source emits a specific wavelength of light (often ultraviolet) that triggers the photopolymerization reaction in the liquid resin.
- **Build platform**: the build platform is a flat surface where the printed object gradually forms layer by layer. It moves up/down (Z axis) to control the height of each layer and ensure precise printing.
- **Resin vat**: the resin vat contains photopolymer resin. It is transparent, enabling light to pass through and selectively focus on specific areas to solidify each layer during the curing process. The bottom of the vat is typically coated with a transparent film or layer to facilitate adhesion of each layer to the build platform.
- **Optical system**: the optical system consists of mirrors, lenses and other optical components that focus and direct the light from the light source/laser onto the surface of the liquid resin in the vat. This system ensures

that the light is projected with high precision to cure the resin according to the desired pattern of each layer.

- **Control system**: the control system of the 3D printing machine manages the entire printing process. It interprets the digital 3D model of the object to be printed and generates the corresponding layer-by-layer instructions for the SLA/DLP printer. It also monitors and adjusts various parameters such as exposure time, layer thickness and build platform movement to ensure accurate and consistent printing.
- **Cooling system**: some SLA/DLP printers may incorporate a cooling system to regulate the temperature of the resin during printing. This helps prevent overheating and ensures optimal curing and layer adhesion.
- **Post-processing station**: once the 3D printing is complete, the printed object requires additional post-processing steps such as washing, curing (additional UV exposure to fully harden the resin) and support removal. Some DLP printers may include integrated post-processing capabilities, while others may require manual post-processing steps.

7.2.8 Printing Procedure in the SLA Machine

By following these steps, you can successfully use a stereolithography apparatus (SLA) to create high-quality 3D printed objects.

Step 1: Prepare the 3D model-start by designing the 3D model of the object you want to print or obtaining an already prepared model from the websites or scanning a real object. Use CAD software such as NX-CAD, SolidWorks, Autodesk, CATIA, UG NX, ProE, Fusion 360, or any other CAD tool you are comfortable for designing your model with the appropriate dimensions, features and details. (As detailed in Chapter 3.)

Step 2: Convert the above CAD model native file to to .STL or .OBJ or 3MF format. This can be done by saving your CAD model in the .STL (stereolithography) file format, which is the standard file format used for 3D printing. (As detailed in Chapter 4.)

Step 3: Import the .ST/.OBJ/3MF model in 3D Sprint software.

- Orient the model: rotate, scale and position the model within the print bed to ensure optimal printing orientation and minimize supports. This helps improve print quality and reduce material usage.
- Check for errors: utilize the analysis tools provided by the software to examine the model for any potential errors, including non-manifold geometry, intersecting faces or inverted normals. Fix any issues to prevent printing problems.

- Add supports: determine where supports are needed to prevent overhangs and ensure successful printing. Use automatic support generation tools or manually add supports as necessary.
- Customize preferences: set up print preferences like layer height, infill density, printing speed and temperature according to the specifications of your printer and the material being used.
- Slice the model: use the slicing feature to generate the tool paths necessary for printing the model layer by layer. Preview the sliced model to verify that it looks correct and make any final adjustments if needed.

Step 4: Export the print file: once satisfied with the settings and slicing, export the print file in the appropriate format for your 3D printer. Common formats include .gcode or specific file formats supported by your printer.

Step 5: Preparing the SLA/DLP printer: prepare the SLA printer by ensuring it is properly calibrated and filled with the appropriate resin material. Make sure the build platform is clean and levelled.

- Carefully select the appropriate resin material for your specific application, taking into account factors such as strength, toughness and temperature resistance. Before you begin, make sure to shake the resin bottle thoroughly for 15 minutes.
- Check resin level: open the resin tray and inspect the resin level. Make sure there is enough resin to complete your intended print job. If the resin level is low, carefully pour in more resin following the manufacturer's instructions, taking care not to spill any resin outside the tank.
- Inspect resin quality: check the quality of the resin for any signs of contamination or degradation. If the resin appears compromised, replace it with fresh resin from a sealed container.
- Clean the build platform: inspect the build platform for any debris or leftover resin from previous prints. Use a clean, lint-free cloth and isopropyl alcohol to wipe down the build platform thoroughly; ensuring itis free from any residues that could affect print quality.
- Calibrate build platform: follow the printer's calibration procedure to ensure the build platform is properly aligned and levelled.

Step 6: Initiate printing: once all preparations are complete, close the printer's cover or door and power on the machine. Navigate the printer's interface to select the desired print file and start the printing process. Monitor the print job periodically to ensure it progresses smoothly and address any issues that may arise, such as resin spills or failed prints. The build platform gradually moves downward/upward, by a value of layer thickness as each layer is solidified, allowing the next layer of resin to be exposed to the laser. This process continues layer by layer until the entire object is printed. During printing, the SLA printer uses a UV laser to selectively solidify thin layers of liquid resin according to the sliced model. The laser beam traces the pattern of each layer onto the surface of a vat of liquid resin.

Step 7: Post-printing cleanup: after the print job is complete, carefully remove the printed components from the build platform using appropriate tools, such as a scraper. Rinse the printed object in a container of isopropyl alcohol to remove any excess resin. Dispose of any used resin and clean the printer's components, such as the resin tank and build platform, according to the manufacturer's instructions. For certain materials, a post-curing process is needed to enhance strength and stability. Following this, supports are removed from the parts and any remaining marks are sanded down for a smooth surface. SLA parts can then be easily machined, primed, painted and assembled according to specific requirements or desired finishes.

7.2.9 Software and Slicing for SLA Printing

SLA printing requires specialized software to prepare 3D models for printing. Some popular software options include the following:

- **Pre Form**: developed by Form labs, Pre Form is widely used software specifically designed for Form labs SLA printers. It allows users to import, orient, and prepare models for printing. It also includes features for adding supports and optimizing print settings.
- **Blender**: while not exclusively designed for SLA printing, Blender is a powerful open-source 3D modelling software that can be used for creating models to be printed with SLA technology. It offers robust modelling and sculpting tools, along with support for various file formats.
- **Mesh mixer**: Autodesk's Mesh mixer is another popular choice for preparing models for SLA printing. It includes tools for mesh editing, analysis and optimization. Users can also generate custom supports and hollow out models to save resin material.
- **Simplify 3D:** although primarily known for FDM printing, Simplify3D also supports SLA printers. It offers advanced slicing capabilities, allowing users to fine-tune print settings and optimize supports for SLA prints.
- **3D Sprint software**: 3D Sprint Software is seamlessly integrated with the "Figure 4 Standalone 3D Printing hardware," allowing for direct communication and control. This tight integration ensures compatibility and synchronization between the software and hardware

components, resulting in a smooth and reliable printing experience. "3D Sprint Software" is a powerful tool used for preparing and optimizing 3D models for printing on various additive manufacturing platforms. Slicing is a critical step in this process, as it involves breaking down the 3D model into thin layers or slices that can be printed one at a time. Here are some technical details about slicing in 3D Sprint Software:

- ✓ **Layer thickness**: users can specify the desired layer thickness for the print job. This determines the height of each individual slice in the final printed object. Common layer thickness options range from a few microns to several millimeters, depending on the printer's capabilities and the desired level of detail.
- ✓ **Slicing algorithms**: 3D Sprint Software utilizes advanced slicing algorithms to generate optimized toolpaths for each layer of the 3D model. These algorithms take into account factors such as print speed, layer adhesion, support structure placement and surface quality to produce high-quality prints with minimal defects.
- ✓ **Support structures**: in addition to slicing the main body of the model, 3D Sprint Software also generates support structures as needed to prevent overhangs and ensure successful printing of complex geometries. Users can customize parameters such as support density, angle and interface layers to achieve the desired balance between print quality and material usage.
- ✓ **Adaptive slicing**: some versions of 3D Sprint Software may feature adaptive slicing capabilities, where the software automatically adjusts the layer thickness and other slicing parameters based on the geometry of the model. This helps to optimize print times and material usage while maintaining print quality throughout the entire object.
- ✓ **Preview and simulation**: before sending the sliced file to the printer, users can preview each layer and simulate the printing process within 3D Sprint Software. This allows them to identify potential issues such as collisions, overheating or warping and make adjustments as needed to ensure a successful print.

7.2.10 Material Selection for SLA Printing

When selecting materials for SLA 3D printing, it is important to consider several criteria to ensure optimal performance and desired outcomes. Here are some key selection criteria:

- Photopolymerization compatibility: ensure that the material is compatible with photopolymerization and verify that the material undergoes rapid and reliable curing when exposed to the specific wavelengths of light emitted by the printer's laser or UV light source.

- Viscosity and flow characteristics: select a resin with an appropriate viscosity to enable smooth printing without issues such as dripping or running, while still allowing for good surface finish.
- Mechanical properties: consider the mechanical properties of the printed parts, including strength, stiffness, toughness and elongation at break.
- Thermal properties: assess the material's thermal stability and resistance to heat deformation during printing and post-processing.
- Chemical resistance: select a resin with appropriate chemical resistance to ensure the durability and longevity of the printed parts in their intended environment.
- Surface finish and detail resolution: choose a resin that offers good resolution and surface quality, especially for applications requiring high precision and intricate details.
- Biocompatibility and safety: for applications in healthcare, medical devices or consumer products, prioritize the biocompatibility and safety of the resin. Ensure that the selected material complies with relevant regulatory standards and does not pose any health risks to end-users.
- Availability and cost: assess the availability and cost of the resin, including factors such as material availability, lead times and pricing.

7.3.1 Material Jetting Techniques: DOD, CIJ

Material jetting is an innovative additive manufacturing method that achieves precise material deposition, layer by layer, to created intricate three-dimensional objects. In this process, droplets of photopolymer or wax-based materials are jetted onto a build platform (refer to Figure 7.10).

These materials, usually in liquid form, undergo rapid solidification or curing upon exposure to UV light. Material jetting boasts remarkable resolution, typically within the range of tens to hundreds of microns, facilitating the production of finely detailed parts with smooth surfaces and intricate features. Additionally, it supports the simultaneous deposition of multiple resin materials, allowing the creation of complex multi-material and multi-colour objects in a single 3D printing cycle. Mainly it comprises of two techniques viz. DOD and CIJ 3D printing.

7.3.2 Drop-On-Demand 3D Printing

The technique of DOD 3D printing enables precise material deposition by releasing droplets from a nozzle onto a surface or previously printed layers (as depicted in Figure 7.11). Its advantages encompass high resolution, efficient

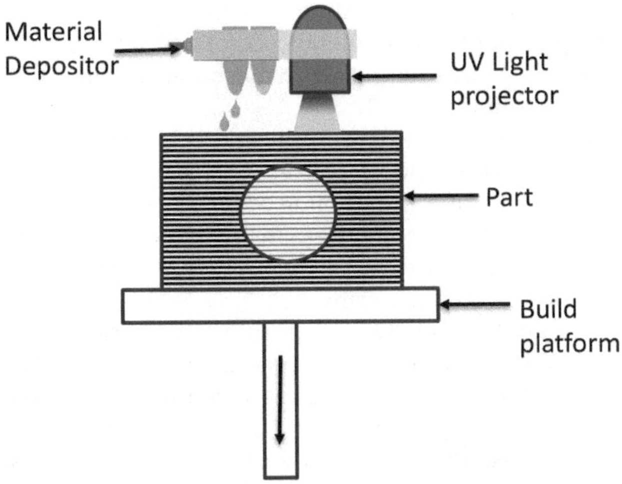

FIGURE 7.10
Schematic diagram of material jetting based 3D printing technique.

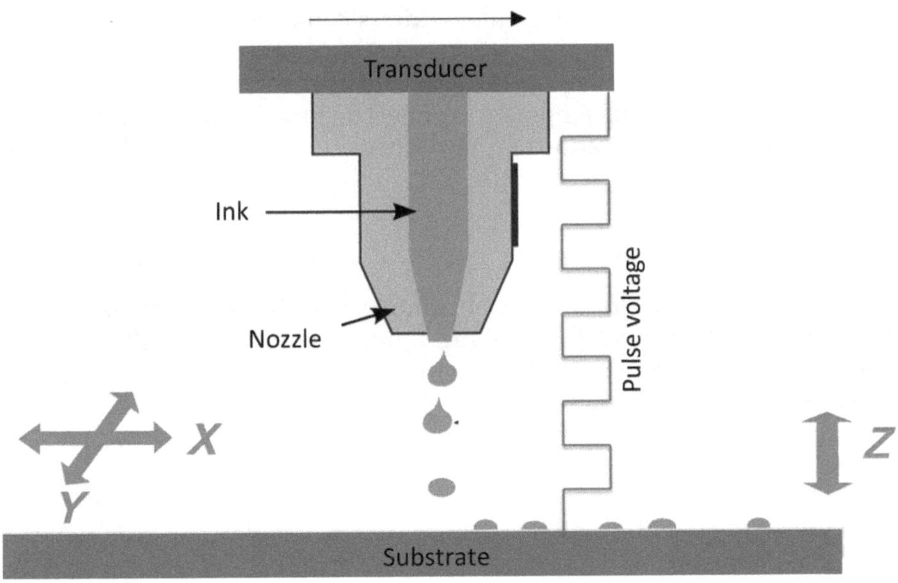

FIGURE 7.11
Schematic diagram of drop-on-demand (DOD) based 3D printing technique.

material usage and compatibility with a wide array of materials. This process relies on the fundamental concept of the layer-by-layer material deposition to construct three-dimensional objects, employing mechanisms akin to inkjet printers. The meticulous control over droplet size, placement and deposition rate facilitates high-resolution printing. Commonly used materials include thermoplastics, photopolymers, metals, ceramics and biomaterials, with ongoing advancements improving their mechanical properties and applicability across various fields.

DOD 3D printing finds extensive applications in industries such as aerospace, automotive, medical field, electronics and miniature parts. In aerospace, it is utilized for fast prototyping, tooling and the production of lightweight components with intricate geometries. In healthcare, DOD 3D printing is employed for crafting customized medical implants, drug delivery systems and framework for tissue engineering.

Recent advancements in DOD 3D printing have concentrated on enhancing printing speed, resolution and material versatility. Upgraded print heads featuring multiple nozzles enable the continuous deposition of resin materials, facilitating the seamless fabrication of multi-material and multifunctional components. Integration of real-time monitoring and feedback systems improves process control and quality assurance. Research into novel deposition techniques, like electro-hydrodynamic jet printing, holds promise for finer resolution and better material compatibility. However, challenges persist, including the need to optimize print speed without compromising quality, reduce material wastage and address compatibility issues with certain materials.

Future research may involve further miniaturization of print heads for finer droplet control, development of novel materials tailored for specific applications, and exploration of new deposition strategies to overcome existing limitations.

7.3.3 Continuous Ink Jetting

It uses traditional continuous inkjet printing principles, like those used in the printing industry. In this method, a liquid material, usually a photopolymer resin, is sprayed onto a platform by a print head nozzle. The material then solidifies, often with UV light, to make the desired shape (refer to Figure 7.12).

Here is how it works:

- A print-head with small nozzles sprays tiny droplets of liquid material onto the platform.
- Computer control ensures precise material deposition.
- Material is sprayed continuously, similar to how inkjet printers work.
- Layer by layer, material is deposited and solidifies, guided by CAD-CAM software.
- Support structures might be added to prevent collapse during printing.
- Curing bonds each layer, resulting in a strong object.

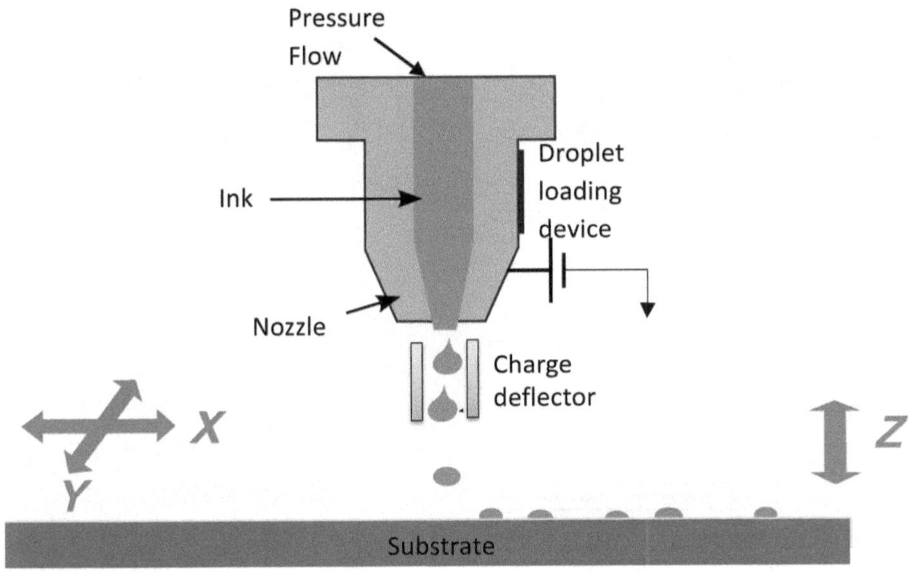

FIGURE 7.12
Schematic diagram of continuous ink jetting (CIJ) based 3D printing technique.

After printing, the object may undergo cleaning, curing or surface finishing for better appearance and properties. This method offers fast printing, detailed results and potential for multi-material printing. However, it may have lower resolution and need support structures for certain designs.

7.4 Direct Ink Writing: DIW

DIW, also called "robocasting," is an advanced form of additive manufacturing that allows for precise creation of complex 3D structures by depositing inks or pastes. Similar to 3D printing but with more material options, DIW uses a syringe or nozzle to dispense inks made of various materials like metals, ceramics, polymers and even biological substances such as cells or biocompatible gels. Following a computer-generated design, the printer builds objects layer by layer (refer to Figure 7.13).

DIW works through the extrusion of liquid or paste-like inks, offering a versatile solution for 3D printing materials beyond solid filament structures. Its capabilities prove valuable for meticulous material management attributes and the fabrication of intricate shapes. A significant innovation lies in the direct ink writing with hydrogels, which has gained attention for its potential in various domains. This technique allows for precise deposition of

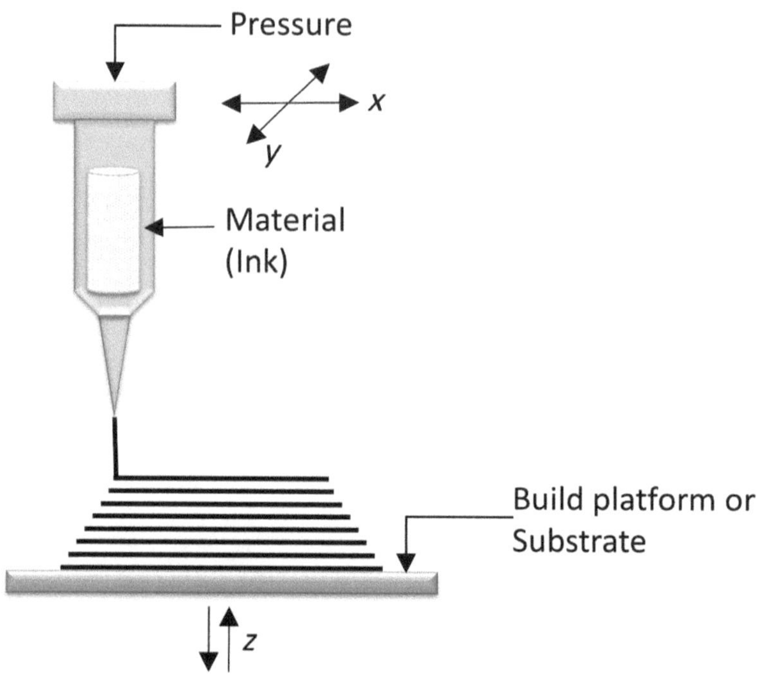

FIGURE 7.13
Schematic diagram of direct ink writing (DIW) based 3D printing technique.

hydrogel-based inks, fabricating complex and functional structures using a combination of polymers, water and sometimes bioactive components like growth factors or cells. Examples of its use include tissue engineering, organ-on-a-chip devices, soft robotics, drug delivery system and bioprinting. DIW's ability to create intricate and precise structures is crucial for making biomimetic tissues and high-fidelity structures.

7.5 Applications of Liquid-Based 3D Printing

- Prototyping: liquid-based 3D printing is widely used for rapid prototyping across various industries, including automotive, aerospace, consumer goods and electronics. It enables engineers and designers to quickly iterate and validate their designs before moving to production, reducing time-to-market and development costs.
- Customization: the ability to produce highly detailed and customized parts makes liquid-based 3D printing suitable for creating personalized

products and one-of-a-kind items. From custom jewellery and fashion accessories to tailored medical implants and prosthetics, this technology empowers designers and manufacturers to cater to individual needs and preferences.

- Production tooling: liquid-based 3D printing is increasingly utilized for producing production tooling, such as injection molds, jigs and fixtures. These tools can be rapidly manufactured with intricate geometries and complex cooling channels, improving efficiency and quality in manufacturing processes.
- Medical and dental applications: in the healthcare industry, liquid-based 3D printing is revolutionizing the way medical devices, surgical guides and patient-specific implants are produced. It enables healthcare professionals to create anatomically accurate models for surgical planning, fabricate custom implants tailored to individual patients, and produce dental restorations with exceptional precision.
- Art and design: artists and designers leverage the capabilities of liquid-based 3D printing to explore new forms, textures and aesthetic possibilities. From sculptural artworks and installations to intricate jewellery and decorative objects, this technology offers a versatile platform for creative expression and experimentation.

7.6 Advantages and Limitations of Liquid-Based 3D Printing

Advantages

- High precision: liquid-based SLA printing enables the creation of highly detailed and intricate structures with exceptional resolution. This is because it uses a laser or other ultraviolet light source to cure photopolymer resin layer by layer, resulting in smooth and precise surfaces.
- Smooth surface finish: SLA printing produces parts with smooth surface finishes, making post-processing requirements minimal. This is particularly advantageous for applications where aesthetics or surface qualities are crucial, such as product prototypes or jewellery.
- Wide material selection: SLA printing supports a wide range of materials, including various types of resins with different properties like transparency, flexibility or toughness. This versatility allows for the production of parts tailored to specific applications, from rigid prototypes to flexible elastomers.

- Support structures: liquid-based SLA printers are capable of generating support structures automatically during the printing process. These supports are easily removable after printing and help in achieving complex geometries and overhangs without sacrificing the quality of the final part.
- Speed: compared to some other 3D printing technologies, SLA printing can be faster, especially for highly detailed parts. The ability to cure entire layers simultaneously contributes to faster print times for certain geometries.

Limitations

- Material limitations: while SLA printing offers a wide range of materials, the selection might not be as extensive as with other 3D printing technologies like FDM or SLS.
- Post-processing requirements: while SLA prints typically have smooth surface finishes, they may require additional post-processing steps such as resin washing, curing of printed object, or sanding to achieve the desired properties or aesthetics. These additional steps can add time and labor to the overall production process.
- Limited build volume: liquid-based SLA printers often have more compact build volumes due to need to store photopolymer resin inside the printer as compared to other 3D printing technologies like FDM, which can restrict the size of parts that can be produced in a single print. This limitation may necessitate splitting larger designs into multiple prints and assembling them afterward.
- Resin handling: liquid resins used in SLA printing can be messy and require careful handling to avoid contamination and spills. Additionally, some resins may emit strong odours or be hazardous if not handled properly, necessitating safety precautions in the workspace.
- UV sensitivity: SLA prints are sensitive to UV light, which can cause them to degrade over time if exposed to sunlight or certain indoor lighting conditions. This limitation requires careful consideration of the intended use and environment of the printed parts.

7.7 Troubleshooting Common Issues

- **Layer adhesion problems:**

  Issue: poor adhesion between layers leading to weak parts or delamination.

Solution: ensure proper resin stirring to prevent settling. Optimize exposure settings for better layer bonding. Increase the build platform temperature slightly to enhance adhesion.

- **Inaccurate dimensions:**

  Issue: printed parts are not matching the intended dimensions.

  Solution: calibrate the printer, including the build platform and laser, regularly. Check for any distortion in the resin tank. Verify the accuracy of the CAD model.

- **Surface imperfections:**

  Issue: surface roughness, pits or dimples on printed parts.

  Solution: adjust the support structures to reduce stress on the part during printing. Optimize exposure settings to minimize overcuring or undercuring. Post-processing techniques like sanding or polishing can help improve surface finish.

- **Warped parts:**

  Issue: printed parts exhibit warping or distortion.

  Solution: ensure proper resin curing by maintaining consistent temperature and humidity in the printing environment. Adjust support structures to minimize stress on the part. Use a heated build platform if available.

- **Resin leakage:**

  Issue: resin leaking from the resin tank or build platform.

  Solution: check for cracks or damage in the resin tank and replace if necessary. Ensure the build platform is properly leveled to prevent resin spillage. Clean the printer regularly to prevent resin buildup.

- **Incomplete curing:**

  Issue: parts are not fully cured, leading to brittleness or softness.

  Solution: verify resin compatibility with the printer. Increase exposure time or power to ensure complete curing. Monitor resin levels to prevent contamination or degradation.

- **Support removal issues:**

  Issue: difficulty in removing support structures without damaging the part.

  Solution: optimize support placement to minimize contact with critical surfaces. Adjust support density and thickness to balance adhesion and ease of removal. Utilize post-processing techniques like sandblasting or chemical baths to aid support removal.

- **Layer registration errors:**

  Issue: misalignment between layers resulting in skewed or distorted prints.

Solution: check for mechanical issues such as loose belts or worn components. Calibrate the printer to ensure accurate layer registration. Monitor resin viscosity and temperature to maintain consistency during printing.

- **Resin contamination:**

  Issue: foreign particles or debris present in printed parts.

  Solution: use filtered resin and ensure a clean printing environment to minimize contamination. Regularly inspect and clean resin tanks and reservoirs. Handle resin cartridges with care to prevent contamination during refilling.

- **Print failure:**

  Issue: prints failing midway or not completing successfully.

  Solution: check for power interruptions or connectivity issues. Ensure adequate resin levels and resin type compatibility with the printer. Monitor print progress and address any errors or warnings promptly. By systematically troubleshooting these common issues, you can optimize the SLA 3D printing process for industrial components and minimize downtime while maintaining quality standards within the SLA.

7.8 Future Trends in Liquid-Based 3D Printing

1. Perspective visualization
   - SLA 3D printing is evolving to incorporate advanced perspective visualization techniques.
   - Real-time rendering of 3D models enables better design evaluation and adjustments.
   - **Example** –the incorporation of augmented reality (AR) technology, allowing engineers to interact with and manipulate their models in a virtual environment.
2. Simulation and artificial organ generation
   - Simulation tools predict printing outcomes, reducing trial and error.
   - Growing focus on printing artificial organs using SLA technology, revolutionizing healthcare. Precise organ replication for surgical practice and medical research.
   - **Example** – Organovo, a biotechnology company, is pioneering the development of 3D bioprinted human tissues for medical research and transplantation (refer to Figure 7.14).

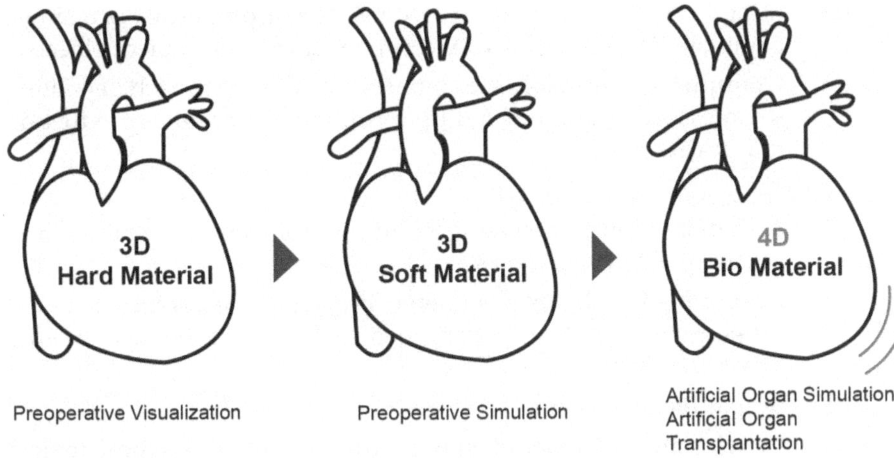

FIGURE 7.14
Evolution of bio printing.

3. 4D printing
    - 4D printing extends the capabilities of 3D printing by introducing time as a fourth dimension. Materials designed to respond to environmental stimuli such as heat, moisture or light.
    - Objects printed with the ability to self-assemble or change shape over time.
    - Potential applications in fields like aerospace, architecture, and biomedicine.
    - **Example** – MIT's Self-Assembly Lab has demonstrated 4Dprinted objects that can change shape in response to environmental triggers.
4. High-speed printing
    - Continuous advancements in SLA 3D printing technology leading to higher printing speeds.
    - Enhanced laser systems and resin formulations enable rapid prototyping and production. Increased efficiency and productivity in manufacturing processes.
    - **Example** – carbons continuous liquid interface production (CLIP) technology enabling rapid resin-based printing with exceptional surface finish and mechanical properties.
5. Multilateral printing
    - Development of SLA printers capable of printing with multiple materials simultaneously.

- Integration of various resins with different properties in a single print job. Enables creation of complex, multi-component objects.
- **Example** – Formlabs are pioneering this approach, enabling seamless integration of multiple printers for parallel production of intricate parts.

6. Post-processing innovations
   - Automated post-processing solutions for cleaning, curing and finishing 3D printed parts.
   - **Example –** Dye Mansions Power fuse system, combines heat and pressure to smooth and strengthen surfaces, enhancing aesthetics and durability.
7. Biocompatible materials
   - Development of materials suitable for implantable medical devices and tissue engineering. Expansion of applications in personalized medicine and prosthetics.
   - **Example –** Formlabs are leading the charge with their biocompatible resins, enabling the fabrication of patient-specific medical devices and implants with unmatched precision and biocompatibility (refer to Figure 7.15).
8. Nanoscale printing
   - Advancements in SLA technology enabling printing at nanoscale resolutions.

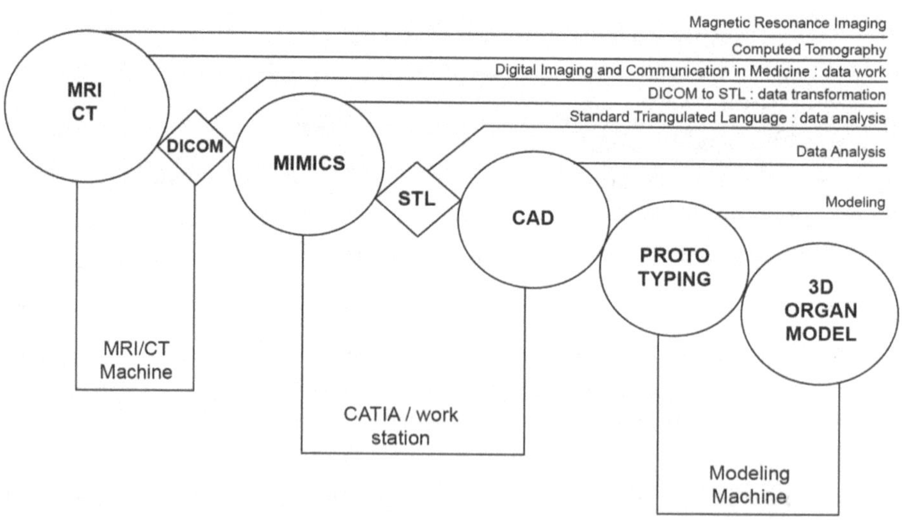

FIGURE 7.15
Technological process of a 3D printing production applied in regenerative medicine.

- Precision printing of intricate structures with unprecedented detail. Applications in microelectronics, photonics and biomedical engineering.
- **Example** – Stratasys digital material allowing for the creation of parts with embedded sensors, actuators and circuitry. Such smart structures find applications in fields like robotics, IoT and aerospace, where adaptive and responsive components are paramount.

9. Environmental sustainability
   - Focus on reducing waste and energy consumption throughout the printing process.
   - Recycling initiatives for used resin and support materials.
   - **Example** – Nano scribes photonic professional GT2 system enables the fabrication of structures at the submicron scale, with applications in optics, microfluidics and nanotechnology.
10. Accessibility and affordability
    - Development of compact desktop SLA printers for home and small-scale use.
    - **Example** – Formlabs sustainable material initiatives (SMI) aims to develop eco-friendly resins derived from renewable sources, minimizing environmental impact without compromising performance.

7.9 Case Study for Prototyping Medical Devices

The implementation of SLA 3D printing for prototyping medical devices, particularly for soft orthodontic aligners, involves a series of sequential steps to ensure accuracy and efficacy. Firstly, the process begins with the scanning of an impression of the patient's teeth from an intra oral scanner, which is then subjected to 3D scanning to create a digital model. Following this, 3D modelling software is utilized to refine and optimize the digital model according to the specific requirements. Next, the orientation of the model is determined, considering factors such as support structures and printing efficiency. Subsequently, the appropriate dental resin is selected based on factors like biocompatibility and mechanical properties. The parameters of the 3D printer machine are then configured to ensure optimal printing conditions, including layer thickness, exposure time and curing settings. The printing itself employs a dental SLA 3D printing machine, where layers of resin are selectively cured by UV light. Post-processing steps such as cleaning support removal and curing may be necessary to finalize the printed part. Finally, the 3D printed part serves as a mould for fabricating soft orthodontic aligners, offering a customized solution for dental applications. This comprehensive process ensures the production of high-quality medical prototypes tailored to individual patient needs, showcasing the potential of SLA 3D printing in advancing healthcare innovation.

Dental 3D SLA printing steps

- Scan impression of patient's teeth with intraoral scanner.
- Utilize 3D scanning to create digital model.
- Refine and optimize digital model using 3D modelling software.
- Determine orientation of model considering support structures and printing efficiency.
- Select appropriate dental resin based on biocompatibility and mechanical properties.
- Configure parameters of 3D printer machine for optimal printing conditions.
- Print using Dental SLA 3D printing machine with selective UV curing.
- Perform post-processing steps such as cleaning, support removal and curing.
- Use 3D printed part as mould for fabricating soft orthodontic aligners.

Self-Check Exercises

1. Classify various liquid based 3D printing techniques.
2. Explain principle of photopolymerization and the processes involved.
3. Summarize the key differences between, FDM and SLA 3D printing technologies.
4. Differentiate SLA, DLP and CLIP vat polymerization 3D printing processes.
5. List the components of a typical SLA/DLP based 3D printing machine.
6. Explain the working of SLA/DLP based 3D printing machine using schematic diagrams.
7. List the software used for slicing in a SLA based 3D printing
8. Explain the different types of slicing options available.
9. Outline the criteria of selecting materials for SLA printing.
10. Explain in brief different material jetting techniques.
11. Make a list of items prepared by vat polymerization and material jetting 3D printing technologies.

12. Describe the key components of an SLA machine, including their functions and importance in the printing process.
13. Developed a detailed step-by-step guide outlining the entire printing process using stereolithography printing machine (SLA), from file preparation to post-processing.
14. Evaluate different software options available for SLA 3D printing, compare their features, and provide details for slicing software best suited for various printing needs.
15. Explain the printing procedure in drop-on-demand based 3D printing.
16. Explain the printing procedure in continuous ink jetting based 3D printing.
17. Explain the printing procedure in direct ink writing based 3D printing.
18. Outline the applications of liquid-based 3D printing.
19. Comprehend the advantages and limitations of liquid-based 3D printing.
20. Explain the common issues and probable troubleshooting methods in liquid-based 3D printing.
21. List emerging trends in SLA printing technology for medical device manufacturing.
22. List emerging trends in CLIP and DIW printing technology.
23. Write down the procedure of dental 3D SLA printing.

Discussion Topics

1. Explore the latest developments in resin formulations, including bio-compatible, high-temperature and engineering-grade resins, and their impact on achieving desired mechanical properties and surface finishes.
2. Examine best practices for designing parts optimized for SLA/DLP/MJ processes, including orientation, support structures, hollowing and lattice structures, to minimize build failures and maximize print accuracy.
3. Dive into the different types of resin materials used in SLA printing, their properties, and considerations for selecting the right resin for specific applications.
4. Explore the slicing process in SLA printing, including layer thickness, exposure settings and support generation, and how these parameters impact print quality and performance.

5. Identify common issues and challenges faced in SLA printing, such as failed prints, resin leaks and print distortions, and discuss strategies for diagnosing and resolving these problems.
6. Outline calibration and optimization of SLA printer settings for achieving high-quality prints (e.g., layer height, exposure time, laser power).
7. Comprehend environmental considerations and sustainability aspects of SLA printing, including material recycling and waste management.
8. How does employing SLA 3D printing contribute to advancing the development of medical devices, specifically in terms of enhancing printing speed, ensuring high quality and optimizing cost- effectiveness.
9. What are some of the most common issues faced during the liquid based 3D printing of components, and what are the best practices for troubleshooting and resolving these issues effectively to ensure high-quality prints.
10. What emerging trends do you foresee in the field of SLA 3D printing, and how might these advancements reshape the future of manufacturing, product development and various industries based on additive manufacturing technologies.

Multiple Choice Questions

1. What is the primary technique used in stereolithography 3D printing?
   (a) Material jetting
   (b) Vat polymerization
   (c) Binder jetting
   (d) Direct energy deposition
2. In the digital light processing 3D printing process, what is used to selectively cure liquid photopolymer resin layer by layer?
   (a) UV light
   (b) Infrared light
   (c) Laser beams
   (d) Visible light
3. Continuous liquid interface printing differs from traditional layer-by-layer methods primarily by:
   (a) Using different types of materials
   (b) Eliminating the need for layering
   (c) Utilizing a different light source
   (d) Slower print times

4. Liquid crystal display 3D printing differs from SLA primarily in:
   (a) Its use of a different light source
   (b) The method by which layers are cured
   (c) The materials it can use
   (d) Its print speed
5. What step immediately follows the design phase in the process of creating objects using 3D printing?
   (a) Slicing
   (b) 3D printing
   (c) Post-processing
   (d) Material selection
6. Material jetting is known for
   (a) Producing parts with the highest printing speed.
   (b) Offering exceptional resolution and the ability to deposit multiple materials simultaneously.
   (c) Being suitable for printing with a wide range of materials except photopolymers.
   (d) Utilizing inkjet-like mechanisms to cure liquid resin layer by layer.
7. Which 3D printing technique employs a nozzle to dispense droplets of material onto a substrate or previously deposited layers?
   (a) Continuous liquid interface printing
   (b) Drop-on-demand 3D printing
   (c) Continuous ink jetting 3D printing
   (d) Direct ink writing
8. Choose the correct sequence to generate prototype.
   (a) 3D CAD data - CAD solid model - STL file - RP prototype
   (b) CAD solid model - 3D CAD data - RP prototype - STL file
   (c) STL file - 3D CAD data - CAD solid model - RP prototype
   (d) 3D CAD data - STL file - CAD solid model - RP prototype
9. STL file format is represented by interaction of ______.
   (a) Lines and hexagons
   (b) Lines and rectangles
   (c) Lines and triangles
   (d) Lines and circles
10. Which file format cannot be converted into STL file format?
   (a). prt
   (b). sld
   (c). dwg
   (d). stp
11. Full form of STL is _____.
   (a) Standard tessellation language
   (b) Streto tessellation lithography
   (c) Stereo tessellation lithography
   (d) Straight tessellation language

12. Polymerization happens ______.
    (a) when the light of appropriate wavelength falls on liquid photopolymer
    (b) when the light of short wavelength falls on liquid photopolymer
    (c) when the light of appropriate wavelength falls on solid photopolymer
    (d) all the above mentioned
13. SLA printer's package material is in a______.
    (a) Chain
    (b) Spool
    (c) Cartridge
    (d) None of the above
14. Which of the following is the process of the pre-processing stage?
    (a) Remove support
    (b) Checking 3D CAD data
    (c) De-powdering loose material
    (d) Dip in binder to strengthen the part
15. Computer model of a part design on a CAD system is called as
    (a) Computer prototype
    (b) Geometric prototype
    (c) Solid prototype
    (d) Virtual prototype
16. Support structure are required for
    (a) Overhang portion
    (b) Thin portion
    (c) Thick portion
    (d) All of the above
17. STL file converts convert curve surface in to ______.
    (a) Polygons
    (b) Surface
    (c) Lines
    (d) All of the above
18. Software that stores lines and shapes rather than individual pixel is known as
    (a) Raster graphic software
    (b) Bitmap graphic software
    (c) Vector graphic software
    (d) Ripped image
19. What is a crucial step to ensure proper print quality before initiating printing?
    (a) Checking the resin level
    (b) Calibrating the build platform
    (c) Cleaning the build platform
    (d) Ambient temperature

20. SLA 3D printer prints object
(a) In pieces
(b) Whole part
(c) Layers
(d) Tetrahedral solids

21. Which should be considered when orienting the part on the build plate in the slicing software
(a) Holes should always be printed horizontally
(b) The footprint of part printed horizontally
(c) The footprint of the part should be small as possible
(d) Minimize the number of overhangs

22. Which software is commonly used for preparing 3D models for printing according to the provided passage?
(a) Solid Edge
(b) AutoCAD
(c) 3D Sprint
(d) CATIA

23. What is the primary consideration for material selection in SLA 3D printing according to the passage?
(a) Material availability
(b) Viscosity and flow characteristics
(c) Cost-effectiveness
(d) Availability and cost

24. What is the main purpose of shaking the resin bottle thoroughly before printing in SLA?
(a) To mix different types of resins
(b) To remove air bubbles
(c) To ensure proper curing
(d) To prevent resin degradation

25. What is a crucial step to ensure proper print quality before initiating printing?
(a) Checking the resin level
(b) Calibrating the build platform
(c) Cleaning the build platform
(d) Generating supports

26. What post-processing technique is commonly used to remove support structures in SLA printing?
(a) Chemical treatment
(b) Ultrasonic cleaning
(c) Sanding and polishing
(d) Manual removal

27. What is the purpose of curing in SLA post-processing?
(a) To remove excess resin
(b) To strengthen the parts

(c) To enhance surface finish
(d) To dissolve support structures

28. What important information does an AM G-code file contain?
(a) AM parts divided into layers and machine tool paths.
(b) Variable part tolerance.
(c) Support structures
(d) Computer-aided design

29. What post-processing technique is commonly used to remove support structures in SLA printing?
(a) Chemical treatment
(b) Ultrasonic cleaning
(c) Sanding and polishing
(d) Manual removal

30. What type of AM layer construction method involves creating an entire layer at once?
(a) 2D channel array
(b) Part orientation
(c) 4D channel array
(d) 1D channel array

31. What is the purpose of calibrating the build platform in SLA printing preparation?
(a) To align the model
(b) To level the platform
(c) To adjust print settings
(d) To generate support structures

32. What measurement tools are recommended for verifying dimensional accuracy of printed objects?
(a) Profile meters
(b) Calipers
(c) Coordinate measuring machines
(d) Both (b) and (c)

33. Which file format is commonly used for 3D printing?
(a). DWG
(b). OBJ
(c). DXF
(d). JPEG

34. What is the main consideration for material selection in SLA printing?
(a) Mechanical properties
(b) Cost-effectiveness
(c) Surface finish
(d) Thermal stability

35. Why is shaking the resin bottle necessary before printing in SLA?
    (a) To remove air bubbles
    (b) To mix different types of resins
    (c) To ensure proper curing
    (d) To prevent resin degradation
36. Which post-processing technique aims to strengthen the printed parts?
    (a) Support removal
    (b) Washing/cleaning
    (c) Curing
    (d) Sanding and polishing
37. What is the significance of visual inspection in inspecting printed objects?
    (a) To detect dimensional inaccuracies
    (b) To identify surface defects
    (c) To assess mechanical properties
    (d) To ensure complete curing
38. What is the primary objective of inspecting and testing printed objects?
    (a) To verify printing parameters
    (b) To assess material properties
    (c) To ensure functional requirements
    (d) To improve post-processing techniques
39. What is the recommended procedure for cleaning the build platform before printing?
    (a) Use a dry cloth
    (b) Rinse with water
    (c) Wipe with isopropyl alcohol
    (d) Apply a cleaning solution
40. Which additive manufacturing technology is commonly used for printing dental teeth?
    (a) Fused deposition modelling
    (b) Selective laser sintering
    (c) Stereolithography
    (d) Binder jetting

Answers

1. (b) Vat polymerization
2. (a) UV light
3. (b) Eliminating the need for layering
4. (a) Its use of a different light source
5. (a) Slicing
6. (b) Offering exceptional resolution and the ability to deposit multiple materials simultaneously

7. (b) Drop-on-demand 3D printing
8. (a) 3D CAD data - CAD solid model - STL file - RP prototype
9. (c) Lines and triangles
10. (c) .dwg
11. (a) Standard tessellation language
12. (a) When the light of appropriate wavelength falls on liquid photopolymer
13. (c) Cartridge
14. (b) Checking 3D CAD data
15. (d) Virtual prototype
16. (a) Overhang portion
17. (a) Polygons
18. (c) Vector graphic software
19. (b) Calibrating the build platform
20. (c) Layers
21. (d) Minimize the number of overhangs
22. (c) 3D Sprint
23. (b) Viscosity and flow characteristic
24. (b) To remove air bubbles
25. (b) Calibrating the build platform
26. (d) Manual removal
27. (b) To strengthen the parts
28. (a) AM parts divided into layers and machine tool paths
29. (d) Manual removal
30. (d) 1D channel array
31. (b) To level the platform
32. (d) Both (b) and (c)
33. (b) .OBJ
34. (d) Thermal stability
35. (a) to remove air bubbles
36. (c) Curing
37. (b) to identify surface defects
38. (c) To ensure functional requirements
39. (c) Wipe with isopropyl alcohol
40. (c) Stereolithography

8

Printing Using Powder-Based Raw Material

In the field of additive manufacturing, lasers or electron beams are utilized as energy sources to treat powdered materials and transform them into solid 3D objects. In powder-based procedures, an energy source such as a laser or electron beam is utilized to selectively melt and fuse powdered particles. There are two primary techniques: one involves fusing the powder in place, while the other involves injecting the powder through nozzles. These machines have the capability to utilize either powders or wires as feeding materials, with metals being the predominant choice. This chapter covers the different variants of powder-based 3D printing processes with their field of application, advancements and future trends. The chapter comprises of following main sections:

8.1 Overview of Powder-Based 3D Printing Processes
8.2 Powder Fusion Mechanisms
8.3 Powder Bed Fusion Processes: SLS, DMLS, EBM
8.4 Binder Jetting
8.5 Direct Energy Deposition Processes: DED, LENS
8.6 Applications of Powder-Based 3D Printing Processes
8.7 Advanced Concepts and Future Trends

After reading through the material, self-check exercises, objective questions and discussion topics presented in this chapter, the learner will be able to do the following:

(a) Explain the powder fusion mechanism used in powder-based 3D printing processes.
(b) Classify the variants of powder-based 3D printing processes.
(c) Explain the working of powder bed fusion (PBF) processes: SLS, DMLS, EBM, BJ.
(d) Describe laser powder fed (LPF)/direct energy deposition (DED) processes: LENS, DED

DOI: 10.1201/9781003562344-8

(e) Apply powder-based 3D printing processes.
(f) Outline the advanced concepts and future trends related to powder-based 3D printing processes.

8.1 Overview of Powder-Based 3D Printing Processes

In the field of additive manufacturing, lasers or electron beams are utilized as energy sources to treat powdered materials and transform them into solid 3D objects. In powder-based procedures, an energy source such as a laser or electron beam is utilized to selectively melt and fuse powdered particles. The powder is evenly distributed in a thin layer, measuring approximately 30–100 μm, throughout a designated construction area. There are two primary techniques: one involves fusing the powder in place, while the other involves injecting the powder through nozzles. As shown in Figure 8.1, the powder-based techniques can be categorized into two types namely, powder bed fusion (PBF) processes and directed energy deposition (DED) processes. The PBF technique employs either a moving blade or a counter rotating cylinder to distribute the powder uniformly. The energy source, be it lasers, electron beams or general electromagnetic radiation, concentrates on the surface, causing the powder particles to melt and fuse with neighbouring particles and layers.

In PBF, two primary forms of energy are utilized: laser, which is the most prevalent and widely employed, and electron beams. Low power lasers are appropriate for manufacturing polymeric components, whereas electron and laser beam methods are favoured for metal components. Practically any powdered substance can be employed in PBF as long as the suitable binder is used. Post-processing processes may be required for metal and ceramic parts

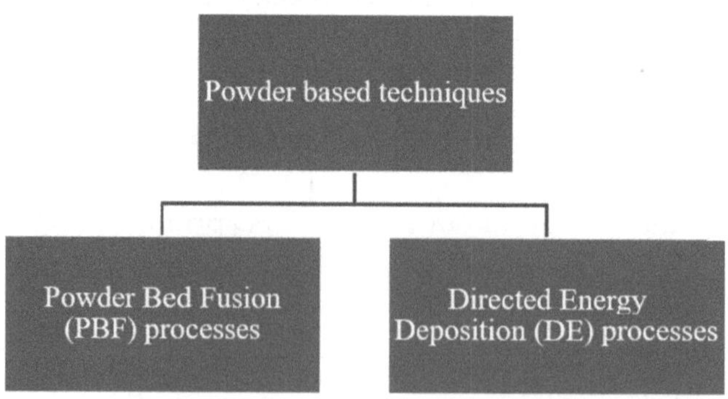

FIGURE 8.1
Classification of powder-based techniques.

to eliminate the binder, and subsequently, sintering can be done to enhance the density of the part.

The polymer feed stocks used in PBF consist of different types of semi-crystalline thermoplastics. In laser-based PBF, the material powder is subjected to intense heat, raising its temperature to a level that is very near its melting point. Cooling after building the parts requires the same amount of time to prevent distortion and oxidation.

The metallic feedstock used in PBF includes aluminium alloys, cobalt-chromium alloys, nickel alloys, gold, silver, stainless steel, tool steel and titanium alloys. Lasers and electron beams are commonly used as power sources. The particle sizes of powders used in PBF techniques generally vary depending on the specific technique being used. In laser PBF, the size of particles usually falls within the range of 20 to 40 µm, while in electron beam PBF, the particle size is slightly bigger, ranging from 45 to 100 µm. The variations in particle sizes can have an impact on other aspects of the printing process, such as the resolution, surface finish and mechanical qualities of the final printed items.

In DED machines, material is deposited through a nozzle, and a laser or electron beam is used to melt it layer by layer until the final layer is properly printed. These machines have the capability to utilize either powders or wires as feeding materials, with metals being the predominant choice. The use of DED is becoming increasingly popular for repair and feature enhancement since it allows for the creation of massive structures that are very close to the final shape, in a faster and more cost-effective manner compared to older approaches. The DED is attracting attention for its capacity to efficiently and cost-effectively produce enormous, nearly finished structures, surpassing the capabilities of conventional technologies.

Powder-based 3D printing techniques have revolutionized manufacturing by enabling the rapid and accurate creation of intricate, high-quality items. PBF, which combines multiple processes, is a significant form of these approaches. The SLS procedure is the foundation of PBF. It uses a powerful laser to selectively liquefy powdered substances, such as polymers or metals, in a sequential manner in order to construct three-dimensional entities.

Within the realm of metal manufacturing, there are several advanced techniques available, including direct metal laser sintering (DMLS), selective laser melting (SLM), electron-beam melting (EBM) and binder jetting. These techniques improve SLS and provide many advantages such as more material possibilities and stronger products. Because it is designed to sinter metal powders instead of polymers, DMLS makes it easier to fabricate complex metal parts that are needed in industries like aerospace, automotive and medicine. A closely comparable method to DMLS is SLM, wherein the metal powder is melted completely by the laser rather than sintering it. This produces totally dense metal components with superior mechanical qualities and material densities. Maintaining precise control over laser parameters is essential to preventing flaws in parts generated by SLM.

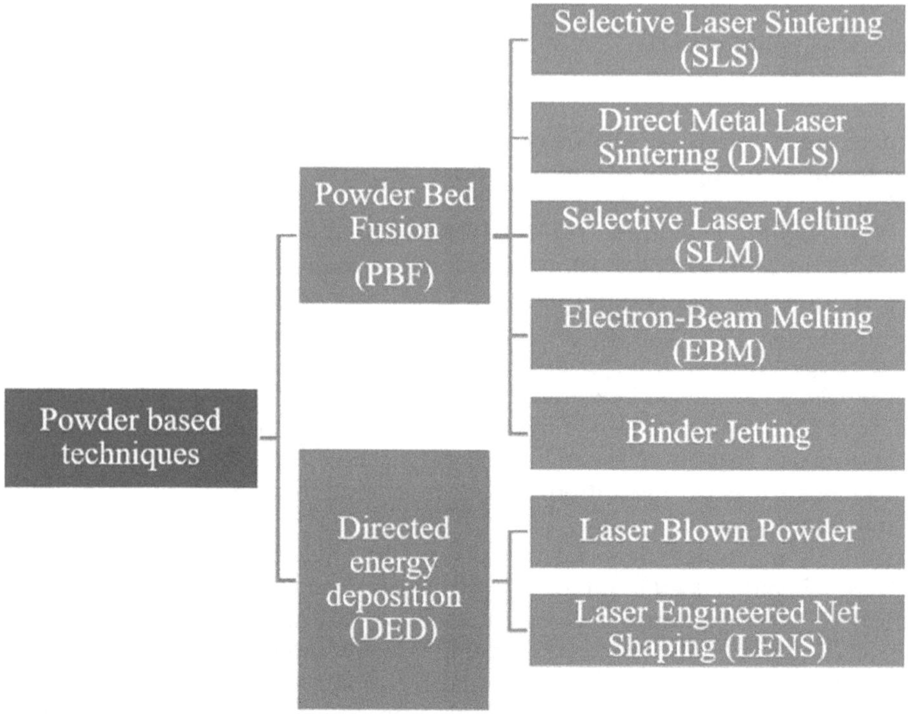

FIGURE 8.2
Powder-based 3D printing techniques.

Metal powder is melted onto a surface using a laser in laser powder-fed (LPF) or directed energy processes. This enables precise metalworking, coating and repair. This is further enhanced by laser engineered net shaping (LENS), which creates entirely solid metal components with intricate geometries.

These powder-based techniques demonstrate the vast array of manufacturing options and are facilitating the advancement in sectors such as aerospace, automotive and healthcare. Figure 8.2 shows various powder-based 3D printing techniques.

8.2 Powder Fusion Mechanisms

We will examine material fusion processes used in powder-based 3D printing in this part. The evolution of different procedures is influenced by differences in energy sources, powder delivery and layer preparation mechanisms, and powder fusion mechanisms. Electron beam melting (EBM), SLS, DMLS, SLM and 3D laser cladding are a few examples.

In these procedures, a powder bed is created by dispersing powder throughout the work area; the average thickness of the powder layers is approximately 100 μm. The thin powder coating is created using a recoater mechanism, which is often a blade or counter-rotating cylinder. In order to obtain the desired shape, the laser or electron beam is subsequently focused over the powder layer with the use of scanning galvanometer mirrors as per the coordinates provided by the .gcode file of the digital object to be printed in layer by layer fashion. The powder is melted or sintered at the points where the laser or electron beam intersects the powdered bed surface due to the application of energy. Particles of powder adhere to both the previously deposited layer and the present layer as a result of this process. Then, excess powder is strewn all over the work area, and the process is repeated in order to create the next layer and finally the solid 3D part.

In most powder-based techniques, each layer of a part undergoes melting in two steps:

1. Contouring: the laser irradiates the outer boundary of the part.
2. Melting or sintering: the laser irradiates the powder within the contour/perimeter.

The main design guideline for this system is to make sure that any pieces with internal features have an exit point where powder can be removed. This makes sure that extra powder is properly evacuated from interior cavities or channels during the post-processing phase, which makes it easier to produce high-quality and functioning components.

Sintering of materials*: sintering is a manufacturing process that involves the heating of powdered materials to the point where their particles adhere and bond together, forming a coherent mass. This process is commonly used in the production of ceramics, metals, and other materials to create finished products with specific properties.*

Contouring*: in the context of 3D printing, "contouring" typically refers to the process of defining the outer boundaries or external shape of a printed object.*

Among the many benefits this system promises is the capacity to create internal pathways and high-resolution features while maintaining accurate dimensional control. In contrast to many other additive manufacturing (AM) techniques, this one allows for the fabrication of components without the need for supports because the powder bed serves as the part's support

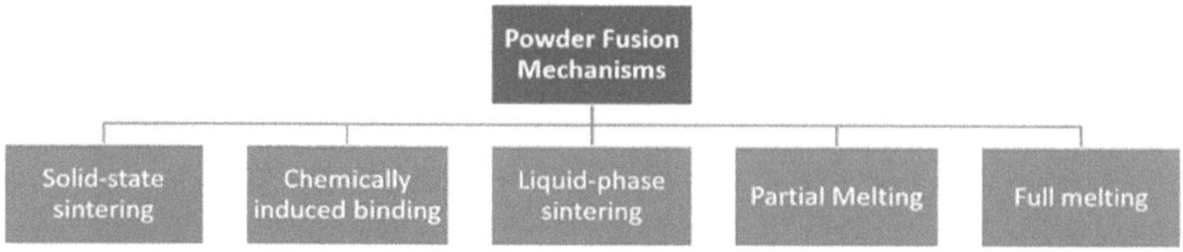

FIGURE 8.3
Powder fusion mechanisms.

structure. Since support removal is not required, the removal of support structures results in material and post-processing cost reductions. In addition to being cost-effective, the lack of support structures promotes geometric design freedom. Supportless manufacturing also makes it simple to nest parts within a single build, which increases part-packing density and speeds up the build process. Polymer PBF, with its faster build speed and higher build rate from the removal of supports, is more cost-competitive than classic "low-cost" AM methods like material extrusion, especially for larger production runs.

The PBF processes encompass five distinct fusion mechanisms: solid-state sintering, chemically induced binding, liquid-phase sintering, partial melting and full melting. Figure 8.3 shows the mechanisms used to fuse the powder in various powder-based 3D printing processes.

The predominant mechanisms employed in 3D printing processes available in market are liquid-phase sintering and full melting. Following sections deal with each mechanism in detail and its relevance to the AM.

8.2.1 Solid-State Sintering

This procedure existed prior to the AM's evolution. To fuse powder particles below their melting point, solid-state sintering makes use of thermal energy. Thus, the development of a liquid is prevented. Typically, metal particles are joined through sintering without melting; this allows them to maintain their "solid state" at elevated temperatures. During this process, the sintering temperature usually ranges from half of the material's absolute melting point to its actual melting point.

The goal of the solid-state sintering process is to reduce the powder particles' total free surface energy (E_s). Both the particle's surface area (S_a) and the surface energy per unit area (γ_s)for a given material, environment, and temperature have an impact on the total free surface energy (E_s).The total free surface energy (E_s) is shown by the equation.

$$E_s = S_a \times \gamma_s$$

When particles undergo fusion at high temperatures, as depicted in Figure 8.4, the overall surface area reduces, resulting in a decrease in surface energy.

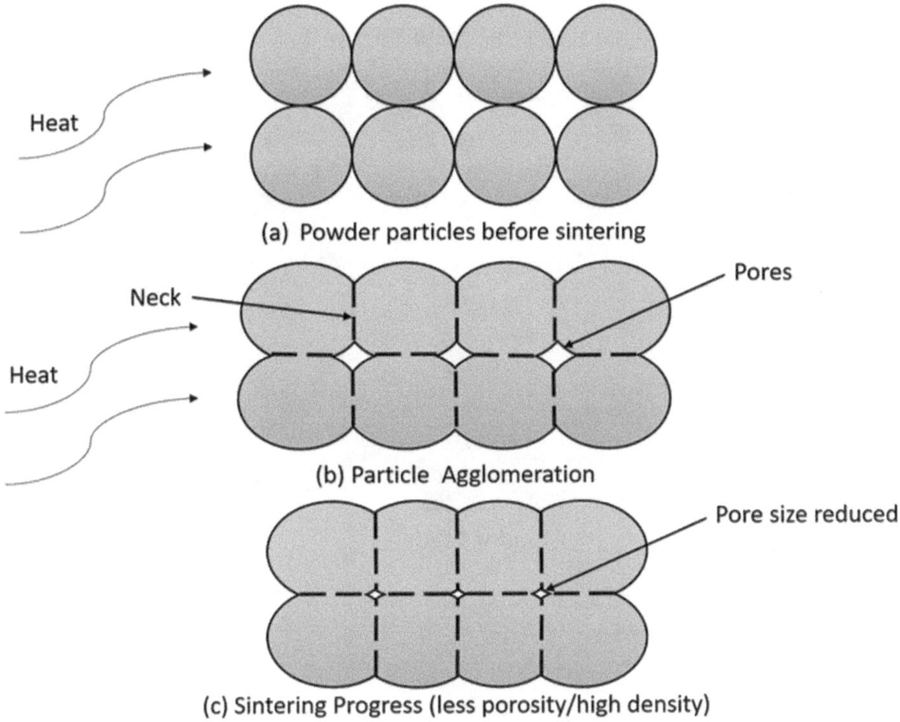

FIGURE 8.4
Solid-state sintering.

The primary process involved is the diffusion of powder particles across a solid-state medium. At first, the particles are densely arranged as depicted in Figure 8.4(a). Diffusion facilitates the creation of a neck at the border between the particles of the powder, as shown in Figure 8.4(b). Through additional diffusion, the neck elongates in order to decrease the porosity, leading to an elevation in density (refer to Figure 8.4(c)).

Consequently, when the size of particles reduces, the surface area increases, resulting in a more powerful stimulus for necking and consolidation processes. Therefore, smaller particles sinter more rapidly and at lower temperatures compared to bigger particles. The rate of particle diffusion is exponentially dependent on temperature, leading to an acceleration of the sintering process as the temperature approaches the material's melting point.

8.2.2 Chemically Induced Sintering

This approach is comparatively infrequently utilized in additive manufacturing operations. Its purpose was to create ceramic matrix composites. Chemical sintering occurs when powder particles react with gases that are

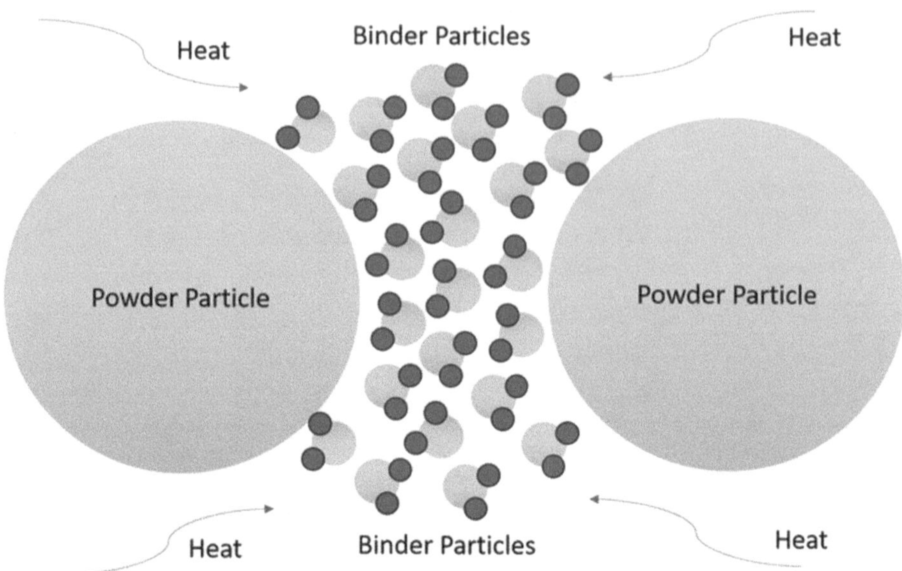

FIGURE 8.5
Schematic of chemical sintering.

triggered by external heating. The chemical reaction produces a byproduct that aids in the cohesion of the powder particles. This is mainly used in additive manufacturing of ceramics materials. Figure 8.5 depicts the chemical sintering process.

A few examples of this are as follows:

- When SiC particles are subjected to high temperatures, the SiC material partially breaks down, resulting in the separation of Si and C. As a result, the unbound silicon combines with oxygen in the atmosphere, forming silicon dioxide (SiO_2). The SiO_2 functions as an adhesive for the SiC particles.
- Another use of chemical sintering involves the creation of chemically bound Si_3N_4. The silicon powder is subjected to temperatures ranging from 1200°C to 1250°C while being exposed to a mixture of nitrogen/hydrogen or nitrogen/helium. Crucially, heat is utilized at a temperature lower than the melting point of silicon to prevent excessive energy usage. The reaction produces Si_3N_4 as a byproduct, which functions as a binder.
- Similarly, laser processing of ZrB_2 with oxygen leads to ZrO_2 formation, binding ZrB_2 and ZrO_2.
- Moreover, when aluminium is subjected to laser processing in the presence of nitrogen (N_2), it undergoes a reaction that results in the

creation of aluminium nitride (AlN) and the joining together of aluminium (Al) and aluminium nitride (AlN) particles.

An inherent limitation of chemical sintering is the presence of increased porosity in the end product. Therefore, it is necessary to use additional post-processing techniques, such as infiltration or high-temperature furnace sintering, in order to get the desired characteristics for various applications. Post-processing infiltration may involve the addition of additional reactive components, resulting in the creation of new compounds after the infiltration process. Nevertheless, the expenses and duration linked to these supplementary procedures have constrained the extensive use of chemically induced sintering in commercially accessible powder-based 3D printers.

8.2.3 Liquid-Phase Sintering

Liquid-phase sintering (LPS) is widely regarded as one of the most adaptable mechanisms in PBF. In the powder processing industry, this term is frequently employed to characterize the process by which powder particles fuse. The classification of the liquid phase sintering processes is illustrated in Figure 8.6. This category encompasses numerous procedures.

The binder material melts and turns into a liquid in most LPS methods, but the structural material stays solid. The solid pieces are held together by the molten parts, which work like glue. Sometimes it is hard to tell the difference between the binder material and the structure grains.

8.2.3.1 Separate Grains

These processes utilize distinct binder and structural grains, as implied by their names. The majority of binders utilized in this context are metallic, while the structural components might consist of both metals and ceramics. The size of the binder material particles is comparatively less than that of the structural material particles. Consequently, this leads to a significant ratio of surface area to volume for the grains. When subjected to heat, binder grains

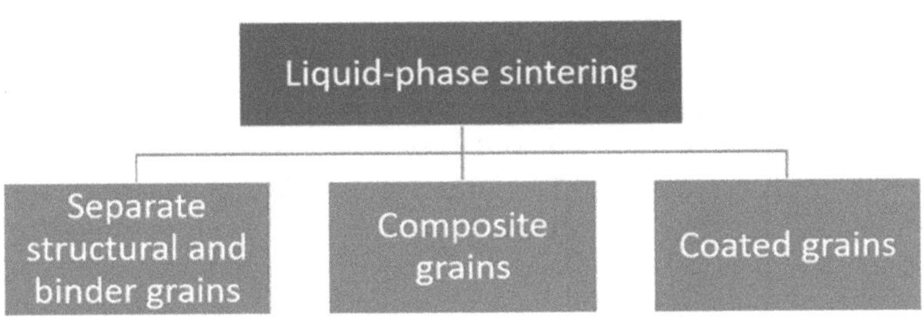

FIGURE 8.6
Classification of liquid phase sintering.

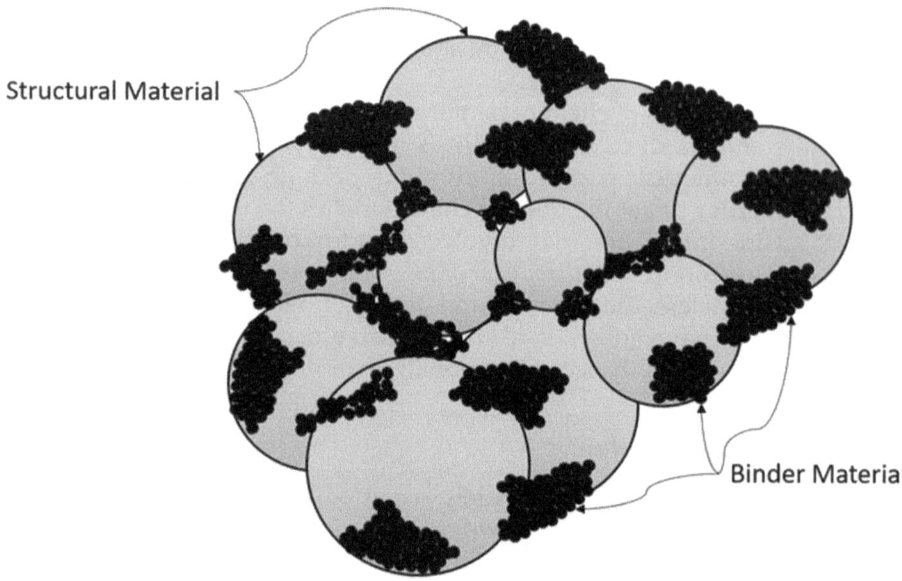

FIGURE 8.7
Illustrates of separate particles based LPS used in additive manufacturing (AM) processes.

are prone to melting due to their lower melting point compared to structural grains. Figure 8.7 displays the diagram of liquid phase sintering that involves individual particles.

Smaller binder grains have the benefit of enhancing packing efficiency. The capillary force acts as a driving engine, enabling the liquid binder material to spread into tiny pores and encouraging the reorganization of particles. Nevertheless, despite these advantages, the resulting "green" component usually maintains a notable degree of porosity and brittleness. Consequently, these components receive further treatments such as furnace post-sintering, hot isostatic pressing (HIP) or infiltration with a low melting point substance to improve their characteristics.

8.2.3.2 Composite Grains

Each particle of powder in this case consists of both structural and binder material (Figure 8.8). These composite particles are obtained by mechanically processing a mixture of, two different powders. This process guarantees that the powder particles undergo fracturing and welding as a result of recurrent milling. This process produces a powder containing micro-grain composite particles. It is easy to distinguish between the binder and structural material phases in these composite grain particles. These particles contribute to a higher SLS green density than a mixture of distinct binder and structural material powders.

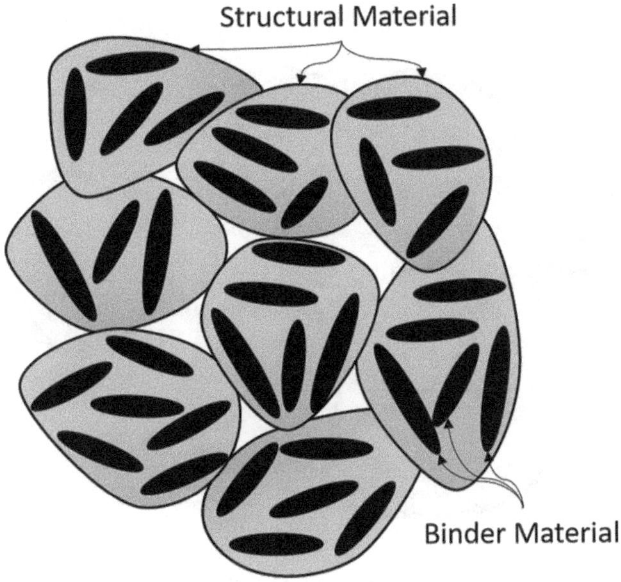

FIGURE 8.8
Composite particles: this involves particles that are a mix of molten and solid components.

Alongside metals, composite powder granules are also present in polymer-based powders, including glass-filled polymer powder. This highlights the adaptability of composite grains in improving the qualities of manufactured parts.

> ***Green part**: in various manufacturing processes, especially in powder metallurgy or additive manufacturing like 3D printing, the initial formed object is often referred to as the "green part" before undergoing further processing steps such as sintering or post-treatments to achieve its final characteristics. The green part is not yet fully densified or solidified according to the intended properties and needs additional steps for completion.*
>
> ***Hot isostatic pressing**: hot isostatic pressing (HIP) is a manufacturing process used to improve the density and properties of materials, particularly metal and ceramic components.*

8.2.3.3 Coated Grains

Coated grains refer to the process of adding a layer of binder onto the structural material. The aim is to achieve optimal heat absorption by the binder components throughout the processing phase. When binder and structural materials are separate, the structural material has a higher likelihood

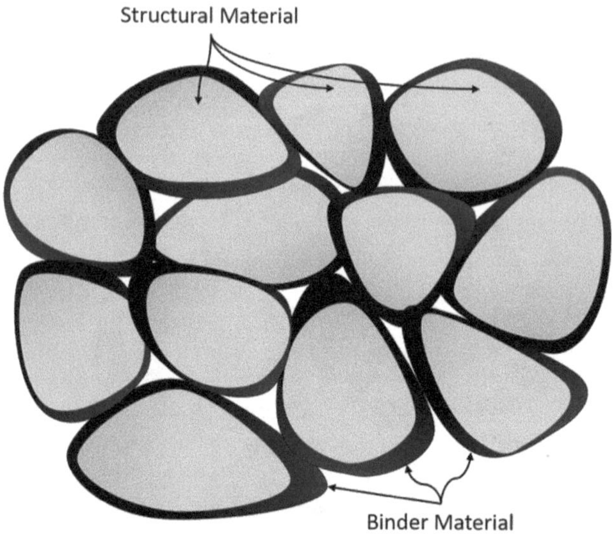

FIGURE 8.9
Coated particles: structural particles (yellow) coated with binder particles in dark (black).

of absorbing more heat and melting before the binder material. Applying a layer of binder onto the structural material also guarantees the efficient adhesion of the structural particles. Figure 8.9 shows the coated powder particles.

There are two types of coated powder available: polymer binder-coated powder and metal binder-coated powder. The market offers polymer-coated steel powders known as "direct steel" and "laser form." In order to enhance the infiltration of components made with polymer-coated binders, an extra de-binding process is required. It is crucial to eliminate the polymer coating in order to remove it completely and create a suitable surface for future infiltration procedures.

Applying a coating is beneficial when dealing with materials that are particularly susceptible to oxidation, such as aluminium particles. Due to the presence of an oxidation layer on the Al particle, direct sintering of Al powder is hindered. To prevent oxidation of the particles, the LPS method can be employed to fuse polymer-coated Al particles.

Essentially, the LPS is a technique that employs heat to liquefy a material and use it as an adhesive to bond solid material particles. It possesses versatility and can be employed in various manners.

8.2.4 Partial Melting

In some techniques, it is impossible to distinguish between the binder particles and structural particles. However, there is a distinction between melted and non-melted material areas. Thus, these techniques are referred as "partial melting" techniques.

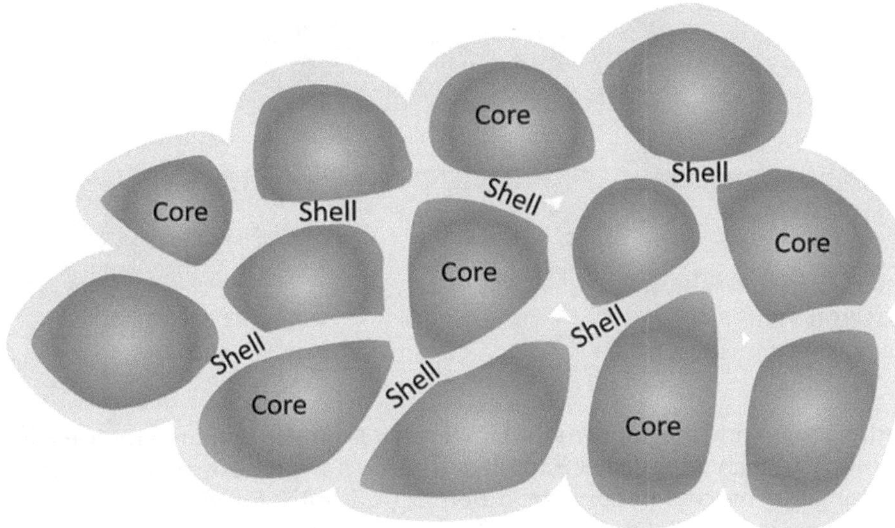

FIGURE 8.10
Schematic of the powder particles with melted shell and unaffected core.

Within partially molten materials, only the external layer of the powder particles undergoes melting and functions as an adhesive to unite the unmelted core powder particles. The central nucleus of the particle remains unaltered. This is accomplished by supplying a level of heat that is inadequate to completely liquefy the entire particle. This method is effective for bonding both metal and polymer particles. Figure 8.10 shows the schematic of the process.

In order to do this, the thermal model is utilized to ascertain the temperature of both the shell and inner core of the powder grains. Additionally, the amount of energy needed to fully transform the powder particle into a liquid state is calculated. The partial melting is accomplished by maintaining the heat energy below this threshold.

8.2.5 Full Melting

A recurring limitation found in all the previously mentioned approaches is the permeability of the components. The components fabricated using these techniques exhibit a lower density in comparison to bulk materials. Once again, in order to obtain denser components, many post-processing operations are required. In order to overcome this constraint, the technique of SLM was devised to create components that have the same density as bulk materials.

The SLM process utilizes the full melting technique, in which the laser is employed to completely liquefy the powder particles. It is compatible with

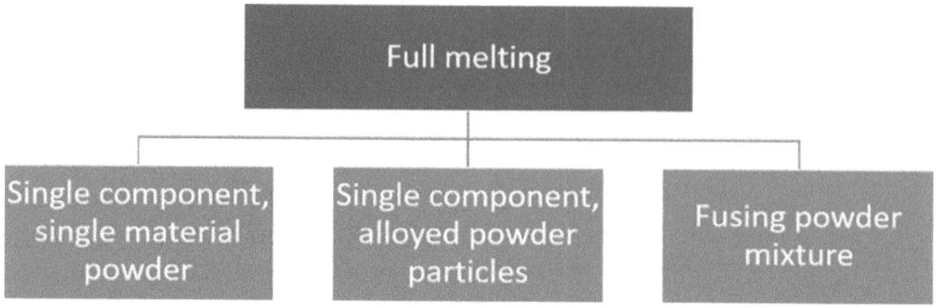

FIGURE 8.11
Categories of full melting.

both polymer and metal powders. However, it is mostly favoured for use with metals. This technology can be applied to process any type of metal. However, the range of process parameters differs significantly for each material due to changes in material properties. The process parameter window could be determined experimentally for each material to prevent scan track instabilities, such as the sphereodization of the liquid melt pool, commonly referred to as "balling," and to minimize part porosity. As a result of the high cost involved, there is currently a restricted availability of materials in the market. The full melting methodology can be categorized into three categories as shown in Figure 8.11.

Pure metal powders are utilized in single-component, single-material powder. It is feasible to produce the pieces with a density of 100 percent. The second technique involves the utilization of powdered alloy materials, such as steel alloys (SS316L, AISI H11 tool steel) and titanium alloys (TiAl6V4). The third process involves the fusion of a heterogeneous mixture of powdered particles composed of various elements. Each particle of every material is expected to completely melt here.

8.3 Powder Bed Fusion Processes: SLS, DMLS, EBM

8.3.1 Selective Laser Sintering

SLS is the predominant method of 3D printing that utilizes powder bed fusion. The invention was credited to Carl Deckard, an American inventor and educator, who developed it during the mid-1980s at the University of Texas in Austin. The SLS method employs a high-intensity carbon dioxide laser to merge powder particles. The SLS utilizes lasers to selectively fuse specific regions inside a powdery bed, layer by layer, resulting in the creation of superior, solid models. The technique involves a repetitive sequence of steps, which includes depositing powder and solidifying a specific area using a laser, followed by depositing the next layer of powder. This operation is continued until the deposition is completed, followed by the sintering of the final layer. The SLS technique is versatile and may be used with a wide range of materials, including polymers, metals and ceramics. This is the first commercially available metal 3D printing technique. The following sections deal with the detailed process and mechanism, followed by process parameters, advantages, limitations and application areas of the SLS technique.

8.3.1.1 Working Mechanism of SLS

In the SLS process, seen in Figure 8.12, a carbon dioxide (CO_2) laser is employed as the primary source of energy. The laser beam fuses the thin layer (with a thickness ranging from 0.075 to 0.1 mm) of powder particles that are distributed throughout the working surface in the construction chamber (see Figure 8.12). The recoater mechanism is utilized to evenly distribute a fine

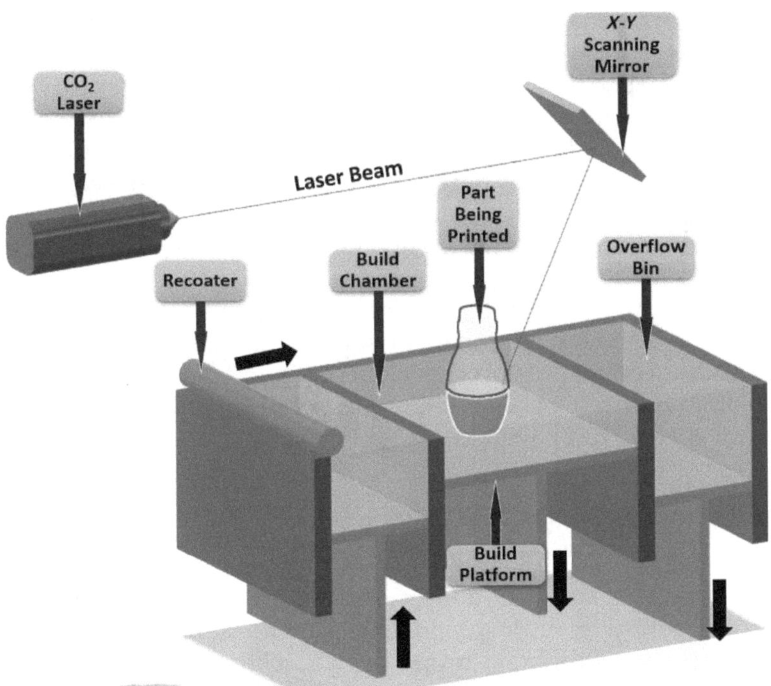

FIGURE 8.12
Schematic of selective laser sintering (SLS).

layer of powder over the designated working area. The mechanism comprises a roller that oscillates in a reciprocating motion to apply the powder layer in the designated area. Typically, in order to prevent contamination from the external environment and the oxidation of powder particles, inert gases like as argon or nitrogen are used to fill all the chambers.

The laser selectively traces the cross section of a thin layer of the powder bed, based on a specific slice obtained from a CAD model (.gcode file). An *XY* mirror system is utilized to manipulate the laser beam, allowing for precise scanning of specific locations and subsequent fusion of each layer. Following the fusion of a layer, the powder bed descends to create room for the recoater mechanism to deposit a fresh coating of powder. The CAD model is replicated by bonding each subsequent layer together, resulting in the formation of a shape. This process is iterated until the final product is fully printed.

It is crucial to emphasize that throughout the entire process, the building chamber is kept at a higher temperature, which is precisely regulated to stay below the threshold at which the polymer powder becomes soft. Increasing the temperature of the chamber accelerates the process and minimizes the thermal stress and distortions that may arise during the solidification of each layer. The temperature management plays a crucial role in attaining enhanced efficiency and stability in the SLS technology. Infrared heaters are utilized to heat the specific region where the component is being shaped. In addition, heaters are positioned above the powder chambers to raise the temperature of the powder prior to its distribution within the designated work zone. Occasionally, resistive heaters are also employed surrounding the powder bed. Preheating and keeping a consistent temperature are essential for minimizing the laser power required to melt the powder and preventing the item from warping during the process.

The unsintered powder in the working area serves as a support material, hence eliminating the need for extra support. Additionally, this approach eliminates the need for post-processing in order to remove support structures from the final printed component. Once the printing process is complete, the unsintered powder can be reused.

8.3.1.2 Process Parameters of SLS

One big benefit of the SLS system is that it can be changed to fit different needs by changing the laser and process settings. Being flexible lets you have precise control over the manufacturing process and makes it possible to make things with different properties and characteristics based on specific needs.

The process parameters for SLS can be categorized into four domains. Figure 8.13 shows the classification of the SLS process parameters.

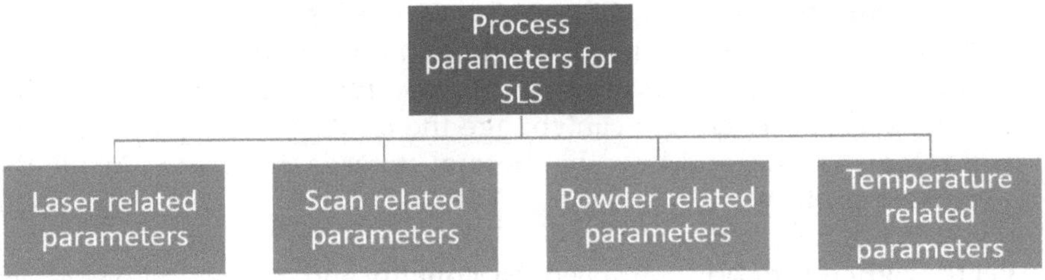

FIGURE 8.13
Classification of the SLS process parameters.

Laser-related process parameters: laser-related process parameters refer to factors that affect how the laser interacts with the material during the SLS process. These parameters include the following:

- **Laser power:** this parameter controls how strong the laser beam is, which in turn controls how much energy is sent to the powder to melt and fuse it together.
- **Spot size:** this measure tells you how big the laser beam spot is on the powder bed's surface. The printed part may have finer details and a better resolution if the spot sizes are smaller.
- **Pulse duration:** the amount of energy sent to the powder depends on how long each laser pulse lasts. The melting is more precise when the pulse lasts less time, and less heat is transferred to other places.
- **Pulse frequency:** the number of light pulses sent out in a given amount of time. Higher pulse rates can make the printing process go faster without lowering the quality.

By changing these settings, you can fine-tune the SLS process and get printed parts with better surface finish, accuracy in measurements and mechanical qualities.

Scan-related process parameters:scan-related parameters in SLS determine how the laser scanning occurs to fuse the material. These parameters include the following:

- **Scan speed:** this setting tells you how fast the laser moves across the powder bed. While faster scan speeds can speed up printing generally, they may lower the quality of the part that is printed.
- **Scanspacing:** the space between two scan lines that are next to each other. When the spaces between scans are closer together, they meet more, which makes the fusion between layers better but may also make the printing take longer.

- **Scan pattern:** the scan pattern shows the laser's path as it moves across the powder bed. Raster (back and forth) and vector (continuous curve) patterns are two common types of scan patterns. The scan design you choose can change the quality of the part, the time it takes to make, and how the thermal stress is distributed inside the printed part.

To get the right mix of printing speed, part quality and mechanical properties in SLS-printed parts, these scan-related parameters must be optimized.

Powder-related process parameters:powder-related process parameters in the SLS pertain to the specific attributes of the powdered material employed. The parameters encompass the following:

- **Particle shape:** the physical structure of the powder particles, which might impact their arrangement and fusion when undergoing the sintering process. Particle forms commonly fall into three categories: spherical, irregular and angular.
- **Particle size and distribution:** refers to the dimensions of the individual powder particles and how they are spread out inside the powder bed. Reducing the size of particles typically results in more intricate details and a more polished surface texture in the printed object.
- **Powder bed density:** this term describes the density of the powder particles when they are tightly packed together within the powder bed. Attaining a consistent density of the powder bed is crucial in order to ensure uniform sintering and precise dimensions in the printed component.
- **Layer thickness:** this parameter indicates the thickness of each layer of powder that is deposited during the printing process. Reducing the thickness of the layers can lead to improved resolution and more intricate features, but it may also result in longer printing durations.
- **Material properties:** the material properties encompass the mechanical and chemical characteristics of the powder material, such as its melting point, thermal conductivity and mechanical strength, which have a vital impact. The qualities of the material have an impact on its behaviour during the sintering process and eventually determine the final properties of the printed part.

It is essential to optimize certain parameters linked to powder in order to achieve superior and uniform outcomes in SLS printing. Certain parameters have a direct impact on the sintering process and the characteristics of the printed components.

Temperature-related process parameters: the term "temperature-related SLS process" pertains to the factors that govern the temperature conditions during the printing process. The parameters encompass the following:

- **Bed temperature:** this parameter represents the temperature of the powder bed that is maintained during the sintering process. By keeping the powder bed at a temperature just below the melting point of the powder material, it promotes the fusing of particles, hence improving the overall quality of the part.
- **Supply chamber temperature:** the temperature within the chamber housing the powder supply cartridges. Applying heat to the powder prior to spreading it across the build area enhances its capacity to flow and guarantees consistent melting while sintering.
- **Temperature homogeneity:** it refers to the evenness of temperature both within the powder bed and inside the supply chamber. Ensuring a steady and even distribution of temperature throughout the entire build volume is crucial for achieving uniform sintering. This helps to minimize the occurrence of flaws, such as warping or cracking, in the printed parts.

It is crucial to optimize these temperature-related factors in order to get dependable and superior outcomes in SLS printing, as they have a direct impact on the sintering process and the characteristics of the printed components.

8.3.1.3 Materials Processed by SLS

An essential feature of SLS is its ability to handle a broad range of materials, meeting various requirements in different industries. The materials encompass a wide range of categories, such as metals, ceramics, composites and polymers, each possessing distinct qualities and applications.

Metals: metals play a crucial role in various sectors because of their robustness, longevity and ability to conduct electricity. During the process of SLS, metals are subjected to a rigorous procedure in order to attain accurate and complex designs. Several frequently utilized metals in SLS including the following:

- **Aluminium**: aluminium is a highly demanded material due to its lightweight properties and ability to resist corrosion. It is extensively used in industries, such as aircraft, automotive and consumer electronics.
- **Titanium**: titanium is renowned for its exceptional strength-to-weight ratio and biocompatibility, which makes it very desirable for a wide range of applications. It is frequently employed in the aerospace industry,

for medical implants, and in the production of high-performance sports equipment.

- **Copper**: copper is widely utilized in electrical and electronic components, as well as in heat exchangers and plumbing systems, due to its exceptional electrical conductivity.
- **Nickel alloys**: nickel alloys are used in gas turbines, chemical processing, and aerospace components due to their high-temperature resistance and superior mechanical qualities.
- **Cobalt chrome**: cobalt chrome is frequently used in medical implants, dental prosthesis and aerospace components due to its excellent wear resistance and biocompatibility.

Ceramics: ceramics provide exceptional hardness, heat resistance and chemical stability, rendering them well-suited for applications that require superior performance in challenging circumstances. Ceramics in SLS undergo processing to produce complicated and long-lasting components. Important ceramic materials consist of:

- **Silicon carbide (SiC)**: silicon carbide is widely recognized for its remarkable hardness and heat conductivity, making it suitable for several high-temperature applications, including ceramic brake discs, armour plating, and semiconductor components.
- **Aluminum oxide (Al_2O_3)**: aluminium oxide is widely known for its exceptional strength, hardness and ability to resist chemicals. It is commonly used in various industries for cutting tools, coatings that resist wear and materials that can withstand high temperatures.
- **Zirconia (ZrO_2)**: zirconia is widely used in dental ceramics, structural ceramics and medical implants due to its notable mechanical strength, toughness and biocompatibility.

Composites: composites possess a distinctive amalgamation of characteristics obtained from their component materials, rendering them adaptable solutions for several applications. Composite materials in SLS undergo processing to attain components that are both lightweight and possess high strength. Some examples of notable composite materials are as follows:

- **Nylon-based composites:** these composites combine nylon with reinforcing materials like carbon fibre or glass fibre, resulting in improved strength, rigidity and resistance to impact. These materials are frequently utilized in the manufacturing of automobile parts, sporting items and aerospace components.

Polymers: polymers are extensively used in SLS due to their versatility and ease of processing. They are employed to create complicated and functional parts in several sectors. Polymers can be categorized into two primary classifications according to their structure:

- **Amorphous polymers:** these polymers exhibit an amorphous structure, providing transparency, resistance to impact, and favourable thermal stability. Polycarbonate (PC), acrylonitrile styrene acrylate (ASA), and polystyrene (PS) are often utilized amorphous polymers in SLS.
- **Semi-crystalline polymers:** these polymers have both amorphous and crystalline areas, which provide them a combination of mechanical strength, chemical resistance and dimensional stability. The significant semi-crystalline polymers used in SLS are polyetheretherketone (PEEK), polyetherketone (PEK), polyaryletherketone (PAEK), nylons/polyamides (PA), polycaprolactone (PCL) and thermoplastic elastomers (TPEs).

The wide range of materials that can be used in SLS allows designers, engineers and manufacturers to create unique and high-performance components in many industries. Each material group, including metals, ceramics, composites and polymers, possesses unique benefits that cater to a diverse array of application requirements and performance standards.

8.3.1.4 Advantages and Disadvantages of the SLS

Advantages: the SLS offers numerous advantages across various industries. Here are some key advantages of SLS:

- **No supports needed**: SLS printing is a process where a 3D object is built by adding layers of powdered material in a progressive manner. This eliminates the need for additional supports to maintain overhangs or complex areas of the model. The neighbouring powder acts as a form of strengthening, enabling the printing of complex designs without the requirement of later removing support structures.
- **Detailed printing**: SLS printers utilize laser technology to fuse nylon powder, resulting in prints of remarkable accuracy. This approach can generate layer heights that range from 50 to 200 microns. Support-free printing enables the easy creation of complex and irregular shapes, guaranteeing consistent and excellent results.
- **Batch printing**: several smaller models can be printed at the same time with SLS printers because of their relatively large printing capacity. Because the laser tightly bonds nylon particles together across a wide surface, ensuring strong bonding between layers, and because supports are unnecessary, this is conceivable.

- **Fast printing**: although not the most rapid in the field, SLS printers nonetheless provide satisfactory printing speeds ranging from 20 to 48 mm per hour. The utilization of a high-power laser and preheating of the bed accelerates the printing process, facilitating the efficient production of parts within a short timeframe.
- **Robust prints**: SLS prints are renowned for their robustness and resilience, making them highly suitable for experimentation and prototyping purposes. Utilizing nylon powder produces prints that possess mechanical attributes at the level of the industry, including notable tensile strength and elastic modulus. This guarantees that the printed components can endure rigorous testing and function consistently in diverse applications.

In summary, SLS printing has numerous benefits such as support-free printing, intricate printing capabilities, the flexibility to print in batches, rapid printing rates and strong print quality. These advantages make it a versatile and dependable choice for creating top-notch 3D prints in various industries.

Disadvantages: while SLS offers numerous advantages, like any manufacturing process, it also has its drawbacks. Here are some disadvantages of SLS:

- **Limitation in raw-material:** at now, SLS printers offer a restricted range of raw materials, with only a handful of commonly used powders, like PA-12, PA-11, and PEEK, being available. Out of these options, PA-12 is the most frequently utilized choice. Furthermore, the range of colour options is restricted, usually limited to a metallic shade of grey, which hampers the potential for creative design.
- **Learning curve is different:** adapting from other 3D printing technologies such as FDM or SLA to SLS requires a period of adjustment due to variations in print orientation, support structure utilization, and post-processing techniques. The cleaning and finishing procedures for SLS prints vary considerably compared to those used for other types of printers, necessitating supplementary training and calibration.
- **No recycling option:** currently, there is no documented recycling solution available for SLS prints, unlike certain other 3D printing technologies. Although the quantity of trash produced during SLS printing is modest, there are continuous efforts to recycle materials. Nevertheless, combining unsintered powder with fresh powder for recycling can undermine the quality of the print and is not advisable.
- **Health factors:** SLS printers employ nylon powder as its raw material, which consists of small particles that can provide health hazards if breathed in. It is advisable to use appropriate protective equipment, such as gloves and an air-filter mask, when using SLS printers in order to reduce the risk of inhaling airborne particles. Specialized

workstations, such as Sinterit's power handling station, have the ability to significantly improve safety and efficiency during the printing and post-processing stages.

- **Cost of SLS printing:** the upfront cost of SLS printers, particularly benchtop versions, can be substantial, ranging from $15,000 to $18,000. Although the cost per print may be more cost-effective in comparison to SLA printers, it is important to consider the price of raw ingredients, such as nylon powder. Furthermore, there may be additional costs associated with post-processing techniques, such as sandblasting and colouring.
- **Post-processing methods:** because SLS prints usually have rough surfaces and solid colours, they need to be processed after they are made to make them look good for presentations. Some methods, like sandblasting and colouring, can make something look better. Other methods may be needed for specific useful or decorative reasons, which can make the production process more complicated and cost more.

Even with these problems, selective laser sintering is still a useful additive manufacturing method because it has special benefits like being flexible, fast and letting you design anything you want. By keeping up with these problems through research and development, SLS can keep finding new uses and getting around problems in many fields.

8.3.2 Direct Metal Laser Sintering and Selective Laser Melting

The working principle of the DMLS and SLM is like the SLS. DMLS is specifically designed for sintering metal powders, not polymers. It enables the production of complex metal components that are well-suited for the aerospace, automotive, and medical industries, where strong metal parts are necessary.

SLM is a method similar to DMLS, where the metal powder is totally melted by the laser instead of being partially fused together. This leads to the creation of metal parts that are fully solid with excellent material density and mechanical properties. Accurate manipulation of laser parameters is crucial in order to prevent flaws in parts generated using SLM.

8.3.3 Electron Beam Melting

EBM is a type of 3D printing that is identical to SLM. Both methods involve the fusion of metal powder to create solid 3D objects. Even so, they vary in the technique employed to liquefy the powder. EBM, or electron beam melting, utilizes an electron beam rather than a laser to induce the melting of powder particles.

EBM, which was initially created at Chalmers University of Technology in Sweden, was later brought to the commercial market by Arcam AB in 2001.

This technique primarily concentrates on the utilization of titanium alloys, however cobalt chrome can also be employed. EBM is widely utilized in the aircraft industry because of its capability to handle strong materials such as titanium. This is crucial in meeting the industry's requirement for lightweight and long-lasting components.

During the EBM process, an electron beam traverses a prespread layer of powder. This process results in the powder undergoing melting and subsequent solidification, thereby forming individual layers of the printed product. Nevertheless, due to the utilization of electrons instead of a laser, there exist certain distinctions between the two procedures in EBM (Figure 8.14).

Some distinctions between EBM and SLM come from the engineering choices made in how they are carried out, rather than being inherent to the processes themselves. These differences can affect factors like the final product quality and the types of materials that can be used.

Electron beams and laser beams have fundamental differences in their composition and behaviour. Electron beams involve electrons that move at a

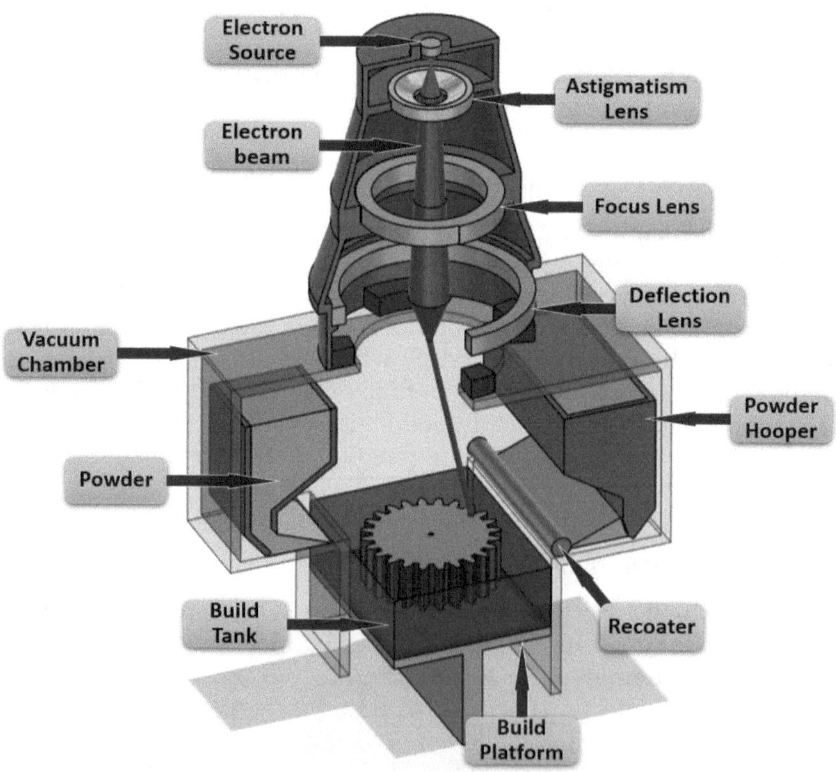

FIGURE 8.14
Schematic of electron beam melting (EBM).

speed comparable to the speed of light, whereas laser beams are composed of photons that move at the speed of light.

When an electron beam encounters a gas at normal atmospheric pressure, the electrons interact with the atoms of the gas, causing them to change direction. On the other hand, laser beams can pass through gas without any interruption, as long as the gas allows the laser wavelength to pass through without obstruction. The difference in the interaction of gases with electrons and photons is due to their distinct features.

EBM is performed in a low-pressure vacuum setting to ensure the removal of gaseous byproducts and oxygen. On the other hand, SLM occurs in a gas environment with no chemical reactions and at the same pressure as the surrounding air.

Electrons, possessing a negative charge, are controlled and concentrated through the utilization of magnetic fields in EBM. This feature facilitates prompt reaction to variations in input conditions, enabling the electron beams to be scanned at both slow and fast speeds. However, laser beams utilize mirrors connected to motors to precisely concentrate and regulate the photons. The mobility of these mirrors dictates the motion of the laser beam focus point, making instantaneous movement impractical. The scan speed in SLM is influenced by variables such as the mirrors' mass, motor parameters, and the distance between the mirrors and the powder bed.

In laser-based procedures, the heating of the powder is a result of the absorption of photons by the particles of the powder. Nevertheless, with EBM, the heating procedure varies. The powder is heated through the transfer of kinetic energy from incoming electrons to the particles of the powder.

In EBM, powder particles become negatively charged after absorbing the electrons. This can potentially lead to two issues:

- Powder particles can be expelled from the powder bed when the repulsive force between negatively charged particles surpasses the gravitational and frictional forces that keep the particles together.
- As the powder particles accumulate negative charges, they have a tendency to resist incoming electrons. This can result in a broader electron beam, which can have an impact on the accuracy and manipulation of the melting procedure.

These phenomena do not occur with photons, making the heating process different between the two techniques.

High conductivity is crucial in EBM to prevent excessive negative charging of powder particles in the powder bed. In order to solve this issue, scan procedures are employed to limit the formation of areas with significant negative charges. Furthermore, intentional modifications are implemented to the electron beam energy in order to sustain diffusion and avoid excessive accumulation of negative charge in a specific location. As a result, this causes the melt pool size in EBM to increase, leading to a bigger heat-affected zone. Consequently, EBM parts often have larger minimum feature size, resolution, and surface polish compared to SLM parts.

EBM can only be utilized with electrically conductive materials, such as metals, due to the necessity of a conductive powder bed. Conversely, the lasers utilized in SLM have the capability to function with any substance that can absorb energy at the specific wavelength of the laser, encompassing metals, polymers and ceramics.

Generating an electron beam is more efficient than generating laser beams. The process involves applying a voltage differential to the heated filament, which turns electrical energy into an electron beam. In addition, beam energies above 1 kW can be obtained at reduced costs. In contrast, laser systems convert only a small portion (up to 20%) of the entire input electrical energy into beam energy. The residual electrical energy is dissipated as heat during the process. Laser beam systems with a capacity above 1 kW are more costly than electron beam systems.

Electron beams are a more cost-effective high-energy source compared to laser beams. Nevertheless, more recent fibre lasers are developing as uncomplicated, more dependable, and more effective substitutes, with reported conversion efficiencies ranging from 70% to 80%. This tendency has the potential to diminish the cost benefit of electron beams in the future. Within the field of EBM, powder beds are kept at elevated temperatures in contrast to SLM. There are other factors contributing to this disparity.

Electron beams are employed in EBM to preheat the build platform prior to the application of the initial powder layer. This guarantees a uniform temperature throughout the whole construction process. This is accomplished by intentionally blurring the electron beam and swiftly moving it across the surface of the substrate. This process is likewise executed using a powder bed that is rapidly and evenly warmed to the necessary temperature. This obviates the necessity for further heating components such as radiative and resistive heaters frequently encountered in SLM systems.

Nevertheless, the higher temperature of the powder bed in EBM leads to a notably distinct microstructure when compared to SLM components. SLM typically exhibits visible laser scan lines, but rapid cooling leads to reduced grain sizes. However, in EBM, the heat is evenly distributed, resulting in a continuous grain pattern that resembles the structures found in cast metal. This design exhibits a lower number of apertures in comparison to SLM structures. The disparity arises due to the higher temperatures and increased heat employed by EBM, resulting in the elimination of the layering phenomenon observed in SLM. EBM has various advantages, such as the capability to create intricate shapes, high rates of material usage, and exceptional mechanical qualities resulting from the

TABLE 8.1

Differences Between EBM and SLM

	EBM	SLM
Energy source	Electron beam	Laser
Atmosphere	Vacuum	Inert gas
Scanning mechanism	Deflection coils	Galvanometers
Energy absorption	Limited by conductivity	Limited by absorbance
Preheating mechanism	Electron beam for preheating	Infrared heaters
Scanning speeds	Very fast due to magnetic-driven	Limited by inertia
Costs of energy production	Moderate	High
Surface finish obtained	Moderate to poor finish	Excellent to moderate finish
Feature resolution	Moderate resolution	Excellent resolution
Materials	Works only with metals	Works with metals, polymers, ceramics

production of fully dense parts. Table 8.1 depicts some key differences between EBM and SLM.

These differences highlight the distinct characteristics and capabilities of each additive manufacturing method, informing material selection and process optimization for various applications.

8.4 Binder Jetting

In this process the liquid binder is selectively deposited on the powder material layer to construct a part layer by layer. The processes uses liquid binder material "glue" to fuse material particles without any heat or light source.

FIGURE 8.15 shows the schematic of the binder jetting process. The process needs a thin layer of the powder material such as metal, sand, or ceramics on a platform. On which the print head moving across selectively deposits the binder material on the powder. It cements the material powder particles together where it is deposited. The platform then moves in a downward direction to make space for deposition of the new powder layer to bind the next layer on the previously deposited layer. This process is repeated to complete the printing of the full part. Depending upon the powder and binder material the further post processing is needed.

For metal parts, sintering of the printed part is needed where the component is heated up to temperature specific to the material. Curing, polishing, and coating are the few more steps in the process.

TABLE 8.2

Summary of Various Binders

Binder Type	Typical Materials Used	Key Applications	Advantages	Disadvantages
Water-Based	Sand, ceramics, some metals	Sand casting, ceramics, metal parts	Environmentally friendly, low toxicity	Can cause warping or cracking
Polymeric	Metals, polymers, composites	Metal parts (stainless steel, titanium)	Better strength after sintering	Requires post-processing (curing, sintering)
Silicate-Based	Sand, ceramics	Molds for casting, sand parts	High thermal stability, heat resistance	Rougher surface finish
Solvent-Based	Ceramics, some metals, polymers	Specialized materials	Faster drying, tailored properties	Toxicity, flammability concerns
Resin-Based	Polymers, specialized metals	Fine-feature parts	Fast curing, precision	Requires UV light, expensive
Metallic	Metals (titanium, stainless steel)	Metal parts	Strong parts after sintering	Requires extensive post-processing
Hybrid	Various powders	Mixed material applications	Balanced strength and detail	More complex processing

8.4.1 Types of Binders Used

The strength, finish, and various mechanical as well as chemical properties of the final parts are depending on the types of binder material used. The selection of the binders is based on the powder material being used and the proposed use of the component. The main types of the binders are water-based binders, polymer binders, silicate-based binders, solvent based binders, resin-based binders (UV curable binders), metal binders, and hybrid binders. The accurate selection of the binder plays an important role in achieving the anticipated quality and functionality of the printed parts.

The summary of various binders is given in Table 8.2.

8.4.2 Materials Processed by Binder Jetting

The powder types processed by the binder jetting 3D printing process are metals, ceramics, sand, and polymers.

- Polymers: in some cases, polymer powders with suitable binders are used for rapid prototyping,
- Metal: Stainless steel, titanium, and various alloy.
- Ceramics: Used for producing parts with specific mechanical properties.
- Sand: Commonly used for molds production in metal castings.

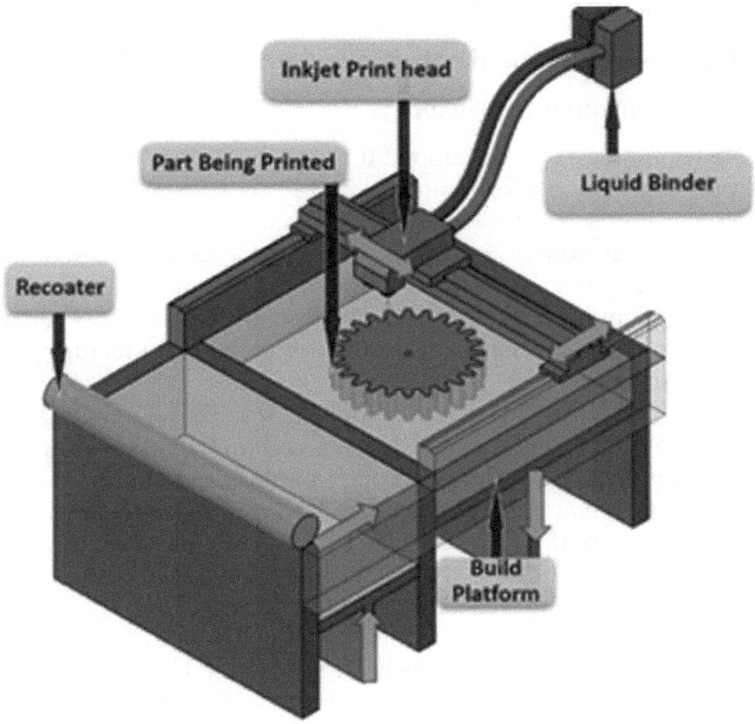

FIGURE 8.15
Schematic of the binder jetting process.

Advantages of Binder Jetting

- For metal 3D printing, binder jetting is a cost-effective option than other 3D printing processes.
- It can be used with a variety of materials as mentioned above, from polymer, metals to sand and ceramics.
- Due to availability of the powder surrounding the layer being printed, the process does not required modeling of the supports.

Disadvantages of Binder Jetting

- The strength of the parts printed with binder jetting may be less compared with the powder bed fusion or traditional casting without post-processing.
- Like strength of the part, to obtain the better surface, finish post-processing steps are required.
- Depending the complexity and material, the process can be slower compared to other techniques for same part.

8.5 Direct Energy Deposition Processes: DED, LENS

8.5.1 Laser Engineered Net Shaping

The LENS technology was developed through a collaborative research effort between Sandia National Laboratories and Pratt & Whitney. Subsequently, in 1997, it was licensed to Optomec, Inc. This cutting-edge technique utilizes lasers as an energy source to fuse air-blown powder particles, often metal particles, enabling the rapid production of tailor-made components and manufacturing moulds. LENS technology obviates the necessity for substantial machining and is particularly suitable for batch manufacturing of high-density parts or moulds.

The LENS functions via a meticulous, computer-regulated procedure. The process combines high-powered lasers and metal powder deposition to effectively manufacture customized components and moulds.

During the process, metal particles are driven through nozzles in the direction of a central point, where they interact with a concentrated laser beam. The laser's high temperature melts the metal particles, resulting in their fusion onto a substrate or previously applied layers.

The laser and powder jets are stationary, while the model and substrate are systematically moved to expose fresh areas for metal deposition. This process is repeated layer by layer until the desired 3D shape is achieved. This method enables the creation of intricate forms with remarkable accuracy and intricacy, hence minimizing the need for substantial further refinement (refer to Figure 8.16).

LENS technology is versatile and may be used with a range of materials, such as titanium, steel, copper and aluminium. This makes it well-suited for a variety of industrial applications, including aerospace and biomedical engineering.

Although it has its benefits, the adoption of this technology must take into account obstacles such as the necessity for post-processing, component deformation caused by residual stresses, and limits in fabricating overhangs compared to other methods. However, LENS technology continues to be flexible and suitable for a wide range of materials, providing the capability to create intricate shapes with accuracy and details.

8.5.2 Advantages and Disadvantages of LENS Technology

Advantages

- **Rapid fabrication of custom parts and molds:** LENS technology facilitates rapid production of personalized components and moulds, resulting in shorter production cycles compared to conventional manufacturing techniques.

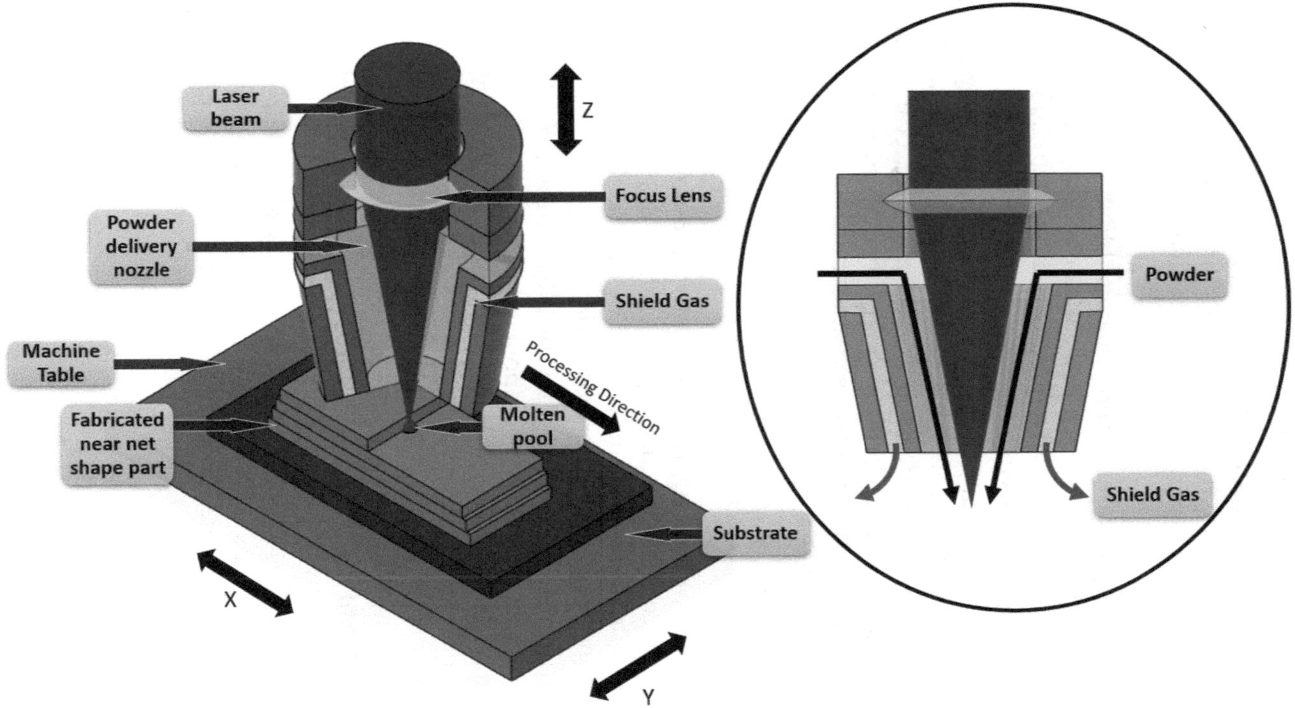

FIGURE 8.16
Schematic of laser engineered net shaping (LENS).

- **High-density parts production:** LENS has the ability to manufacture components with a high level of density, which guarantees their structural integrity and endurance.
- **Capability to produce shapes close to the final product:** LENS technology enables the production of shapes that closely resemble the final result, reducing the requirement for extensive post-processing or rough machining.
- **Versatility in processing various metals:** LENS has the capability to handle a diverse array of metals, such as titanium, steels, copper and aluminium, providing the advantage of being able to choose from a variety of materials for varied purposes.
- **Ability to selectively apply metal to existing parts or repair worn or broken parts:** LENS technology allows for targeted metal deposition onto pre-existing components, making it possible to carry out repairs and alterations without compromising the integrity of the original material. This function improves cost-efficiency and extends the durability of components.

Disadvantages

- **Need for post-processing due to suboptimal component surface finish:** LENS-produced components might require polishing or machining in order to attain the desired surface qualities, particularly for applications that demand enhanced aesthetic criteria.
- **Component distortion caused by residual stresses:** residual stresses may accumulate within the component throughout the LENS process, resulting in distortion or warping of the finished part. This distortion might impact the precision of dimensions and the strength of the structure, requiring the implementation of correction actions.
- **Limitations in fabricating overhangs compared to other techniques like SLS/SLM:**LENS technology may be less capable than SLS and SLM in producing intricate features or complex geometries with overhangs.
- **Exclusion of plastics due to challenges in nozzle flow, restricting certain applications:** the LENS technology is best suited for the processing of metals and ceramics, but it may not be ideal for applications that involve the use of plastics or polymers due to difficulties in nozzle flow and material compatibility. This constraint limits its suitability in specific sectors or product categories.

8.5.3 Application of LENS Technology

The applications of LENS technology encompass a wide range of industries and fields, including the following:

- **Fabrication of complex geometrical components:** the LENS technology is employed to manufacture complicated and sophisticated components with exceptional accuracy and details, especially in sectors like aerospace, automotive, and consumer electronics.
- **Repair and refurbishment of damaged components and structures:** LENS technology facilitates the repair and restoration of worn or broken parts, hence prolonging their lifespan and decreasing the necessity for expensive replacements. This application is widely used in industries such as aviation, marine and manufacturing.
- **Biomedical applications, such as porous titanium implants for bone replacement:** LENS is used in the biomedical field to make unique patient-specific implants and prosthetics with exact shapes and geometries, such as porous structures that help bone tissue integrate and grow back.
- **Manufacturing moulds and tooling for various industries:** moulds, dies and other tools for different manufacturing processes, like injection moulding, metal forming and die casting, are made with LENS technology. This allows for quick testing and customization.
- **Research and development of new materials and metallurgical phenomena:** LENS serves as a valuable tool in materials science research, facilitating the exploration of novel materials, alloy compositions, and metallurgical phenomena. It enables the development of functionally graded materials (FGMs) and the investigation of composition variation to enhance material properties and performance.

8.6 Applications of Powder-Based 3D Printing Processes

The powder-based processes are a cutting edge form of additive manufacturing that use a strong energy source, like an electron beam or laser, to fuse powdered materials, usually metals or plastics, together, building complex three-dimensional things layer by layer. Different businesses use these processes in many different ways because they are flexible, accurate and can make complex shapes without the need for support structures. Some of them are as follows:

- **Prototyping excellence:** SLS is often used for quick and cheap prototyping in the automobile, aerospace, consumer goods and electronics industries. This method makes it easier to make prototypes that have properties like the final goods.
- **Customized part production for biomedical applications:** SLS makes it possible to make parts and goods that are exactly what each person

needs. This feature can be used for custom medical devices, orthotics and prosthetics. In the medical field, SLS is used to make surgical guides, implants that are custom made for each patient, oral prosthetics and anatomical models that are needed for planning surgeries and teaching students how to do them.

- **Strong functional parts for rapid tooling and other uses:** SLS technology is very good at printing functional parts that are strong and last a long time. Because of this, it is a popular choice for making parts for machines, tools and other industry equipment. When you need to make precise tooling parts, jigs and fittings, SLS is the way to go. These enhance speed, accuracy, and consistency in manufacturing processes, leading to improved efficiency.
- **Direct manufacturing of end-use goods:** as materials get better, powder-based 3D printing processes can now be used to make end-use goods directly. These include a wide range of consumer goods, fashion items and models for buildings.
- **Innovative ideas in aerospace and defence:** powder-based printing is a key part of making lightweight, high-performance parts that are used in aerospace and defense. These include frames, housings, structural parts and drone parts.
- **Driving car advancements:** powder-based methods are used to make cabin parts, engine parts, prototypes, and custom tools in the car industry. This makes it easier to quickly test and improve design ideas.
- **Creative expression and ingenuity in design:** SLS technology lets artists and designers use their imaginations to make sculptures, building models, jewellery, and decorative items with lots of small details and complicated shapes.
- **Enhancement of education and research:** SLS technology helps schools and research labs learn more about materials science, additive manufacturing and cutting edge new uses in many different fields.

Powder-based methods are becoming a powerful force in many fields, leading to new ideas, better efficiency and better manufacturing. As technology keeps improving and changing, so will the range of things it can be used for. This means that the future will be full of possibilities that have never been seen before.

8.7 Advanced Concepts and Future Trends

The SLS method is a new way to combine printing materials that is being used successfully in many fields, including art, aerospace and the car industry. As new additives have been added, it has become useful in more and more areas. One way that SLS is changed is by using a substrate coated in photosensitive ink instead of the powder bed. This makes it possible to print micro-PCBs and brings SLS uses to the electronics industry. SLS also has a lot of potential in biology, especially when it comes to tissue engineering and medical implants. PCL/Hap composites and SLS technology can be utilized to fabricate tissue engineering scaffolds. These scaffolds have stable mechanical qualities and a lot of metabolic activity from cells.

Self-Check Exercises

1. Provide a concise overview of powder-based additive manufacturing techniques.
2. Provide examples of polymer and metallic feedstocks for solid-state sintering, including their characteristics and the preheating process for laser-based PBF.
3. Find out the prominent applications of the materials discussed in chemically-induced sintering.
4. Explain the concept of liquid-phase sintering in the context of powder-based 3D printing processes.
5. Provide examples of materials suitable for liquid-phase sintering in additive manufacturing.
6. Define the concept of partial melting in the context of powder-based 3D printing processes.
7. Compare partial melting with other fusion mechanisms, such as liquid-phase sintering and solid-state sintering, in terms of applications and advantages.
8. Differentiate between partial melting and full melting in terms of their impact on the final properties of 3D printed parts.
9. Highlight the advantages of employing full melting in the production of metal and polymer parts through additive manufacturing.
10. Describe the working principle of SLS.
11. Explain the role of a CO_2 laser in the sintering process.
12. Discuss the recoater mechanism and its function.
13. Illustrate the temperature control methods used in SLS.
14. How does the laser-related process parameters affect the end part in the SLS process?
15. How does the layer thickness influence the quality of SLS printed parts?

16. How do variations in recoating speed affect the SLS printing process?
17. What types of materials are commonly used in SLS printing?
18. What are the challenges and limitations associated with using metal powders in SLS?
19. How does the ability to print complex geometries without the need for support structures benefit manufacturers in SLS?
20. How does the cost of SLS equipment and materials compare to other additive manufacturing technologies?
21. What environmental and health considerations should be taken into account when using SLS due to the handling of powdered materials and emissions from the process?
22. What emerging applications are being explored for SLS in fields such as automotive, consumer goods and electronics manufacturing?
23. What industries commonly utilize SLS technology for rapid prototyping and product development?
24. Differentiate between EBM and SLS processes.
25. Find some more applications of LENS those are not listed here.

Discussion Topics

1. How the SLS technique differ from other 3D printing techniques? Discuss from working, materials used, advantages, limitations and applications perspective.
2. Which properties of LASER make it most relevant and useful to additive manufacturing? Which aspects of the LASER is not friendly with widespread use of AM?

Multiple Choice Questions

1. In powder-based additive manufacturing processes, what is the primary role of lasers/electron beams?
   (a) To spread the powdered material
   (b) To sinter the powdered material
   (c) To cool the melted material
   (d) To apply pressure to the powder bed

2. Which of the following is NOT a characteristic of powder-based additive manufacturing processes?
   (a) Use of lasers/electron beams
   (b) Creation of 3D structures from powdered materials
   (c) Requirement of a solid substrate for printing
   (d) Melting or sintering of powdered particles
3. What is the primary difference between powder bed fusion and laser powder fed/direct energy depositionprocesses?
   (a) PBF uses lasers, while LPF/DED uses electron beams
   (b) PBF works in a vacuum, while LPF/DED uses inert gas
   (c) PBF involves fusing powdered particles in place, while LPF/DED deposits powder through nozzles
   (d) PBF requires post-processing to remove the binder, while LPF/DED does not
4. What is the purpose of the recoater mechanism in powder-based additive manufacturing processes?
   (a) To spread the powder thinly over the build area
   (b) To heat the powdered material
   (c) To cool the melted material
   (d) To remove excess powder
5. Which of the following is a common energy source used in powder bed fusion processes?
   (a) Microwave
   (b) Electron/laser beam
   (c) Hydraulic pressure
   (d) Magnetic field
6. Which type of powder fusion mechanism is based on the concept of minimizing total free surface energy?
   (a) Solid-state sintering
   (b) Chemically-induced sintering
   (c) Liquid-phase sintering
   (d) Partial melting
7. What is the primary advantage of solid-state sintering in additive manufacturing?
   (a) Low energy consumption
   (b) High-density parts production
   (c) Compatibility with a wide range of materials
   (d) Ability to create parts with internal features
8. Which powder fusion mechanism involves the use of separate grains of binder and structural materials?
   (a) Solid-state sintering
   (b) Chemically-induced sintering
   (c) Liquid-phase sintering
   (d) Partial melting

9. In partial melting techniques, what remains unaffected in the powder particles?
   (a) The entire particle
   (b) The outer shell
   (c) The inner core
   (d) The surrounding environment
10. Which full melting methodology is preferred for producing components with density equal to bulk materials?
    (a) Single component, single material powder
    (b) Alloyed materials powder
    (c) Mixture of powder particles of different materials
    (d) Polymer-coated powder
11. What type of laser is used in selective laser sintering technology?
    (a) Nitrogen laser
    (b) Carbon dioxide laser
    (c) Fiber laser
    (d) Diode laser
12. What mechanism is used to spread the thin layer of powder across the working area in SLS?
    (a) Inkjet printing
    (b) Extrusion
    (c) Recoater mechanism
    (d) Stereolithography
13. Which gas is commonly used to fill the chambers in SLS to avoid contamination and oxidation?
    (a) Oxygen
    (b) Carbon dioxide
    (c) Argon
    (d) Helium
14. What is the primary function of infrared heaters in the SLS process?
    (a) To cool down the printing chamber
    (b) To heat the powder before spreading it
    (c) To provide light for visibility
    (d) To control humidity in the chamber
15. Which parameter determines the intensity of the laser beam in SLS?
    (a) Spot size
    (b) Pulse duration
    (c) Laser power
    (d) Pulse frequency
16. What is the primary advantage of SLS regarding supports needed for printing?
    (a) No supports are needed
    (b) Supports are automatically generated
    (c) Supports are easy to remove
    (d) Supports improve printing speed

17. Which type of polymer is commonly used in SLS printing due to its strength and robustness?
(a) Polyethylene
(b) Polycarbonate
(c) Polypropylene
(d) Polyamide (nylon)

18. What is a limitation of SLS regarding raw materials?
(a) Limited colour options
(b) Limited availability of metals
(c) Limited size of prints
(d) Limited recycling options

19. What protective gear is recommended when working with SLS printers due to health concerns?
(a) Gloves and apron
(b) Goggles and earmuffs
(c) Air filter mask and gloves
(d) Safety boots and helmet

20. In which industries is SLS commonly used for rapid prototyping?
(a) Construction and agriculture
(b) Healthcare and fashion
(c) Automotive and aerospace
(d) Food and hospitality

21. What is the primary energy source used in electron beam melting for melting metal powders?
(a) Laser
(b) Electron beam
(c) Infrared radiation
(d) Ultraviolet light

22. Which of the following metals is mainly used in electron beam melting?
(a) Copper
(b) Aluminum
(c) Titanium
(d) Silver

23. How is the atmosphere during electron beam melting compared to selective laser melting?
(a) Both are conducted in a vacuum
(b) Both are conducted in an inert gas atmosphere
(c) EBM is conducted in a vacuum, while SLM is conducted in an inert gas atmosphere
(d) EBM is conducted in an inert gas atmosphere, while SLM is conducted in a vacuum

24. Whatcontrols the movement of the electron beam in electron beam melting?
 (a) Mirrors attached to motors
 (b) Magnetic fields
 (c) Infrared sensors
 (d) Deflection coils
25. Which heating mechanism is employed in electron beam melting to heat the powder particles?
 (a) Absorption of photons
 (b) Infrared radiation
 (c) Transfer of kinetic energy from electrons
 (d) Conduction from the substrate
26. What is the primary challenge regarding the powder bed in electron beam melting that is not present in selective laser melting?
 (a) High temperatures
 (b) Low conductivity
 (c) Excessive negative charge
 (d) Rapid cooling
27. Which method is used to preheat the powder bed in electron beam melting systems?
 (a) Infrared heaters
 (b) Radiative heaters
 (c) Electron beam defocusing
 (d) Laser preheating
28. What is one advantage of laser engineered net shaping technology?
 (a) Excellent surface finish
 (b) Limited material compatibility
 (c) Slow fabrication process
 (d) High cost
29. In which industries is laser engineered net shaping technology commonly used?
 (a) Textile
 (b) Agriculture
 (c) Aerospace
 (d) Hospitality
30. What is one limitation of laser engineered net shaping technology compared to other additive manufacturing techniques?
 (a) High-density parts production
 (b) Rapid fabrication of custom parts
 (c) Fabrication of complex geometries with overhangs
 (d) Versatility in processing various metals

Answers

1. (b) To sinter the powdered material
2. (c) Requirement of a solid substrate for printing
3. (c) PBF involves fusing powdered particles in place, while LPF/DED deposits powder through nozzles
4. (a) To spread the powder thinly over the build area
5. (b) Electron/laser beam
6. (a) Solid-state sintering
7. (c) Compatibility with a wide range of materials
8. (c) Liquid-phase sintering
9. (c) The inner core
10. (a) Single component, single material powder
11. (b) Carbon dioxide laser
12. (c) Recoater mechanism
13. (c) Argon
14. (b) To heat the powder before spreading it
15. (c) Laser power
16. (a) No supports are needed
17. (d) Polyamide (nylon)
18. (d) Limited recycling options
19. (c) Air filter mask and gloves
20. (c) Automotive and aerospace
21. (b) Electron beam
22. (c) Titanium
23. (c) EBM is conducted in a vacuum, while SLM is conducted in an inert gas atmosphere
24. (b) Magnetic fields
25. (c) Transfer of kinetic energy from electrons
26. (c) Excessive negative charge
27. (c) Electron beam defocusing
28. (a) Excellent surface finish
29. (c) Aerospace
30. (c) Fabrication of complex geometries with overhangs

9

Post-Processing Techniques and Future Trends

Post-processing techniques play a crucial role in enhancing the quality and functionality of 3D printed components. After the printing process, the next step involved is removing the printed part from the build platform. Printed parts are often inspected for dimensional accuracy, surface defects or other imperfections that may have occurred during printing. This can involve visual inspection, measurements with calipers or other precision tools, or even non-destructive testing methods like X-ray or CT scanning for complex geometries. This chapter explains different post-processing, inspection and testing techniques used in solid-based, liquid-based and powder-based 3D printing processes. It also outlines the advancements and future trends in FDM based 3D printing. Further it provides information about the sustainability in additive manufacturing by understanding its environmental benefits, resource efficiency and potential to reduce waste in production processes. At the end the role of AI and machine learning in optimizing processes, enhancing design, and improving quality control in additive manufacturing is also covered.The chapter comprises of following main sections:

9.1 Post-Processing Techniques
9.2 Other Future Trends in Additive Manufacturing
9.3 Sustainability in Additive Manufacturing
9.4 Integration of AI and Machine Learning in Additive Manufacturing

After reading through the material, self-check exercises, objective questions and discussion topics presented in this chapter, the learner will be able to do the following:

(a) Explain different post-processing techniques used in different 3D printing processes.
(b) Explain inspection and testing techniques used for 3D printed object.
(c) Outline the advancements and future trends in FDM-based 3D printing.

DOI: 10.1201/9781003562344-9

(d) Explain sustainability in additive manufacturing by understanding its environmental benefits, resource efficiency and potential to reduce waste in production processes.

(e) Explain the role of AI and machine learning in optimizing processes, enhancing design and improving quality control in additive manufacturing.

9.1 Post-Processing Techniques

9.1.1 Solid-Based 3D Printing Processes

Post-processing techniques play a crucial role in enhancing the quality and functionality of FDM 3D printed components. After the printing process, the next step involved is removing the printed part from the build platform.

1. *Support removal* is typically the first stage of post-processing for any 3D printing technologies that require support to accurately produce parts. Support can generally be separated into two categories viz. standard and dissolvable.
   - Standard support material can generally be removed manually from the print with little effort, and cleaning of support material in hard to reach places (like holes or hollows) can be achieved with dental picks and needle-nose pliers.
   - Dissolvable support materials are removed from a print by placing the print in a bath of the appropriate solvent until the support material dissolves. The support is typically printed in:

     HIPS (usually associated with ABS)

     PVA (usually associated with PLA)

     HydroFill

For HIPS/ABS prints, a bath in a 1:1 ratio of (R)-(+)-limonene and isopropyl alcohol is good for rapid support removal. Other support materials, such as PVA (used with PLA) and HydroFill (PLA and ABS), simply dissolve in plain water.

2. *Sanding* –after supports are removed or dissolved, sanding can be done to smooth the part and remove any obvious blemishes, such as blobs or support marks. The starting grit of sandpaper depends on the layer height and print quality. For layer heights of 200 microns and lower, or prints without blemishes, sanding can be started with

150 grit. If obvious blemishes are present or the object was printed at a layer height of 300 microns or higher, start sanding with 100 grit sand paper.

3. *Cold welding* – when the size of a print exceeds the maximum volume of the printer, the design is often broken down into smaller sections and assembled together after printing. For PLA and other materials, assembly can be done using Bond-O or an appropriate glue (glue selection will depend upon plastic). For ABS, multi-part assemblies can be "welded" together using acetone. The mating surfaces need to be brushed lightly with acetone, and firmly held together, or clamped if possible, until the majority of the acetone evaporates. At this point, the two parts are chemically bonded to one another.
4. *Gap filling*– after sanding a print, or dissolving soluble supports, it is not uncommon for gaps to emerge on the print. During printing, gaps are formed when layers are incomplete due to tool-path constraints and are often inevitable. Small gaps and voids can easily be filled with epoxy (like XTC-3D), and may not require additional processing. Large gaps, or hollows left from joining a multi-print assembly, can be successfully filled with autobody filler, which will require additional sanding once dry. Gaps in an ABS print can also be filled by creating a slurry of ABS filament and acetone, which chemically reacts with the ABS print and seeps into any voids in the surface.
5. *Polishing* – after sanding a print, a plastic polish can be applied to give standard thermoplastics, like ABS and PLA, a mirror-like surface finish. Once the print is sanded up, wipe excess dust off the print with a tack cloth then clean the print in warm water bath with a toothbrush. Allow the print to dry fully, and buff using a buffing wheel or by hand with a microfibre cloth and plastic polishing compound, such as Blue Rouge.
6. *Priming andpainting* – Once the print is properly sanded (only need to go up to 600 grit for painting), the print can be primed. Priming should be done in two coats, using an aerosol primer. Once priming is complete, painting can begin. Painting can be done with artist acrylic paints and brushes, but the use of an airbrush or aerosol can provide a smoother surface finish.
7. There are other post processes also existing like vapour smoothing; dipping; epoxy coating; metal plating. (For more details refer-www.hubs.com/knowledge-base/post-processing-fdm-printed-parts/.)
8. Additionally, post-processing may involve functional enhancements. For instance, heat treatment processes like annealing can be used to improve the mechanical properties of certain materials, increasing their strength or reducing internal stresses.

Printed parts are often inspected for dimensional accuracy, surface defects or other imperfections that may have occurred during printing. This can involve visual inspection, measurements with calipers or other precision tools, or even non-destructive testing methods like X-ray or CT scanning for complex geometries.

9.1.2 Sheet-Based 3D Printing Processes

Post-processing techniques in sheet-based 3D printing processes are essential to refine the final printed objects, improving their mechanical properties, aesthetics, dimensional accuracy, and overall quality.

9.1.2.1 Extraction from Surrounding Sheet Material

After the printing process, the part needs to be extracted from the surrounding sheet material. In LOM, cross hatching is employed to facilitate this extraction process. While paper is commonly used as the sheet material, the extraction does not necessitate specialist tools, making it a cost-effective and time-efficient step.

9.1.2.2 Support Removal

*Purpose:*during the printing process, support structures may be added to provide stability to overhanging or complex geometries. After printing, these supports need to be removed.

*Manual removal:*simple supports can often be removed manually using tools like pliers, knives or files. This requires careful handling to avoid damaging the printed part.

*Automated processes:*in industrial settings, automated systems like laser cutting or water jet cutting can be used to remove supports with high precision.

*Considerations:*careful removal of supports is critical to maintaining the integrity of the surface finish and dimensions of the printed part.

9.1.2.3 Trimming and Edge Smoothing

*Purpose:*sheet-based 3D printing processes often leave edges that may need to be smoothed or trimmed for aesthetic or functional reasons.

*Manual techniques:*this is commonly done using abrasive tools like sandpaper or files to ensure clean edges. It is a particularly important step when layers are not perfectly aligned during the printing process.

Automated techniques: CNC milling or laser cutting can be used to accurately trim and smooth edges in large-scale or industrial processes.

*Challenges:*achieving a balance between removing excess material and preserving the intended design requires skill, especially when working with complex geometries.

9.1.2.4 Heat Treatment (Annealing)

*Purpose:*heat treatment is used to improve mechanical properties like strength and durability by relieving internal stresses that arise during the printing process.

*Process:*the printed part is heated to a specific temperature and then allowed to cool gradually. This helps in achieving uniformity in material properties.

*Materials:*the process varies depending on the material used. For example, thermoplastics require a lower annealing temperature than metals or composites.

*Benefits:*it enhances the overall strength, dimensional stability, and resistance to deformation.

9.1.2.5 Surface Finishing (Polishing and Smoothing)

*Purpose:*surface finishing improves the appearance of the printed part and reduces roughness, often resulting in a smoother texture and improved aesthetics.

*Mechanical polishing:*for metallic sheets, mechanical polishing using abrasive wheels, belts or media tumbling can be employed.

*Chemical smoothing:*solvents or chemicals are applied to smooth the surface without physically contacting the part. This is often used for plastic or composite materials.

*Electro-polishing:*a technique used for metallic parts where an electric current is applied to dissolve surface imperfections, creating a polished and smooth finish.

*Challenges:*achieving a uniform surface finish across complex geometries can be challenging. Over-polishing may lead to reduced part thickness and altered dimensions.

9.1.2.6 Bonding and Lamination

*Purpose:*sheet-based 3D printing involves the layer-by-layer stacking of sheets. Post-processing bonding ensures that these sheets are fully adhered to one another for enhanced strength and durability.

*Heat bonding:*applying heat can melt layers slightly, fusing them together. This is commonly used for thermoplastics.

*Adhesive bonding:*for certain materials, adhesive bonding using specialized resins or glues ensures a strong inter-layer bond.

*Mechanical pressing:*for certain sheet-based processes, mechanical pressing ensures layers are compressed and adhered properly.

*Challenges:*ensuring uniform bonding across large surfaces and avoiding air bubbles or delamination is critical for part integrity.

9.1.2.7 Painting and Coating

*Purpose:*painting and coating improve both aesthetics and functional properties such as resistance to corrosion, UV radiation or wear.

*Painting:*the printed part can be painted with traditional methods such as spray painting, hand-painting or powder coating for better appearance or color uniformity.

*Coating:*protective coatings such as polyurethane, epoxy or metal plating (like nickel or chrome) may be applied to enhance wear resistance or provide surface protection against environmental factors.

*Electroplating:*for metallic parts, electroplating may be used to add a thin layer of another metal, improving appearance and resistance to corrosion.

*Challenges:*careful selection of paints and coatings is important as some materials may negatively react, causing warping or peeling.

9.1.2.8 Dimensional Accuracy Adjustments

Purpose: After printing, parts often require fine adjustments to meet strict tolerances, especially for functional parts where high precision is critical.

Machining: CNC milling, lathing or grinding are used to refine part dimensions and ensure they meet required specifications.

Laser trimming: for highly precise adjustments, laser cutting or trimming can remove small amounts of material without damaging surrounding areas.

Challenges: adjusting dimensions post-printing requires high precision. Over-correction can lead to undersized parts or compromise structural integrity.

9.1.2.9 Joining Multiple Parts

*Purpose:*in cases where the final object is too large or complex to be printed as a single part, multiple smaller components may need to be joined post-processing.

*Mechanical fasteners:*bolts, screws or rivets may be used to join printed parts. This is common in large structures where disassembly may be required later.

*Adhesive joining:*high-strength adhesives like epoxy or UV-curable adhesives may be used to permanently bond parts together.

*Thermal welding:*for thermoplastics or metals, welding processes can be used to fuse parts together.

*Challenges:*ensuring strong, reliable joints without introducing weak points or compromising the part's overall design is crucial.

9.1.2.10 Inspection and Quality Control

*Purpose:*this step ensures that the final part meets all functional, dimensional and aesthetic requirements.

*Visual inspection:*the simplest form of quality control involves inspecting the surface finish, support removal quality and overall appearance of the part.

*Dimensional inspection:*tools like calipers, micrometers or coordinate measuring machines (CMM) are used to ensure the printed part meets precise dimensional tolerances.

*Non-destructive testing (NDT):*techniques such as ultrasonic testing, X-ray, or CT scanning are used for internal inspection, especially for critical parts where internal defects are unacceptable.

*Challenges:*ensuring the accuracy of large and complex parts may require multiple inspection techniques to avoid missing potential issues.

9.1.2.11 Specialized Post-Processing Techniques

*Electromagnetic shielding coating:*for parts intended to be used in electronics, a conductive coating may be applied to provide electromagnetic shielding.

*Surface texturing:*to improve grip or aesthetics, specialized texturing can be added after the part is printed. This can be done with laser etching or mechanical engraving.

*Surface hardening:*for mechanical parts that require a hardened surface, surface hardening techniques like carburizing or nitriding can be applied post-printing.

9.1.2.12 Cleaning and Final Touches

*Purpose:*before final delivery, cleaning is necessary to remove any residual material, dust or contaminants.

*Cleaning methods:*simple soap and water cleaning, air blasting or ultrasonic cleaning methods can be employed based on the material and size of the part.

Polishing and buffing: a final round of polishing or buffing may be performed to ensure the surface is clean and smooth.

*Packaging:*proper packaging is important, especially for delicate parts, to prevent damage during transportation or handling.

9.1.3 Liquid-Based 3D Printing Processes

Resin-based 3D printing, such as SLA (stereo lithography), DLP (digital light processing), and MSLA (masked stereolithography), produces high-resolution prints. However, the raw prints require several post-processing steps to enhance their mechanical properties, aesthetics, and functionality. These steps include washing, curing, support removal, sanding, and finishing to ensure optimal part quality.

9.1.3.1 Initial Cleaning: Removing Uncured Resin

The first step in post-processing resin-based 3D prints is the removal of uncured resin from the surface of the printed part. Uncured resin is not only sticky and messy but can also be toxic if left on the model. The cleaning process typically involves the following steps:

- **Isopropyl Alcohol (IPA) Bath**: The printed part is submerged in a bath of isopropyl alcohol (usually 90% or higher concentration) to dissolve and remove excess resin. Ultrasonic cleaners can be used to enhance the cleaning process, especially for intricate geometries.
- **Agitation and Brushing**: For parts with complex details, gentle agitation or brushing with a soft-bristle brush may be necessary to ensure all residual resin is removed.
- **Secondary Rinse**: After the initial cleaning, the part is often rinsed in a second, cleaner IPA bath to remove any remaining contaminants.
- **Safety:** Proper personal protective equipment (PPE), such as gloves and safety goggles, should always be worn during this process to avoid skin contact with uncured resin and IPA.

9.1.3.2 Support Removal: Refining the Surface

Resin-based 3D prints often require support structures to prevent deformation during printing. Once the part is cleaned and cured, these supports must be carefully removed.

- **Manual Removal**: Supports can be removed using precision tools such as flush cutters, tweezers, or hobby knives. Care must be taken to avoid damaging the part.

9.1.3.3 Post-Curing: Enhancing Mechanical Properties

Resin-based 3D prints are not fully cured when they come off the printer. Post-curing is a critical step that ensures the resin achieves its maximum mechanical strength and stability. This process involves exposing the part to ultraviolet (UV) light to complete the polymerization process.

- **UV Curing Chambers**: Dedicated UV curing chambers are commonly used for post-curing. These devices provide uniform UV exposure, ensuring consistent curing across the entire part.
- **Sunlight Curing**: In the absence of a UV chamber, natural sunlight can be used as an alternative. However, this method is less controlled and may result in uneven curing.
- **Curing Time**: The duration of post-curing depends on the resin type and the size of the part. Over-curing can lead to brittleness, while under-curing may result in weak or tacky surfaces.

9.1.3.4 Surface Finishing: Achieving Aesthetic Perfection

The surface finish of a resin-based 3D print can be further enhanced through various techniques, depending on the desired outcome.

- **Sanding and Polishing**: Sequential sanding with progressively finer grits, followed by polishing compounds, can produce a glossy, mirror-like finish.
- **Vapour Smoothing**: Some resins can be smoothed using chemical vapour, such as acetone or specialized smoothing agents, to achieve a uniform surface.
- **Painting and Coating**: Resin prints can be painted with acrylics, enamels, or spray paints to add colour and detail. A clear coat can be applied to protect the finish and add durability.

9.1.3.5 Advanced Techniques: Specialized Applications

For specialized applications, additional post-processing techniques may be employed:

- **Electroplating**: Resin prints can be electroplated with metals such as copper, nickel, or gold to enhance their appearance and conductivity.
- **Dyeing**: Transparent or semi-transparent resin prints can be dyed to achieve vibrant colours.
- **Annealing**: Some high-performance resins may benefit from thermal annealing to improve their mechanical properties.

9.1.3.6 Assembly and Finishing

Assembling parts may require additional post-processing steps such as machining, tapping, threading, or bonding to achieve the desired final product. These processes may be performed manually or using specialized equipment depending on the complexity of the assembly.

9.1.3.7 Safety and Environmental Considerations

Post-processing resin-based 3D prints involves handling chemicals and UV light, which pose potential health and environmental risks. Proper disposal of uncured resin, IPA, and other waste materials is essential to minimize environmental impact. Additionally, working in a well-ventilated area and using appropriate PPE is crucial to ensure safety.

9.1.4 Powder-Based 3D Printing Processes

In powder-based 3D printing processes such as selective laser sintering (SLS), direct metal laser sintering (DMLS), and binder jetting, post-processing is essential for improving the mechanical properties, surface finish, dimensional accuracy, and overall aesthetic of the printed part. These techniques can be broadly categorized into mechanical, thermal and chemical methods, each serving different purposes based on the requirements of the final product.

9.1.4.1 Depowdering

Purpose: depowdering is the first post-processing step in powder-based 3D printing. After printing, parts are surrounded by loose, unsintered powder that must be removed.

Methods:

Manual brushing: involves using soft brushes to remove excess powder from the surface of printed parts.

Compressed air : a blast of compressed air is commonly used for more thorough powder removal, especially for intricate geometries.

Vibration: for more complex internal geometries, vibrating platforms are used to shake off loose powder.

Challenges:

Hard-to-reach areas: complex internal channels and cavities may retain powder, requiring specialized techniques like vacuuming or ultrasonic cleaning.

Powder reclamation: reusing unsintered powder is common but requires careful screening to ensure the particle size and quality meet printing standards.

9.1.4.2 Heat Treatment (Annealing, Stress Relief)

Purpose: to relieve internal stresses and improve the mechanical properties of the printed part.

Methods:

Annealing: a controlled heat treatment that alters the microstructure of the printed material to reduce residual stress caused by the rapid heating and cooling cycles in the printing process.

Hotisostatic pressing (HIP): applied mostly to metal parts, HIP involves heating the part in a high-pressure inert gas environment to remove porosity and improve density, leading to better mechanical properties.

Benefits:

Enhances ductility and toughness of parts.

Reduces the risk of cracking or warping in service.

Challenges: precise control over temperature and time is critical to avoid undesirable changes in material properties.

9.1.4.3 Surface Finishing

Purpose: powder-based 3D printing often leaves a rough surface finish due to the nature of sintering or binding powders, necessitating additional surface finishing processes.

Methods:

Media blasting: using materials such as glass beads, sand or aluminum oxide to remove surface imperfections and smooth the surface. It is often used for aesthetic and tactile improvement.

Vibratory finishing: involves placing the printed part in a vibratory bowl with abrasive media that smooths the surface through mechanical action.

Polishing: for achieving a high-gloss finish, mechanical or chemical polishing techniques are used, especially for metallic parts.

Challenges: over-processing during finishing can lead to changes in part dimensions and removal of critical details.

9.1.4.4 Infiltration

Purpose: infiltration improves the strength and mechanical integrity of porous parts, particularly in binder jetting processes where the green part lacks adequate strength.

Methods:

Polymer infiltration: the part is submerged in or coated with a liquid polymer that infiltrates the porous structure, enhancing the part's strength, density, and surface quality.

Metal infiltration: a low-melting metal like bronze can be infiltrated into a porous metal part to increase its strength and improve load-bearing capabilities.

Benefits: infiltration greatly enhances part density and reduces porosity, making it suitable for functional applications.

Challenges: the process can alter the thermal and mechanical properties of the material, and achieving uniform infiltration may be difficult, especially for complex geometries.

9.1.4.5 Machining

Purpose: to improve dimensional accuracy or add features that cannot be achieved in the 3D printing process alone.

Methods:

CNC machining: common for metal parts, CNC milling, or turning removes excess material to meet precise tolerances.

Drilling and tapping: post-printing, additional threaded holes or precise features may be added.

Benefits:

High-dimensional accuracy and precision.

Allows integration with traditional manufacturing techniques.

Challenges: since 3D printed parts may have internal stresses, machining can sometimes lead to part deformation. Careful planning is required to avoid such issues.

9.1.4.6 Sintering

Purpose: in some processes, like binder jetting, the green part is only held together by a binder. Sintering is necessary to fuse the powder particles and provide strength to the part.

Process: the part is heated in a furnace at a temperature below its melting point to fuse the powder particles together. This causes densification and enhances the mechanical properties of the part.

Challenges: sintering can lead to shrinkage, so the dimensions of the green part must account for this. Additionally, controlling the sintering atmosphere is crucial to prevent oxidation or contamination.

9.1.4.7 Surface Coating

Purpose: surface coating is used to enhance the corrosion resistance, wear resistance or aesthetic appearance of the part.

Methods:

Electroplating: a common method for metallic parts, electroplating involves coating the part with a thin layer of another metal, such as nickel, chrome, or gold, to improve surface properties.

Powder coating: this involves applying a dry powder to the part and heating it to form a protective, durable coating.

Painting: especially useful for aesthetic improvements, painting allows customization of color and surface texture.

Challenges: coating processes can change the surface geometry, and care must be taken not to clog fine features or obscure important details.

9.1.4.8 Hot Isostatic Pressing (HIP)

Purpose: for metal parts, HIP reduces internal porosity and improves overall material density and mechanical properties.

Process: parts are placed in a high-pressure, high-temperature chamber filled with an inert gas (usually argon), which applies uniform pressure from all sides to remove porosity.

Benefits:

Improves fatigue life and tensile strength.

Reduces the likelihood of defects such as cracks or voids.

Challenges: HIP is energy-intensive and requires specialized equipment, making it costly for smaller production runs.

9.1.4.9 Sealing (For Porosity Reduction)

Purpose: sealing is used to address inherent porosity in certain powder-based 3D printed parts, especially in ceramics and some metals, where porosity can negatively impact performance.

Methods:

Resin sealing: parts are immersed in a liquid resin that seeps into the pores. Upon curing, the resin solidifies, reducing porosity.

Vacuum impregnation: in this process, parts are placed in a vacuum chamber to draw out air from the pores, then a sealing resin or polymer is introduced to fill the voids.

Benefits: sealing improves the part's resistance to leakage, corrosion, and enhances structural integrity.

Challenges: sealing adds weight to the part and may alter some of its thermal or mechanical properties.

9.1.4.10 Colouring and Dyeing

Purpose: to improve the aesthetic appeal of parts, especially for consumer products or prototypes.

Methods:

Dyeing: particularly common for polymer-based powder parts (like nylon), dyeing the part in a heated solution provides a uniform colour throughout.

Painting and coatings: can provide custom colours and surface finishes.

Challenges: dyeing is limited to certain materials and can fade over time if exposed to harsh environments or UV light.

9.1.5 Inspecting and Testing the Printed Object

When inspecting and testing a 3D printed object, it is essential to follow a systematic approach to ensure quality, functionality and aesthetic appeal. Here is a detailed guide covering various aspects of inspecting and testing 3D printed objects:

9.1.5.1 Visual Inspection

Surface quality: examine the surface for any visible defects such as layer lines, blobs or stringing. Good-quality prints should have smooth surfaces without significant imperfections.

Color consistency: check for uniform colour throughout the object. Inconsistent colocring may indicate issues during the printing process, such as material inconsistency or temperature fluctuations.

Dimensional accuracy: measure critical dimensions using calipers or a ruler. Compare these measurements with the original CAD model specifications to identify any discrepancies.

9.1.5.2 Geometric Integrity

Check for deformation: inspect for any warping or bending, especially in larger prints. Warping can affect fit and function, particularly for parts intended to fit together.

Overhangs and bridges: examine how well the printer managed overhangs and bridging sections. Poor handling of these features can lead to sagging or unsupported areas.

Holes and fittings: verify that holes and fittings are present and of the correct size. This is especially critical for mechanical parts where precision is essential.

9.1.5.3 Functional Testing

Fit testing: for parts that are meant to fit together (like interlocking components), test the fit. Components should assemble smoothly without excessive force or gaps.

Load testing: if applicable, test the object under expected load conditions. This is crucial for functional parts such as brackets or supports to ensure they can withstand operational stresses.

Movement testing: for moving parts, ensure they operate as intended. Check for smooth motion and that there are no binding or sticking points.

9.1.5.4 Mechanical Properties Testing

Tensile testing: conduct tensile tests to assess the strength and flexibility of the printed material. This involves applying force until the material breaks to determine its tensile strength.

Impact testing: assess the impact resistance of the material by dropping a weight onto the object from a specified height and observing whether it deforms or breaks.

Thermal resistance: evaluate the object's thermal properties, especially if it will be used in high-temperature environments. This can include heat deflection temperature tests.

9.1.5.5 Material Inspection

Material composition: ensure that the correct material has been used for printing. Verify the filament type, batch number and expiration date, if applicable.

Moisture content: check for moisture absorption in the filament, which can affect print quality. Filaments like PLA and nylon are particularly sensitive to humidity.

Filament consistency: inspect the filament diameter consistency using a caliper. Variations can lead to issues during printing.

9.1.5.6 Post-Processing Evaluation

Cleaning: assess the effectiveness of any post-processing steps taken, such as removing support material or sanding surfaces. Ensure that the post-processing does not compromise the integrity of the part.

Coating and painting: if the object is painted or coated, evaluate the uniformity and adherence of the finish. Look for bubbles, peeling or uneven application.

9.1.5.7 Documentation and Reporting

Inspection records: maintain detailed records of the inspection and testing process, including photographs and measurements. This documentation is valuable for quality control and future reference.

Defect reporting: if any defects or issues are identified, document them with detailed descriptions and photographs. Include potential causes and suggested remedies.

9.1.5.8 Feedback Loop

Iterative improvement: use the findings from inspections and tests to improve future prints. This might involve adjusting printer settings, selecting different materials, or modifying designs to enhance printability.

9.1.5.9 Compliance and Standards

Industry standards: familiarize yourself with relevant industry standards for 3D printed objects. Compliance may be necessary for specific applications, particularly in regulated industries such as aerospace or medical devices.

Certification testing: for critical applications, consider submitting parts for certification testing to validate their performance against industry standards.

9.2 Other Future Trends in Additive Manufacturing

9.2.1 Hybrid Additive Manufacturing

This is an advanced manufacturing approach that combines the best aspects of traditional subtractive manufacturing methods (such as CNC machining) with modern additive manufacturing (AM) technologies (like 3D printing). The integration of these two processes creates a powerful tool capable of producing high-quality, complex parts with increased efficiency, better material utilization, and greater design freedom. hybrid additive manufacturing (hybrid AM) refers to the combination of additive and subtractive processes in a single setup, allowing manufacturers to both add material and remove excess material within the same production cycle. It leverages the precision of subtractive techniques like milling and the flexibility of additive techniques like 3D printing.

9.2.1.1 Components of Hybrid AM (refer to Figure 9.1)

Additive component: the additive process involves the layer-by-layer building of material, often using metal powders, resins or other materials. Common techniques include direct energy deposition (DED), laser metal deposition (LMD) or powder bed fusion (PBF). These

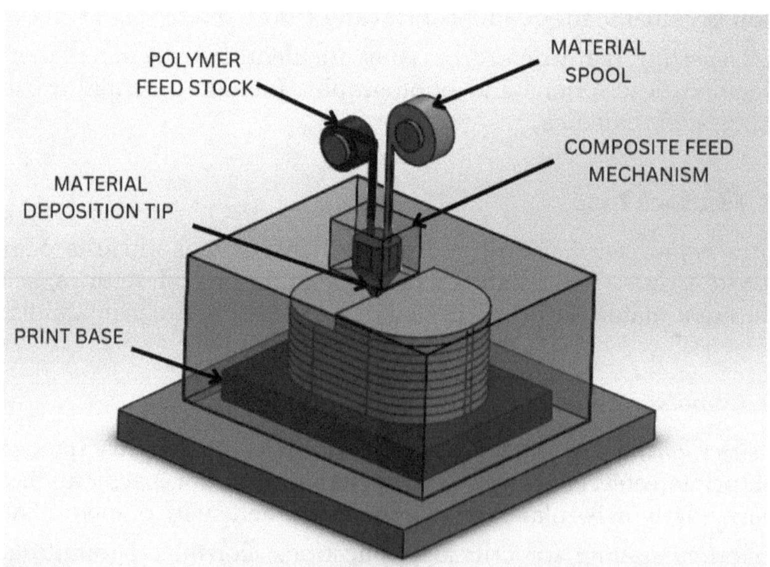

FIGURE 9.1
Hybrid 3D printing.

processes allow complex geometries, intricate internal structures and the ability to add material in specific regions.

Subtractive component: traditional CNC machining or milling is the subtractive part of hybrid AM. This component allows for precision finishing, removing excess material, improving tolerances and creating smooth surfaces in areas where additive processes may fall short.

Integrated control system: a key aspect of hybrid AM is the software that manages both additive and subtractive operations. This system coordinates toolpaths, material deposition and machining instructions seamlessly, ensuring that the two processes are in sync.

9.2.1.2 Advantages of Hybrid Additive Manufacturing

Improved surface finish: while additive manufacturing excels in producing complex geometries, it often results in rough surface finishes. Integrating CNC machining into the process enables post-processing during or after material deposition, leading to smoother surfaces and tighter tolerances.

Enhanced design flexibility: hybrid AM allows for greater design freedom by taking advantage of the capabilities of both additive and subtractive processes. For instance, designers can build internal channels and cavities using additive methods, while maintaining critical tolerances on external features through subtractive techniques.

Reduced production time: in many cases, integrating additive and subtractive processes into a single machine can reduce lead times. Rather than moving parts between different machines for additive production and post-processing, the hybrid system handles both in one workflow, minimizing downtime and transportation.

Material efficiency: additive processes generally generate less waste compared to traditional subtractive methods that remove material from a block. Hybrid systems optimize material use by depositing material only where it is needed, reducing waste and lowering costs.

Complex geometries with high accuracy: components that have intricate internal structures (e.g., lattice structures or conformal cooling channels) can be built using additive techniques, while machining ensures that external features are dimensionally accurate and meet specifications.

Repair and remanufacturing: hybrid AM is ideal for repairing high-value parts, such as turbine blades, moulds or tools. The process can deposit new material onto worn or damaged areas and then machine it back to its original specifications. This capability opens up new possibilities for sustainable manufacturing and extending the life of critical components.

9.2.1.3 Key Technologies in Hybrid AM

Directed energy deposition (DED): DED is a common technology used in hybrid AM. It involves depositing material onto a substrate via a focused energy source such as a laser or an electron beam. Metal powder or wire is fed into the energy source to form the part layer by layer. Once material is added, subtractive processes like CNC milling refine the shape and dimensions.

Laser cladding: a process where a laser is used to melt metal powder or wire onto a substrate, creating a new surface layer. It is often used for coating and repairing parts. Hybrid systems can add this material and then machine it to final dimensions.

Powder bed fusion (PBF): PBF processes, including selective laser melting (SLM) and electron beam melting (EBM), use a laser or electron beam to selectively fuse powdered material together, layer by layer. In hybrid systems, PBF is used for building complex shapes, followed by CNC machining to enhance surface finish and dimensional accuracy.

9.2.1.4 Applications of Hybrid Additive Manufacturing

Aerospace: hybrid AM is highly advantageous in the aerospace sector, where parts often have complex geometries and must meet stringent performance standards. The ability to produce lightweight, high-strength components with intricate internal structures makes it an ideal solution for manufacturing parts like turbine blades, engine components and structural elements.

Medical implants: the medical field benefits from the customization capabilities of hybrid AM. Complex medical implants, such as hip and knee joints, can be designed with patient-specific geometries, produced through additive processes, and machined to precise tolerances to ensure compatibility and performance.

Tool and die making: hybrid AM is transforming tool and die production by enabling conformal cooling channels within molds, which are difficult to achieve with traditional methods. These complex channels can be additively manufactured and then precisely machined to achieve the final form.

Energy sector: in the energy industry, hybrid AM is utilized for producing and repairing high-value components, such as gas turbines and other critical equipment. The ability to repair worn-out parts by adding material and then machining it back to specifications extends the lifespan of expensive equipment.

9.2.1.5 Challenges and Limitations

Complexity in process integration: the synchronization between additive and subtractive processes requires precise control and robust software. Any misalignment or error can result in defects, such as poor surface quality or dimensional inaccuracies.

Cost of hybrid systems: the upfront cost of hybrid machines, which combine both additive and subtractive capabilities, can be high. Although long-term savings in terms of material efficiency and production speed can offset the initial investment, smaller manufacturers may find it difficult to justify the cost.

Material compatibility: not all materials are suitable for both additive and subtractive processes. For example, certain metals or polymers may work well in 3D printing but pose challenges when machined, or vice versa.

*Post-processing requirements:*while hybrid AM reduces the need for post-processing compared to pure AM processes, many parts still require additional treatments such as heat treatment, surface polishing, or coating to meet final specifications.

9.2.1.6 Future Trends in Hybrid Additive Manufacturing

Automation and AI integration: as more hybrid machines are developed, we can expect increased automation of processes and integration with artificial intelligence (AI) for enhanced quality control. AI can monitor additive and subtractive processes in real-time, ensuring high precision and reducing human error.

Multi-material capabilities: future hybrid systems may offer multi-material printing, allowing manufacturers to combine different materials with distinct properties in a single part. This could lead to new possibilities in fields such as electronics, medical devices, and aerospace.

On-site manufacturing: with advancements in portable hybrid systems, on-site manufacturing and repair, especially in remote or challenging environments, will become more common. For example, repairs on oil rigs, ships or even in space could be handled using hybrid machines.

9.2.2 Multi-Material 3D Printing

Multi-material 3D printing represents an advanced stage of additive manufacturing, allowing the simultaneous use of different materials within a single print (refer to Figure 9.2). This ability has revolutionized industries by enabling the creation of complex, functional parts with various properties.

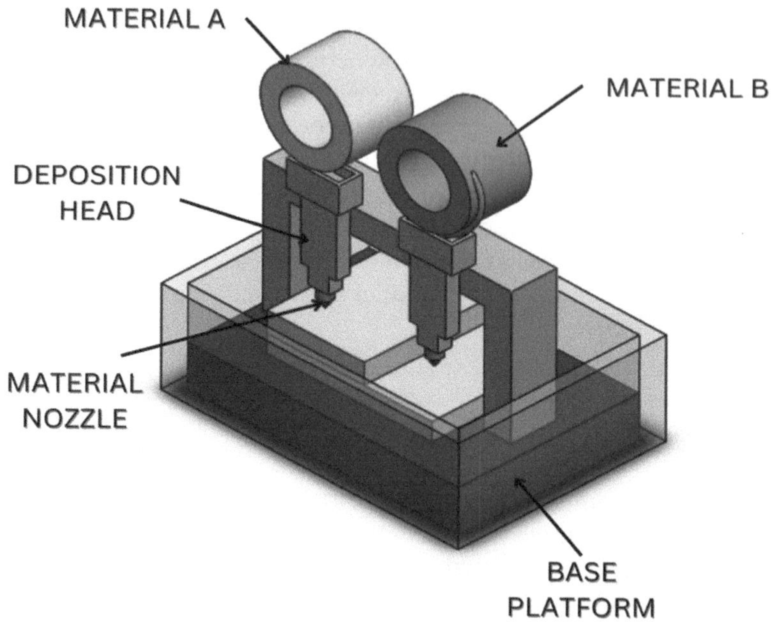

FIGURE 9.2
Multi material 3D printing.

Multi-material 3D printing is particularly valuable in sectors like aerospace, healthcare, automotive and consumer electronics, where components often require a combination of mechanical strength, flexibility and conductivity, among other characteristics.

9.2.2.1 Types of Multi-Material 3D Printing Techniques

There are various methods for multi-material printing, each offering different advantages, depending on the materials and the complexity of the object being printed.

- **Fused deposition modelling (FDM) with multi-material extruders**

  In FDM-based multi-material printing, different filament materials are fed through multiple extruders. This technique is among the most accessible for multi-material printing. By switching between extruders, the printer can deposit various filaments – such as ABS, PLA or flexible thermoplastics – within a single object.

  Advantages: cost-effective, wide material choice, simple setup.

  Limitations: limited resolution when switching between materials, fewer material compatibility options compared to other methods.

- **PolyJet printing**

  PolyJet printers spray multiple types of liquid photopolymers onto a build tray, followed by curing them with UV light. This technique offers extremely high resolution and allows the creation of multi-colour and multi-material objects in a single print run.

  Advantages: high resolution, smooth finishes, multiple material properties, including transparency and flexibility.

  Limitations: expensive equipment and materials, less suitable for large prints.

- **Selective laser sintering (SLS) with multiple powders**

  In multi-material SLS, different powders, such as metal and ceramic, can be sintered layer by layer using lasers. This process is effective for producing parts that combine the properties of different materials, like strength, heat resistance or conductivity.

  Advantages: high mechanical strength, suitable for industrial applications, excellent material fusion.

  Limitations: complex process, high cost, requires post-processing.

- **Multi-material stereolithography (SLA)**

  In SLA-based systems, resins with different properties can be selectively cured by a UV laser. Similar to PolyJet, SLA offers high-resolution capabilities but typically requires post-processing to improve material properties like strength.

  Advantages: high precision, fine details, suitable for intricate parts.

  Limitations: requires post-processing, limited material choices compared to FDM or SLS.

9.2.2.2 Material Classes

Materials for 3D printing generally fall into three categories:

- **Thermoplastics**: ABS, PLA, nylon, and polycarbonate. These are common in FDM printing.
- **Resins**: photosensitive polymers used in SLA and PolyJet printing.
- **Metals and ceramics**: utilized in SLS and direct energy deposition (DED) systems, mainly for industrial applications.

9.2.2.3 Design Considerations for Multi-Material 3D Printing

The potential for multi-material printing opens up new avenues for innovation in product design. However, it also brings unique challenges that designers need to address.

- **Optimizing material layout**

  In multi-material designs, certain areas of the object can be tailored for specific functions:

  Rigid sections for structural support.

  Flexible sections for mechanical movement or shock absorption.

  Conductive materials for electrical pathways.

- **Transition zones**

  Designing smooth transitions between different materials is critical. Sudden changes between rigid and soft materials, for example, can cause stress concentrations, leading to cracks or delamination. Gradients or intermediate layers can be used to soften these transitions.

- **Layer thickness and resolution**

  Each material may have different optimal printing parameters, including layer thickness and print speed. High-resolution materials may be used for aesthetic or fine details, while lower-resolution materials are reserved for bulk areas of the print.

9.2.2.4 Post-Processing in Multi-Material Prints

Post-processing is often necessary for multi-material prints to achieve the desired finish, strength or functionality. Different materials might require unique post-processing techniques, which can complicate the process.

- **Post-curing**

  In SLA and PolyJet printing, UV post-curing solidifies any unreacted photopolymers. Care must be taken to ensure that the curing process does not over-expose certain materials, which can lead to brittleness.

- **Smoothing and polishing**

  For aesthetic applications, multi-material prints might need to be smoothed or polished. FDM prints, for example, can be treated with acetone vapour to give them a glossy finish. However, this can degrade the bond between different materials, requiring caution.

- **Machining and drilling**

  Some multi-material parts, especially in industrial applications, may require additional machining. This poses a challenge if different sections of the part have different hardness levels. Specialized tools or methods might be needed to handle varying material properties.

9.2.2.5 *Applications of Multi-Material 3D Printing*

The versatility of multi-material 3D printing has opened doors to a variety of applications, from prototyping to end-use parts in highly specialized fields.

- **Medical devices and prosthetics**

  Multi-material 3D printing is ideal for producing medical devices that need to combine flexibility and strength. Prosthetics, for example, can be custom-designed to fit a patient's anatomy while using flexible materials for comfort and rigid ones for support.

- **Wearable technology**

  Wearable devices often need to be lightweight, flexible and durable. Multi-material 3D printing allows for the integration of electronics into wearable devices, combining conductive inks with flexible and rigid polymers to create functional, lightweight gadgets.

- **Aerospace and automotive components**

  The aerospace and automotive industries have adopted multi-material printing for lightweight, strong and heat-resistant components. The ability to combine materials like metals and composites reduces the number of parts, simplifies assembly and improves performance.

9.2.2.6 *Challenges and Future Prospects*

Despite the rapid development of multi-material 3D printing, there are several challenges that must be overcome.

- **Material development**

  The need for materials specifically tailored to multi-material 3D printing remains high. This includes materials that can bond well during the printing process, maintain compatibility with others, and retain their properties under stress or in harsh environments.

- **Software and design tools**

  Multi-material design requires more advanced software capable of handling multiple materials and adjusting the printing process parameters for each. Current CAD tools are not fully optimized for this, limiting the complexity of designs that can be easily printed.

- **Cost**

  Multi-material printers, especially those capable of handling metals and high-performance polymers, are significantly more expensive than single-material counterparts. However, as demand increases, these costs are expected to drop, making the technology more accessible.

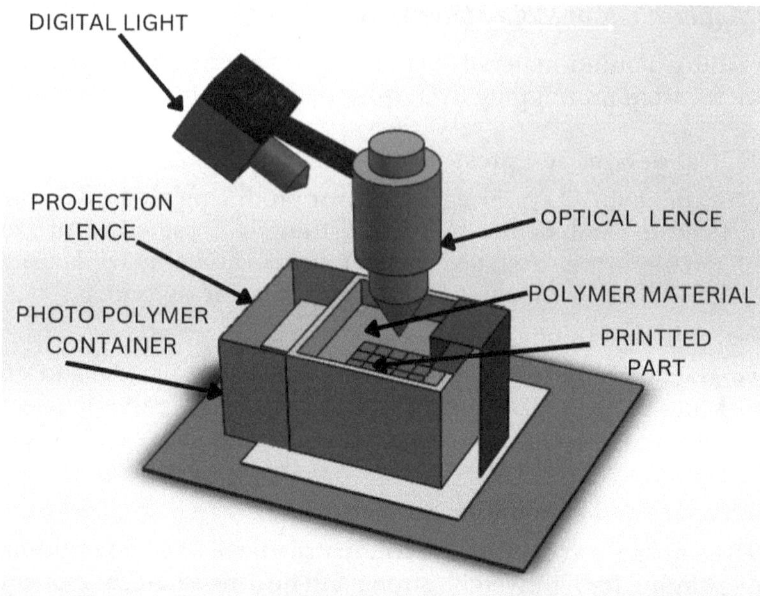

FIGURE 9.3
Micro scale 3D printing.

9.2.3 Micro-Scale 3D Printing

Micro-scale 3D printing, also referred to as micro-additive manufacturing (μ-AM), is a highly precise technology that enables the creation of structures on the scale of micrometers (refer toFigure 9.3). This emerging technology bridges the gap between conventional 3D printing and nanofabrication, offering applications across industries such as electronics, biomedical engineering and materials science.

- **Scale and dimensions**: objects printed at the micro-scale typically range between 1 micron and a few hundred microns. For reference, a human hair is approximately 70 microns thick. This enables the construction of extremely small yet functional parts for various applications.
- **Precision**: micro-scale 3D printing technology relies heavily on precision and control. The ability to fabricate at this level requires highly accurate positioning systems and ultra-fine material deposition techniques.

9.2.3.1 Technologies Involved in Micro-Scale 3D Printing

The key technologies include the following:

- **Two-photon polymerization (TPP)**: this technique uses a laser to polymerize a photosensitive resin. The laser is focused to a small point (a few microns in diameter) and induces polymerization only at the focal point, allowing for highly detailed structures. TPP can achieve resolutions below 100 nm, making it ideal for applications requiring extreme precision, such as biomedical scaffolds or photonic devices.

- **Electrohydrodynamic jet printing (E-Jet):** E-Jet printing is an advanced micro-scale technology that uses an electric field to draw material from a nozzle. By manipulating the electric field and controlling the viscosity of the material, E-Jet can produce structures with resolutions in the range of tens of microns. This is useful for fabricating electronics and sensors.
- **Micro-stereolithography (μ-SLA):** an adaptation of traditional stereolithography, μ-SLA uses a focused UV laser to cure liquid resin layer by layer. The laser's fine focus allows for very small feature sizes, and μ-SLA is frequently used to produce biomedical components, micromechanical devices, and other precision tools.
- **Direct ink writing (DIW):** DIW is a nozzle-based technique where inks, typically composed of polymers, ceramics or even living cells, are extruded to form microstructures. The nozzle diameter controls the resolution of the printed object. DIW allows for multi-material printing at the microscale, making it particularly suitable for applications like tissue engineering and soft robotics.
- **Focused ion beam induced deposition (FIBID):** this process involves the focused deposition of materials using an ion beam, which enables patterning with a resolution of a few nanometers. FIBID is primarily used for prototyping of nanostructures, particularly in electronics.

9.2.3.2 Materials for Micro-Scale 3D Printing

The choice of materials in micro-scale 3D printing is crucial for achieving the desired structural, mechanical and functional properties. Since the precision of the process is directly influenced by the properties of the materials used, the development of suitable materials for micro-scale printing has been an active area of research.

- **Polymers:** polymers are the most common materials used in micro-scale 3D printing due to their versatility and ease of processing. UV-curable resins, used in TPP and μ-SLA, are particularly popular. Advanced polymers with tailored mechanical, thermal or optical properties can be synthesized for specific applications, such as flexible electronics or microfluidics.
- **Metals:** micro-metal printing is often performed using direct ink writing or E-Jet technologies. Metals such as silver, gold, copper and titanium are used to create functional components in electronics, photonics, and mechanical systems.

- **Ceramics**: ceramics offer excellent thermal, electrical and mechanical properties, making them ideal for high-temperature or wear-resistant micro-scale components. Micro-ceramic printing is often achieved using techniques like DIW with specially formulated inks.
- **Bio-Inks**: for biomedical applications, such as tissue engineering or drug delivery, bio-inks composed of hydrogels and living cells are used. These materials are optimized to mimic biological tissues and promote cellular growth.

9.2.3.3 *Applications of Micro-Scale 3D Printing*

The ability to fabricate objects at the micro-scale opens up a wide range of applications across several industries:

- **Biomedical devices**: micro-scale 3D printing is revolutionizing the field of biomedicine. The technology allows for the creation of scaffolds for tissue engineering, micro-needles for drug delivery and even the fabrication of organ-on-a-chip devices. These devices can simulate human organs at a miniature scale, aiding drug development and personalized medicine.
- **Micro-electromechanical systems (MEMS)**: MEMS are small mechanical devices integrated with electronics, such as sensors, actuators and microfluidic systems. Micro-3D printing enables the precise fabrication of these components, leading to innovations in robotics, wearables and medical implants.
- **Optoelectronics and photonics**: the fabrication of optical elements and devices at the micro-scale, such as waveguides, lenses and photonic crystals, has significant applications in communication technologies and sensors. Micro-3D printing enables the customization of these components with high precision.
- **Microfluidics**: micro-scale 3D printing allows for the development of intricate microfluidic devices, used for lab-on-a-chip applications. These devices manipulate tiny amounts of fluids for chemical analysis, biomedical diagnostics, and environmental monitoring.
- **Soft robotics**: micro-scale soft robots can be printed using DIW techniques, using flexible and responsive materials that allow for motion. These robots can be used in minimally invasive surgery or in delicate manipulation tasks in confined spaces.

9.2.3.4 *Challenges in Micro-Scale 3D Printing*

Despite its vast potential, micro-scale 3D printing faces several challenges that need to be addressed:

- **Resolution and accuracy**: as objects become smaller, maintaining precision becomes more difficult. Any small deviation in the printing process can result in substantial errors in the final product, especially at the micron scale.
- **Material limitations**: not all materials are suitable for micro-scale 3D printing, and there are ongoing efforts to develop new materials that can meet the demands of different industries, such as biocompatibility in biomedical applications or conductivity in electronics.
- **Speed**: printing at such a fine resolution is a time-consuming process, especially for larger structures. The need to balance speed with resolution and accuracy is a critical challenge that manufacturers face.
- **Post-processing requirements**: often, additional steps such as curing, sintering or annealing are required after printing to achieve the desired material properties. This adds complexity and cost to the process.
- **Cost**: the equipment required for micro-scale 3D printing is often expensive, and the limited speed of production can make it cost-prohibitive for certain applications. Reducing these costs will be key to wider adoption of the technology.

9.2.3.5 Future Directions and Innovations

- **Multi-material printing**: advances in multi-material printing at the micro-scale are expected to significantly enhance the functionality of printed objects, enabling the integration of different materials with varying properties within a single object. This could lead to new innovations in electronics, soft robotics and biomimetic systems.
- **Increased resolution**: new techniques, such as hybrid additive manufacturing, combine 3D printing with other fabrication methods to achieve even finer resolutions, potentially at the nanoscale.
- **Biofabrication**: the future of micro-3D printing lies in biofabrication, where living tissues, complex biological structures, and even organs can be printed for medical applications. Advances in bio-inks and scaffold materials will drive the development of this field.
- **Scaling up**: while micro-3D printing is focused on small-scale applications, there is potential for scaling the technology to print larger structures with micro-level details. This could revolutionize industries such as aerospace, automotive and energy, where precision components are required.

9.2.4 4D Printing

4D printing builds upon the foundation of 3D printing, but takes it a step further by incorporating smart materials and responsive designs. While 3D

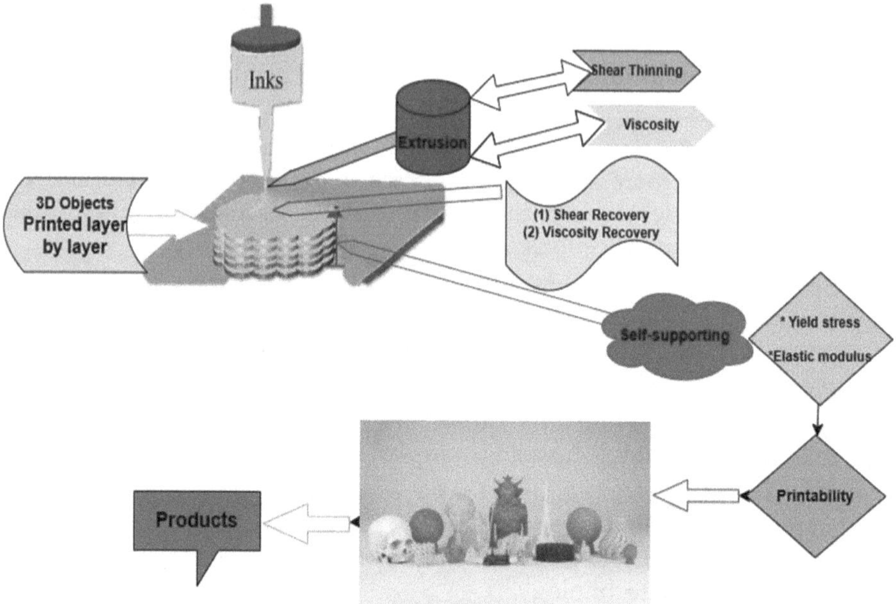

FIGURE 9.4
4D printing process.

printing creates static objects, 4D printing allows these objects to transform their shape and properties over time. This "fourth dimension" refers to the programmed ability of the printed object to change based on external stimuli like temperature, light or moisture (refer to Figure 9.4).

- **Principle:** the core principle of 4D printing lies in using special materials that possess inherent properties to respond to external cues. These materials, including hydrogels, shape-memory polymers and even biomaterials, are programmed during the 3D printing process itself. This programming instructs the material on how to react to specific stimuli, causing it to self-assemble, morph or change functionalities. Imagine printing a flat structure that folds itself into a specific 3D shape when placed in water
- **Smart materials:** the essence of 4D printing lies in the use of materials that can change in response to external factors. These materials are programmable and designed to have certain "memory" properties that allow them to return to or adopt new shapes under specific conditions.
- **Shape-memory polymers (SMPs):** these materials can return to a preprogrammed shape when exposed to heat or light. They have been widely used in medical devices like stents and in structural components in aerospace.

- **Hydrogels:** these are polymers that swell and shrink in response to moisture or temperature. They are commonly used in bioengineering and medical applications.
- **Liquid crystal elastomers (LCEs):**these combine the properties of liquid crystals with rubber-like elasticity, allowing them to respond to temperature changes with a reversible shape transformation.
- **Multi-material printing:** 4D printing often relies on the use of multiple materials with different properties to enable complex transformations. For example, a combination of rigid and flexible materials might allow a flat object to curl into a tube when heated.
- **Software and design:** designing for 4D printing involves complex algorithms that simulate how materials will behave under different stimuli. CAD (computer-aided design) and CAM (computer-aided manufacturing) systems are used to model these transformations accurately, ensuring that the object behaves as intended after printing.

- **Applications of 4D printing**

✓ **Aerospace and defense**

Self-repairing components: 4D printing could allow for the creation of materials that self-repair when damaged. For example, a wing of an aircraft could automatically seal itself after a crack forms.

Deployable structures: satellites and spacecraft could use 4D-printed components that unfold or expand once they reach space, reducing launch volume and weight.

✓ **Medical field**

Biocompatible devices: 4D-printed stents or implants could be made to expand or change shape in response to the body's temperature, improving patient outcomes.

Smart structures: sutures that change tension in response to healing rates or that dissolve over time are an exciting potential application of 4D printing in surgery.

✓ **Textiles and fashion**

Adaptive clothing: clothing that adjusts its properties based on environmental conditions, like a jacket that opens ventilation when the temperature rises, could be created using 4D printing.

Shape-changing footwear:footwear could be designed to change its structure to provide better support or comfort depending on the user's activity.

✓ **Construction**

Self-assembling structures: 4D printing could revolutionize construction by enabling materials that self-assemble into complex structures when triggered by environmental factors such as heat or moisture. This could reduce labour costs and assembly time.

✓ **Robotics**

Soft robotics: the field of robotics could use 4D-printed components for soft robotic arms or hands that can change stiffness or flexibility depending on the task, enabling more adaptable, lifelike machines.

9.2.4.1 Challenges and Limitations

- **Material limitations:** while smart materials have been through a breakthrough, there are still limitations in the range of stimuli-responsive materials available. Researchers are continuously working to develop materials that respond to a wider array of stimuli with greater precision and durability.
- **Complexity in design:** designing for 4D printing requires precise knowledge of how different materials will interact and respond to external factors. This adds complexity to the design process, making it more time-consuming and costly.
- **Cost:** currently, 4D printing is more expensive than traditional manufacturing processes, including 3D printing, due to the advanced materials and specialized software required. This limits its adoption to niche applications.
- **Speed:** the time it takes for 4D-printed objects to transform can vary, depending on the material and stimulus. Some transformations occur in seconds, while others can take hours, limiting the technology's practicality in certain situations.

9.2.5 3D Printing Coloured Objects

3D printing has revolutionized manufacturing by enabling the production of complex, customized objects directly from digital designs. A further advancement in this technology is the ability to produce coloured objects, adding a new dimension to aesthetics and functionality.

9.2.5.1 Methods for 3D Printing Coloured Objects

There are several methods to incorporate colour into 3D printed objects. These vary in complexity and the degree of colour customization available.

- **Filament-based colour mixing (FDM)**

  This method involves using coloured filaments, either pre-coloured or mixed during the printing process. Some advanced printers can switch between multiple filament colours or even blend them in real-time.

Single extruder switching: printers with a single extruder can switch between colours, printing in one colour, then pausing to switch filament to another.

Multi-extruder systems: printers with multiple extruders can use several colours simultaneously, with each extruder assigned a different color filament.

Pros: simple and relatively low-cost for consumer-level printers.

Cons: limited to a few colours, and transitions between colours can be abrupt.

- **Full-colour binder jetting**

  This technique uses powdered materials and injects colour directly into the powder layer by layer, allowing for a full-colour object with a high resolution and vibrant hues.

  Process: a binding agent and coloured ink are sprayed on a layer of powdered material, binding and colouring it simultaneously.

  Pros: enables photo-realistic colour applications and intricate colour gradients.

  Cons: requires post-processing for strength, such as infiltration with a resin or wax. The object's durability can be a concern in some applications.

- **PolyJet technology**

  PolyJet printing uses photopolymer resin, which is cured by ultraviolet light. This method can produce objects in multiple colours and textures by mixing resins during the print process.

  Process: the printer jets layers of liquid photopolymer onto a build tray and cures them with UV light. By mixing different resin colours, it can achieve high colour fidelity and smooth finishes.

  Pros: high detail and smooth surface finishes with accurate colour representation.

  Cons: higher costs and the need for professional-grade equipment.

- **Voxel-based colouring (3D inkjet printing)**

  Voxel-based methods allow each voxel (3D pixel) of an object to be individually coloured, leading to extremely detailed and high-resolution prints.

  Process: the printer jets layers of material, each voxel infused with a color value, similar to how a 2D inkjet printer works.

  Pros: unparalleled colour control and the ability to produce highly detailed, coloured models.

  Cons: high cost and slower printing times.

9.2.5.2 Post-Processing Techniques

Sometimes colour is applied after the object has been printed. This can be done through painting, dyeing or airbrushing.

Pros: flexibility in adding colours to prints and creating custom colour schemes without needing specialized 3D printing equipment.

Cons: labour-intensive and may affect the surface finish or durability of the object.

9.2.5.3 Challenges in 3D Printing Coloured Objects

While technology has advanced significantly, there are still several challenges in 3D printing coloured objects:

Colour fidelity: achieving the exact colour as in the digital model can be difficult, especially when blending multiple filaments or working with complex colour patterns.

Material compatibility: not all materials can be easily coloured, and some require specialized processing.

Resolution vs. colour accuracy: higher colour accuracy often requires trade-offs in print resolution, especially when mixing colours during the process.

*Cost:*many colour 3D printing technologies are expensive, making them inaccessible for small businesses or hobbyists.

9.2.5.4 Applications of Colour 3D Printing

Coloured 3D printing has opened up new possibilities across various industries:

Prototyping: full-colour prototypes allow designers and engineers to visualize products more accurately, spotting potential design issues early.

Medical models: surgeons use coloured anatomical models for presurgical planning, enhancing the understanding of complex procedures.

Architecture: full-colour models give architects the ability to present realistic models of buildings or landscapes, providing better visualization for clients.

Art and fashion: artists and designers are leveraging colour 3D printing to create intricate, multicoloured objects that would not be possible using traditional methods.

9.2.6 3D Bioprinting

9.2.6.1 3D Bioprinters

3D bioprinting is an emerging field that applies 3D printing technology to create biological structures, such as tissues, organs and other medical applications, using biomaterials. Unlike traditional 3D printing, which uses plastics or metals, bioprinting involves cells and biomaterials, presenting unique opportunities and challenges.The process begins with the design of a digital model of the tissue or organ using computer-aided design (CAD). This model serves as a blueprint for printing each layer. Bioinks –materials composed of living cells and nutrients – are then extruded layer by layer to create a 3D structure. Unlike traditional 3D printing, where the printed objects are static and rigid, bioprinting requires living cells that continue to grow and differentiate after the printing process

9.2.6.2 Materials Used in 3D Bioprinting

- **Bioinks**: these are the "ink" in 3D bioprinting, consisting of living cells, growth factors and biomaterials. Bioinks need to be biocompatible, meaning they should not cause adverse reactions when implanted into the body. Hydrogels are a common bioink material because they can encapsulate cells and provide a supportive environment similar to human tissue.
- **Scaffolds**: scaffolds provide mechanical support and help cells to grow in a specific pattern. They need to be porous and biodegradable so that cells can migrate through the structure and eventually replace the scaffold with natural tissue. The scaffold material should degrade at the same rate as the new tissue grows.
- **Natural vs.synthetic materials**: natural materials (such as collagen or alginate) are often used because they closely resemble human tissue, but synthetic materials (like polyethylene glycol) can be designed to have specific mechanical properties that natural materials may lack.

9.2.6.3 Types of 3D Bioprinting Technologies

- **Inkjet bioprinting**: this technology works similarly to a traditional inkjet printer but uses bioinks instead of ink. It deposits tiny droplets of bioink onto a substrate in a controlled manner to build the tissue layer by layer. Inkjet bioprinting is fast and can handle multiple types of bioinks, but it is limited by the type of materials it can print (refer to Figure 9.5).
- **Extrusion-based bioprinting**: in extrusion printing, a continuous flow of bioink is extruded through a nozzle to create patterns. This method

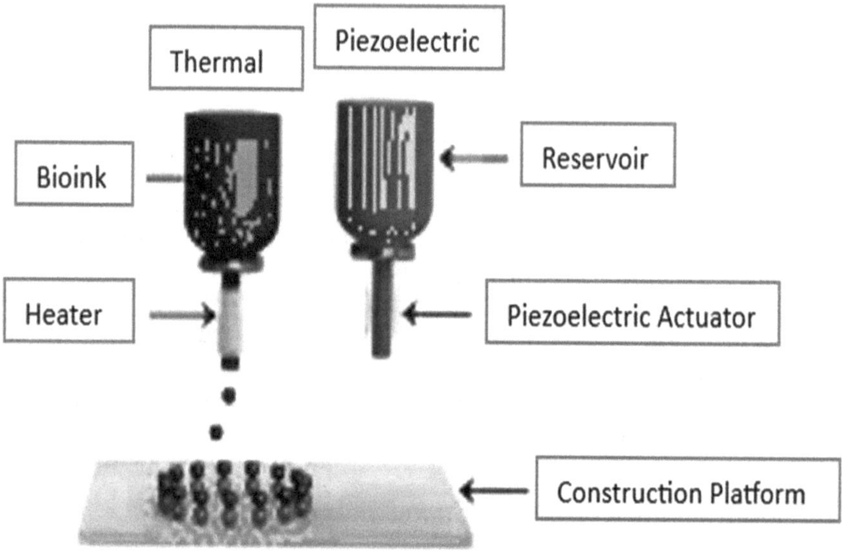

FIGURE 9.5
Ink jet bio 3D printing.

allows for higher cell density and can print more complex structures but is slower than inkjet bioprinting.

- **Laser-assisted bioprinting**: this technique uses a laser to focus on a substrate, which transfers cells and biomaterials onto a surface. It provides a high level of precision and allows for the printing of complex cell patterns, but it is expensive and less commonly used (refer to Figure 9.6).

9.2.6.4 The Process of 3D Bioprinting

- **Preprocessing**: this involves designing the structure to be printed, including the shape, cell types and the positioning of biomaterials. It begins with imaging techniques, such as CT or MRI scans, to obtain data that will be used to construct a CAD model.
- **Processing (printing)**: the 3D bioprinter is programmed to follow the blueprint from the preprocessing stage. The printer deposits the bioink layer by layer, forming the desired 3D structure. The speed, temperature and pressure during the printing process must be carefully controlled to maintain cell viability.
- **Post-processing**: after printing, the structure may need to undergo maturation in a bioreactor, where the cells can proliferate, differentiate and integrate into a functional tissue. This step ensures the bioprinted tissue

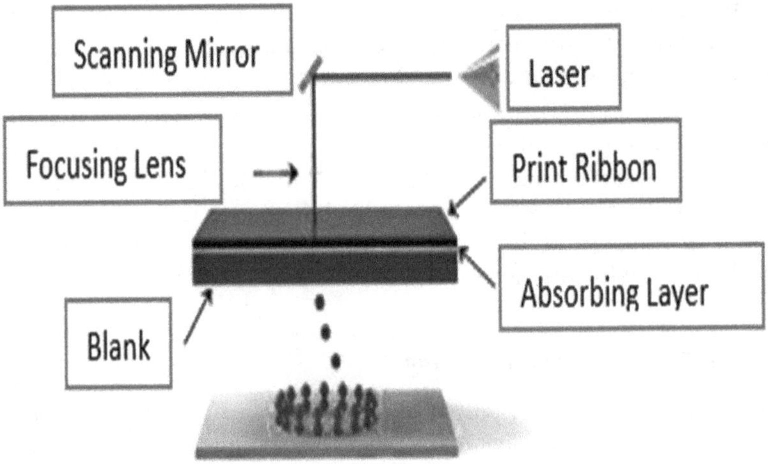

FIGURE 9.6
Laser assisted bio 3D printing.

closely mimics natural tissue in terms of mechanical and biological properties.

9.2.6.5 Applications of 3D Bioprinting

- **Tissue engineering and regenerative medicine**: one of the most exciting applications is the potential to print tissues and organs for transplantation. Bioprinted tissues could be used to replace damaged skin, cartilage or even organs like the liver or kidney.
- **Drug testing and personalized medicine**: pharmaceutical companies are using bioprinted tissues to test the efficacy and safety of new drugs. By printing tissues from a patient's own cells, researchers can tailor drug treatments to individual patients, minimizing side effects and improving outcomes.
- **Cosmetic and skin applications**: companies are exploring the potential for bioprinted skin in wound healing and cosmetic surgery. Bioprinted skin could be used to test cosmetics or for grafting onto patients with severe burns.

9.2.6.6 Challenges and Limitations of 3D Bioprinting

- **Vascularization**: one of the main obstacles is creating tissues with a proper vascular system (blood vessels) to supply nutrients and remove waste. Without blood vessels, thick tissues cannot survive, as cells in the center die from lack of oxygen and nutrients.

- **Cell viability and functionality**: keeping cells alive and functional during and after the printing process is challenging. Cells may undergo stress during extrusion or inkjet printing, leading to a loss of function.
- **Ethical and regulatory issues**: as bioprinting moves closer to the production of organs, ethical concerns arise. How do we ensure equitable access to these technologies? Should there be limits to what types of tissues or organs can be printed? Additionally, regulatory agencies must develop guidelines to ensure that bioprinted products are safe and effective for medical use.

9.2.6.7 The Future of 3D Bioprinting

- **Advances in vascularization**: researchers are working on techniques to bioprint blood vessels or use sacrificial materials that dissolve after printing, leaving behind channels for blood flow.
- **Multi-material printing**: in the future, bioprinters may be able to print with multiple types of bioinks simultaneously, allowing for more complex tissues with different types of cells.
- **Organ printing**: while functional organ printing is still in its infancy, advances in bioprinting technology could eventually lead to the production of entire organs, solving the organ donor shortage.
- **Personalized medicine**: bioprinting could play a significant role in the future of personalized medicine. By using a patient's own cells, it may be possible to create tissues and organs that are a perfect genetic match, reducing the risk of rejection.

9.3 Sustainability in Additive Manufacturing

9.3.1 Material Efficiency

- **Reduced waste generation**: traditional manufacturing processes, such as subtractive machining, often result in significant material waste. AM, on the other hand, builds objects layer by layer from digital models, using only the necessary material for production. This intrinsic efficiency can reduce overall material waste by up to 90%.
- **Use of recycled materials**: many additive manufacturing processes can utilize recycled materials, such as plastics and metals, reducing the need for virgin resources. This can lead to a circular economy, where materials are reused, reprocessed and reintroduced into the production cycle.

- **Biodegradable and sustainable materials**: the development of bio-based materials, such as polylactic acid (PLA), which is derived from renewable resources like corn starch or sugarcane, enhances sustainability in AM. These materials can decompose under the right conditions, minimizing environmental impact.

9.3.2 Energy consumption

- **Lower energy requirements**: compared to traditional manufacturing methods, AM often requires less energy. This is due to its material efficiency and the fact that AM processes can be more flexible, allowing for shorter production runs without extensive setup.
- **Localized production**: AM enables localized manufacturing, which can reduce transportation emissions associated with moving materials and finished products. By producing goods closer to the point of use, energy consumption linked to logistics is minimized.
- **Energy-efficient processes**: emerging technologies in AM, such as selective laser sintering (SLS) and digital light processing (DLP), are being optimized for energy efficiency. Innovations in process control and machine design can further lower energy requirements.

9.3.3 Waste Reduction

- **On-demand production**: AM allows for on-demand production, which helps avoid overproduction and reduces the need for extensive inventory storage. This just-in-time approach can significantly lower waste, especially in industries like fashion and consumer goods.
- **Support structures and reusability**: many AM technologies use support structures to stabilize parts during printing. These supports can often be reused or recycled, further reducing waste. Additionally, innovations in design allow for more complex geometries that require less support material.

9.3.4 Economic Implications

- **Cost savings**: although the initial setup costs for AM can be high, the long-term savings associated with material usage, reduced waste and lower energy consumption can make it a cost-effective choice. This is especially true for small production runs or custom parts.
- **Market flexibility**: AM allows for rapid prototyping and design iteration, which can lead to quicker responses to market changes. This flexibility can be crucial for businesses looking to innovate or adapt to consumer demands without incurring significant costs.

- **Job creation**: the rise of AM technologies may create new jobs in design, machine operation, and material development. As the industry grows, so does the potential for skilled labor opportunities in sustainable manufacturing practices.

9.3.5 Social Impact

- **Customization and accessibility**: AM can enable customized products that meet specific user needs, from prosthetics tailored to individual patients to unique designs in consumer goods. This personalization can enhance the quality of life for users and broaden accessibility to specialized products.
- **Community development**: localized production through AM can bolster community economies by enabling small businesses and entrepreneurs to produce goods without the need for large-scale manufacturing facilities. This can foster innovation and support local job markets.
- **Education and awareness**: the integration of AM in educational institutions promotes awareness about sustainable practices and environmental responsibility. As students learn about and engage with these technologies, they are better equipped to pursue sustainable solutions in their future careers.

9.3.6 Lifecycle Assessment

- **Holistic evaluation of impact**: implementing a lifecycle assessment (LCA) approach can help evaluate the environmental impact of AM from raw material extraction to end-of-life disposal. Understanding the full impact enables manufacturers to make informed decisions that enhance sustainability.
- **End-of-life solutions**: AM technologies can facilitate easier disassembly and recycling of products, enabling more sustainable end-of-life management. This is critical in reducing landfill waste and promoting a circular economy where materials are reused rather than discarded.

9.3.7 Regulatory and Standards Development

- **Industry guidelines**: as AM technologies evolve, so does the need for regulations and standards that address sustainability. Developing and adhering to industry standards can ensure that AM practices remain environmentally friendly and socially responsible.
- **Certifications and labels**: certifications for sustainable materials and processes can help consumers make informed choices about the

products they purchase. Standards for AM sustainability can guide manufacturers in improving their practices and transparency.

9.4 Integration of AI and Machine Learning in Additive Manufacturing

Integrating AI and machine learning (ML) into AM is transforming the industry by enhancing processes, improving product quality, and enabling more efficient operations.

9.4.1 Process Optimization

AI algorithms can analyze vast amounts of data from the additive manufacturing process to identify patterns and correlations that human operators might miss.

- **Data collection**: sensors and IoT devices collect real-time data during the AM process, including temperature, humidity, layer thickness and print speed.
- **Data analysis**: machine learning models can process this data to find optimal settings for specific materials and designs, adjusting parameters dynamically to improve print quality.
- **Feedback loops**: implementing feedback loops allows the system to learn from previous builds and continuously improve, leading to reduced trial and error.

9.4.2 Predictive Maintenance

Machine learning can help predict equipment failures before they occur, reducing downtime and maintenance costs.

- **Data monitoring**: AI systems can monitor the health of AM machines by analyzing performance data over time.
- **Failure prediction**: using historical data, ML algorithms can predict when a machine is likely to fail or require maintenance, allowing proactive measures to be taken.
- **Maintenance scheduling**: by optimizing maintenance schedules based on usage patterns and predictions, companies can reduce operational costs and improve reliability.

9.4.3 Quality Control

AI technologies enhance quality control processes in additive manufacturing by analyzing defects and ensuring product consistency.

- **Defect detection**: machine learning algorithms can analyze images from cameras positioned around the printing area to detect defects in real time, such as warping, layer misalignment or material inconsistencies.
- **Statistical process control**: AI can apply statistical methods to monitor the manufacturing process continuously, ensuring that it stays within predetermined quality parameters.
- **Root cause analysis**: when defects are detected, AI can assist in identifying the root causes by correlating the data from various sensors and logs, helping engineers troubleshoot and improve processes.

9.4.4 Design Optimization

AI can assist in the design process, allowing for the creation of more complex and efficient geometries that take full advantage of additive manufacturing's capabilities.

- **Generative design**: AI algorithms can explore a wide range of design alternatives based on specified constraints and objectives (e.g., weight, strength, material usage), producing innovative designs that may not be intuitive to human designers.
- **Topology optimization**: Machine learning can optimize the material distribution within a given volume, enhancing performance while minimizing material usage.
- **Simulation and validation**: AI can simulate how designs will perform under various conditions, speeding up the validation process before physical prototyping.

9.4.5 Supply Chain and Inventory Management

AI and machine learning can streamline supply chain operations and optimize inventory levels in the context of additive manufacturing.

- **Demand forecasting**: ML algorithms can analyze historical sales data and market trends to forecast demand for specific products, enabling better inventory management.
- **Material optimization**: AI can help determine the optimal materials needed for production based on design requirements, costs, and availability, thereby reducing waste and cost.

- **On-demand production**: integrating AI with AM allows for on-demand production models, reducing the need for large inventories and enabling just-in-time manufacturing.

9.4.6 Customization and Personalization

AI can facilitate the customization of products to meet individual customer needs, a significant advantage of additive manufacturing.

- **Customer data analysis**: by analyzing customer preferences and historical data, AI can suggest personalized designs tailored to specific requirements.
- **Real-time adaptation**: AI systems can adapt designs in real time based on user feedback or changing requirements during the manufacturing process.
- **User-Friendly Interfaces**: AI-driven design tools can provide intuitive interfaces for customers to customize their products, expanding the potential market for AM services.

9.4.7 Material Development

AI can accelerate the development of new materials for additive manufacturing by predicting material properties and performance.

- **Material property prediction**: ML models can analyze existing data on material compositions and their properties to predict how new formulations will perform.
- **Experimentation automation**: AI can automate the experimentation process by determining which material combinations to test, speeding up the R&D phase.
- **Optimizing material usage**: AI can help identify the most efficient use of materials, minimizing waste and improving sustainability in additive manufacturing.

9.4.8 Regulatory Compliance and Certification

Ensuring compliance with regulatory standards in AM is crucial, and AI can streamline this process.

- **Documentation automation**: AI systems can automate the generation of documentation required for compliance, ensuring that all necessary data is collected and recorded accurately.

- **Real-time monitoring for compliance**: AI can monitor manufacturing processes to ensurethey comply with industry standards, automatically flagging deviations from protocols.
- **Data-driven audits**: machine learning can analyze past compliance data to improve audit processes, making them more efficient and less prone to human error.

Self-Check Exercises

1. Explore various post-processing techniques such as sanding, polishing and painting for enhancing SLA printed object aesthetics.
2. Explore various post-processing techniques such as bead blasting, infiltration, and dyeing for enhancing the aesthetics and durability of powder-based 3D printed objects.
3. Develop a checklist and procedure for inspecting and testing SLA/DLP printed objects to ensure quality and functionality.
4. Develop a methodology of hybrid additive manufacturing (HAM), which is an amalgamation of vat photopolymerization (VPP) based 3D printing and pressureless sintering, for the fabrication of Ti6Al4V parts for biomedical applications.
5. What are the different classes of thermoresponsive polymers (TRP) and their applications in multi-material 3D printing, particularly in fields like biosensors, biofabrication and biotechnology?
6. Identify and describe three types of smart materials used in 4D printing. How does each material respond to external stimuli?
7. Outline the three main stages of the 3D bioprinting process and Identify three medical applications of 3D bioprinting and describe one in detail.
8. Identify a biobased material used in AM and discuss its environmental benefits. Provide an example of a customized product made possible by AM and discuss its social benefits.
9. Describe the role of data collection and analysis in optimizing additive manufacturing processes and explain the purpose of feedback loops in process optimization, and how do they contribute to reduced trial and error.
10. What are shape-memory polymers (SMPs), and how are they utilized in medical devices? Provide an example of their application.

Multiple Choice Questions

1. Which post-processing technique gives thermoplastics a mirror-like finish after sanding?
   (a) Gap filling
   (b) Priming and painting
   (c) Polishing
   (d) Cold welding
2. Which solvent is used for dissolving HIPS support material in an ABS print?
   (a) Plain water
   (b) Ethanol
   (c) A 1:1 ratio of (R)-(+)-limonene and isopropyl alcohol
   (d) D. Acetone
3. What is the purpose of support removal in sheet-based 3D printing processes?
   (a) To improve the surface finish of the printed part
   (b) To remove excess material around the edges of the part
   (c) To provide stability to overhanging or complex geometries during printing
   (d) To improve the strength of the final part
4. Which post-processing technique involves heating a printed part to relieve internal stresses?
   (a) Trimming and edge smoothing
   (b) Surface finishing
   (c) Annealing
   (d) Dimensional accuracy adjustments
5. Which method is commonly used to remove uncured resin residue from SLA-printed parts?
   (a) Heating in an oven
   (b) Sandblasting
   (c) Solvent bath or ultrasonic cleaning
   (d) Mechanical cutting
6. What is the purpose of post-curing in SLA-printed parts?
   (a) To reduce the weight of the part
   (b) To fully strengthen the part by promoting additional polymerization
   (c) To remove support structures
   (d) To improve surface adhesion for coatings
7. What is the primary purpose of depowdering in powder-based 3D printing post-processing?
   (a) To improve the surface finish of the part
   (b) To remove loose, unsintered powder from the printed part
   (c) To increase the dimensional accuracy of the part
   (d) To enhance the mechanical properties of the part

8. Which post-processing method is used to relieve internal stresses and enhance mechanical properties by controlled heating?
   (a) Media blasting
   (b) Depowdering
   (c) Annealing
   (d) Infiltration
9. In the context of infiltration, which material is commonly used to improve the strength of porous metal parts?
   (a) Bronze
   (b) Polymer
   (c) Aluminum oxide
   (d) Nickel
10. Which test would be most relevant for a 3D printed part intended to withstand high temperatures?
    (a) Moisture content test
    (b) Impact testing
    (c) Thermal resistance testing
    (d) Geometric integrity check
11. What is the purpose of conducting load testing on a 3D printed object?
    (a) To ensure the object can withstand operational stresses
    (b) To measure dimensional accuracy
    (c) To evaluate thermal properties
    (d) To check for moisture absorption in the filament
12. Which technology is commonly used in hybrid AM for adding material layer by layer?
    (a) Powder bed fusion (PBF)
    (b) Computer numerical control (CNC) milling
    (c) Laser sintering
    (d) Surface grinding
13. What is one of the major advantages of hybrid AM compared to traditional additive manufacturing?
    (a) It eliminates the need for any post-processing
    (b) It allows for the repair of worn or damaged parts
    (c) It only uses metal materials
    (d) It requires less precise software for control
14. Which material category is commonly used in selective laser sintering for multi-material 3D printing?
    (a) Thermoplastics
    (b) Metals and ceramics
    (c) Resins
    (d) Elastomers

15. In multi-material 3D printing, why are gradient or intermediate layers important between different materials?
   (a) To speed up the printing process
   (b) To prevent stress concentrations and reduce cracking or delamination
   (c) To allow for faster cooling
   (d) To minimize the use of UV light during curing

16. What is the typical size range for objects created with micro-scale 3D printing?
   (a) 1 to 10 microns
   (b) 1 micron to a few hundred microns
   (c) 100 microns to 1 mm
   (d) 1 mm to 10 mm

17. Which technology in micro-scale 3D printing uses a laser to polymerize photosensitive resin at a focal point?
   (a) Electrohydrodynamic jet printing (E-Jet)
   (b) Direct ink writing (DIW)
   (c) Two-photon polymerization (TPP)
   (d) Micro-stereolithography (μ-SLA)

18. What is the "fourth dimension" in 4D printing?
   (a) Colour
   (b) Weight
   (c) Transformation over time
   (d) Size

19. Which of the following materials is commonly used in 4D printing due to its ability to change shape in response to moisture?
   (a) Shape-memory polymers (SMPs)
   (b) Liquid crystal elastomers (LCEs)
   (c) Hydrogels
   (d) Thermoplastics

20. In the medical field, a key application of 4D printing is the creation of:
   (a) Coloured surgical models
   (b) Smart sutures that adjust based on healing rates
   (c) Deployable satellites
   (d) Full-colour anatomical models

21. Which 3D printing technique allows for high colour fidelity and smooth finishes by mixing photopolymer resins?
   (a) Filament-based colour mixing (FDM)
   (b) Full-colour binder jetting
   (c) PolyJet technology
   (d) Voxel-based colouring

22. What is a primary difference between traditional 3D printing and 3D bioprinting?
    (a) Traditional 3D printing uses bioinks.
    (b) 3D bioprinting involves living cells and biomaterials.
    (c) Traditional 3D printing cannot use CAD models.
    (d) 3D bioprinting creates rigid, static structures.
23. Which type of bioprinting technology uses a continuous flow of bioink to create complex structures with higher cell density?
    (a) Inkjet bioprinting
    (b) Laser-assisted bioprinting
    (c) Extrusion-based bioprinting
    (d) Scaffold printing
24. Which bioink material is commonly used for its ability to encapsulate cells and mimic human tissue?
    (a) Polyethylene glycol
    (b) Hydrogels
    (c) Polymers
    (d) Ceramics
25. What is the purpose of the post-processing stage in 3D bioprinting?
    (a) Designing the CAD model
    (b) Printing the bioink layers
    (c) Allowing the printed structure to mature in a bioreactor
    (d) Developing ethical and regulatory guidelines
26. What type of material used in additive manufacturing is derived from renewable resources and can decompose under the right conditions?
    (a) Polylactic acid (PLA)
    (b) Polyethylene terephthalate (PET)
    (c) Polyvinyl chloride (PVC)
    (d) Polycarbonate (PC)
27. Which of the following additive manufacturing processes is known for being optimized for energy efficiency?
    (a) Subtractive machining
    (b) Selective laser sintering (SLS)
    (c) Injection molding
    (d) Die casting
28. How does on-demand production in additive manufacturing contribute to waste reduction?
    (a) By increasing inventory storage needs
    (b) By allowing overproduction
    (c) By reducing the need for extensive inventory storage
    (d) By requiring complex support structures

29. In quality control for additive manufacturing, which AI-driven feature helps detect defects in real-time?
    (a) Predictive maintenance
    (b) Generative design
    (c) Defect detection
    (d) Documentation automation
30. Which AI technique is used to create complex geometries in additive manufacturing, optimizing material distribution within a given volume?
    (a) Topology optimization
    (b) Maintenance scheduling
    (c) Demand forecasting
    (d) Root cause analysis
31. What role does AI play in improving regulatory compliance in additive manufacturing?
    (a) It eliminates the need for regulatory audits
    (b) It reduces the documentation required
    (c) It automates documentation and monitors real-time compliance
    (d) It replaces all human operators with AI systems
32. How does AI contribute to defect detection in additive manufacturing?
    (a) By manually inspecting each product
    (b) By monitoring statistical parameters without image analysis
    (c) By analyzing images to detect defects like warping or misalignment in real-time
    (d) By relying on human quality control inspectors

Answers

1. (c) Polishing
2. (c) A 1:1 ratio of (R)-(+)-limonene and isopropyl alcohol
3. (c) To provide stability to overhanging or complex geometries during printing
4. (c) Heat treatment (annealing)
5. (c) Solvent bath or ultrasonic cleaning
6. (b) To fully strengthen the part by promoting additional polymerization
7. (b) To remove loose, unsintered powder from the printed part
8. (c) Annealing
9. (a) Bronze
10. (c) Thermal resistance testing
11. (a) To ensure the object can withstand operational stresses
12. (a) Powder bed fusion (PBF)

13. (b) It allows for the repair of worn or damaged parts
14. (b) Metals and ceramics
15. (b) To prevent stress concentrations and reduce cracking or delamination
16. (b) 1 micron to a few hundred microns
17. (c) Two-photon polymerization (TPP)
18. (c) Transformation over time
19. (c) Hydrogels
20. (b) Smart sutures that adjust based on healing rates
21. (c) PolyJet technology
22. (b) 3D bioprinting involves living cells and biomaterials.
23. (c) Extrusion-based bioprinting
24. (b) Hydrogels
25. (c) Allowing the printed structure to mature in a bioreactor
26. (a) Polylactic acid (PLA)
27. (b) Selective laser sintering (SLS)
28. (c) By reducing the need for extensive inventory storage
29. (c) Defect detection
30. (a) Topology optimization
31. (c) It automates documentation and monitors real-time compliance
32. (c) By analyzing images to detect defects like warping or misalignment in real-time

Bibliography

3D Systems. (n.d.). Selective laser sintering. Retrieved from www.3dsystems.com/selective-laser-sintering

Agarwal, R., et al. (2023). Additive manufacturing: Materials, technologies, and applications. In Pulak Mohan Pandey, Nishant K. Singh, Yashvir Singh (Eds.) *Additive manufacturing* (pp. 77–97). CRC Press.

Alojaly, H. M., Hammouda, A., & Benyounis, K. Y. (2024). Review of recent developments on metal matrix composites with particulate reinforcement. *Comprehensive Materials Processing*, 350–373. https://doi.org/10.1016/B978-0-323-96020-5.00041-8

Ataee, A., Li, Y., Song, G., & Wen, C. (2017). Metal scaffolds processed by electron beam melting for biomedical applications. In *Metallic foam bone: Processing, modification, characterization, and properties* (pp. 83–110). Elsevier. https://doi.org/10.1016/B978-0-08-101289-5.00003-2

Bandyopadhyay, A., et al. (2019). Introduction to additive manufacturing. In Amit Bandyopadhyay, Susmita Bose (Eds.) *Additive manufacturing, second edition* (pp. 1–23). CRC Press.

Chang, K. H. (2015). Rapid prototyping. In Amit Bandyopadhyay, Susmita Bose (Eds.) *e-Design* (pp. 743–786). Elsevier. https://doi.org/10.1016/B978-0-12-382038-9.00014-4

Chua, C. K., Leong, K. F., & An, J. (2021). Additive manufacturing and 3D printing. In Narayan, R. (ed) *Biomedical Materials*, Springer, Cham. https://doi.org/10.1007/978-3-030-49206-9_19.

Dasgupta, N., et al. (2023). *3D printing in biotechnology: Current technologies and applications*. Elsevier.

Deshmukh, K., Muzaffar, A., Kovářík, T., Křenek, T., Ahamed, M. B., & Pasha, S. K. K. (2020). Fundamentals and applications of 3D and 4D printing of polymers: Challenges in polymer processing and prospects of future research. In *3D and 4D printing of polymer nanocomposite materials: Processes, applications, and challenges* (pp. 527–560). Elsevier. https://doi.org/10.1016/B978-0-12-816805-9.00017-X

Doddamani, M., et al. (2023). *3D printing of composites*. Springer.

Elibol, C., Güner, M. B., & Sagir, K. (2023). Processing and characterization of polymeric biomaterials. In *Handbook of polymers in medicine* (pp. 653–701). Elsevier. https://doi.org/10.1016/B978-0-12-823797-7.00024-1

Formlabs. (n.d.). Guide to Selective Laser Sintering (SLS) 3D printing. Retrieved from https://formlabs.com/asia/blog/what-is-selective-laser-sintering/

Gan, X., Fei, G., Wang, J., Wang, Z., Lavorgna, M., & Xia, H. (2020). Powder quality and electrical conductivity of selective laser sintered polymer composite components. In *Structure and properties of additive manufactured polymer components* (pp. 149–185). Elsevier. https://doi.org/10.1016/B978-0-12-819535-2.00006-5

Garcia-Leiner, M., Ghita, O., McKay, R., & Kurtz, S. M. (2019). Additive manufacturing of polyaryletherketones. In *PEEK biomaterials handbook* (2nd ed., pp. 89–103). Elsevier. https://doi.org/10.1016/B978-0-12-812524-3.00007-7

Gebhardt, A., Kessler, J., & Thurn, L. (2018). *3D printing: Understanding additive manufacturing*. Carl Hanser Verlag GmbH Co KG.

Gibson, I., Rosen, D. W., & Stucker, B. (2021). *Additive manufacturing technologies: 3D printing, rapid prototyping, and direct digital manufacturing* (3rd ed.). Springer.

Gupta, V. K., Nayak, A., Thakur, U. S., & Jain, P. K. (2022). Heterogeneous modeling for fabrication of anatomical structure from DICOM image using additive manufacturing. In *Proceedings of the International Conference on Industrial and Manufacturing Systems (CIMS-2020): Optimization in Industrial and Manufacturing Systems and Applications* (pp. 353–370). Springer International Publishing.

Hagedorn, Y. (2017). Laser additive manufacturing of ceramic components: Materials, processes, and mechanisms. In *Laser additive manufacturing: Materials, design, technologies, and applications* (pp. 163–180). Elsevier. https://doi.org/10.1016/B978-0-08-100433-3.00006-3

Horvath, J., & Cameron, R. (2020). *Mastering 3D printing: A guide to modeling, printing, and prototyping*. Apress.

Islam, A., & Rahman, M. Z. (2024). Recent advances in additive manufacturing techniques: An in-depth review. *Comprehensive Materials Processing*, 352–378. https://doi.org/10.1016/B978-0-323-96020-5.00187-4

Kalita, S. J., & Dubey, S. K. (2022). *3D printing in biomedical engineering: Principles and applications*. Academic Press.

Kumar, A., et al. (Eds.). (2024). *3D printing technologies: Digital manufacturing, artificial intelligence, industry 4.0*. Walter de Gruyter GmbH & Co KG.

Kumar, L., & Sharma, M. (2023). *3D printing: Innovations and industrial applications*. Wiley-Blackwell.

Kumar, S. (2014). Selective laser sintering/melting. In *Comprehensive materials processing: Thirteen volume set* (Vol. 10, pp. 93–134). Elsevier. https://doi.org/10.1016/B978-0-08-096532-1.01003-7

Kumar, S. A., & Prasad, R. V. S. (2021). Basic principles of additive manufacturing: Different additive manufacturing technologies. In *Additive manufacturing: A tool for industrial revolution 4.0* (pp. 17–35). Elsevier. https://doi.org/10.1016/B978-0-12-822056-6.00012-6

Luximon, A., & Luximon, Y. (2021). New technologies—3D scanning, 3D design, and 3D printing. In *Handbook of footwear design and manufacture* (pp. 477–503). Elsevier. https://doi.org/10.1016/B978-0-12-821606-4.00019-3

Markforged. (n.d.). What is Selective Laser Sintering (SLS)? Retrieved from https://markforged.com/resources/learn/3d-printing-basics/3d-printing-processes/what-is-selective-laser-sintering-sls

Materialise. (n.d.). Materialise Magics 26 user manual. Retrieved February 20, 2024, from https://static.helpjuice.com/helpjuice_production/uploads/upload/image/5908/direct/1652962888547-Magics_manual.pdf

May, M. (2024, March 18). *ACCURACIES OR TOLERANCES IN 3D PRINTING*. 3Faktur. https://3faktur.com/en/accuracies-or-tolerances-in-3d-printing/

Munir, K. S., Li, Y., & Wen, C. (2017). Metallic scaffolds manufactured by selective laser melting for biomedical applications. In *Metallic foam bone: Processing, modification, characterization, and properties* (pp. 1–23). Elsevier. https://doi.org/10.1016/B978-0-08-101289-5.00001-9

Muralidhara, H. B., & Banerjee, S. (Eds.). (2021). *3D printing technology and its diverse applications*. CRC Press.

Murr, L. E., & Gaytan, S. M. (2014). Electron beam melting. In *Comprehensive materials processing: Thirteen volume set* (Vol. 10, pp. 135–161). Elsevier. https://doi.org/10.1016/B978-0-08-096532-1.01004-9

Najmon, J. C., Raeisi, S., & Tovar, A. (2019). Review of additive manufacturing technologies and applications in the aerospace industry. In *Additive manufacturing for the aerospace industry* (pp. 7–31). Elsevier. https://doi.org/10.1016/B978-0-12-814062-8.00002-9

Noorani, R. (2006). *3D printing technology, applications and solution*. CRC Press.

Pagac, M., Hajnys, J., Ma, Q.-P., Jancar, L., Jansa, J., Stefek, P., & Mesicek, J. (2021). A review of vat photopolymerization technology: Materials, applications, challenges, and future trends of 3D printing. *Polymers*, 13(4), 598. https://doi.org/10.3390/polym13040598

Pandey, P., Krishnanand, Nayak, A., & Taufik, M. (2023). Development of mathematical model for surface roughness estimation in material jetting 3D printed parts. *Proceedings of the Institution of Mechanical Engineers, Part E: Journal of Process Mechanical Engineering*. 2023; 0(0). doi:10.1177/09544089231205960

Pandey, P., Nayak, A., & Taufik, M. (2022). Evaluation of mathematical models for surface roughness prediction of PolyJet 3D printed parts. *Advances in Materials and Processing Technologies*. www.tandfonline.com/doi/full/10.1080/2374068X.2022.2097416

Rahmati, S. (2014). Direct rapid tooling. In *Comprehensive materials processing: Thirteen volume set* (Vol. 10, pp. 303–344). Elsevier. https://doi.org/10.1016/B978-0-08-096532-1.01013-X

Riza, S. H., Masood, S. H., & Wen, C. (2014). Laser-assisted additive manufacturing for metallic biomedical scaffolds. In *Comprehensive materials processing: Thirteen volume set* (Vol. 10, pp. 285–301). Elsevier. https://doi.org/10.1016/B978-0-08-096532-1.01017-7

Riza, S. H., Masood, S. H., Rashid, R. A. R., & Chandra, S. (2020). Selective laser sintering in biomedical manufacturing. In *Metallic biomaterials processing and medical device manufacturing* (pp. 193–233). Elsevier. https://doi.org/10.1016/B978-0-08-102965-7.00006-0

Sahoo, S., & Roy, S. (2021). Additive manufacturing of titanium alloys: Microstructure and texture evolution, defect formation, and mechanical response. In *Additive manufacturing: A tool for industrial revolution 4.0* (pp. 151–182). Elsevier. https://doi.org/10.1016/B978-0-12-822056-6.00013-8

Sheng, R. (2022). *3D printing: A revolutionary process for industry applications*. Woodhead Publishing.

Singh, N. (2021). Additive manufacturing for functionalized nanomaterials breaks limits. In *Additive manufacturing with functionalized nanomaterials* (pp. 1–34). Elsevier. https://doi.org/10.1016/B978-0-12-823152-4.00008-9

Singh, R., Singh, S., & Hashmi, M. S. J. (2016). Implant materials and their processing technologies. In *Reference module in materials science and materials engineering*. https://doi.org/10.1016/B978-0-12-803581-8.04156-4

Soni, S., Nayak, A., & Taufik, M. (2022). Development of graphics user interface for process planning in extrusion-based additive manufacturing. *Materials Today: Proceedings*, 49, 403–411.

Srivastava, S., & Singh, R. (2023). *Advances in 3D printing: Materials, technologies, and applications*. Springer.

Taufik, M., & Jain, P. K. (2014). Volumetric error control in layered manufacturing. *Proceedings of the ASME 2014 International Design Engineering Technical Conferences and Computers and Information in Engineering Conference*, 4, V004T06A017. https://doi.org/10.1115/DETC2014-35099

Wimpenny, D. I., Pandey, P. M., & Kumar, L. J. (Eds.). (2017). *Advances in 3D printing & additive manufacturing technologies* (Vol. 1). Springer.

Yang, L., Hsu, K., Baughman, B., Godfrey, D., Medina, F., Menon, M., & Wiener, S. (2017). *Additive manufacturing of metals: The technology, materials, design and production*. Springer.

Index

Kolibri 360 GmbH
Ferdinand-Jühlke-Straße 7
99095 Erfurt, Deutschland
produktsicherheit@kolibri360.de